THE OHIO RIVER VALLEY SERIES

Rita Kohn and William Lynwood Montell
Series Editors

THE OHIO

R.E. Banta

Illustrated by Edward Shenton
Foreword by Thomas D. Clark

THE UNIVERSITY PRESS
OF KENTUCKY

This edition is reprinted by arrangement with Henry Holt and Company, Inc.

Publication of this volume was made possible in part by a grant from the National Endowment for the Humanities.

Scholarly publisher for the Commonwealth,
serving Bellarmine College, Berea College, Centre College of Kentucky, Eastern Kentucky University, The Filson Club Historical Society, Georgetown College, Kentucky Historical Society, Kentucky State University, Morehead State University, Murray State University, Northern Kentucky University, Transylvania University, University of Kentucky, University of Louisville, and Western Kentucky University.

Editorial and Sales Offices: The University Press of Kentucky
663 South Limestone Street, Lexington, Kentucky 40508-4008

02 01 00 99 98   5 4 3 2 1

Library of Congress Cataloging-in-Publication Data

Banta, R. E. (Richard Elwell), 1904 -
    The Ohio / R.E. Banta.
        p.   cm.   — (Ohio River Valley series)
    Originally published: New York : Rinehart, [1949],
in series : Rivers of America.
    Includes bibliographical references and index.
    ISBN 0-8131-2098-5 (cloth : alk. paper). —
ISBN 0-8131-0959-0 (paper : alk. paper)
    1. Ohio River—History. I. Title. II. Series.
F516.B18   1998
977—dc21   98-30104

This book is printed on acid-free recycled paper meeting the requirements of the American National Standard for Permanence of Paper for Printed Library Materials.

Manufactured in the United States of America

To the ladies
Caroline and Kathleen

Contents

Series Foreword

THE Ohio River Valley Series examines and illuminates the Ohio River and its tributaries, the lands drained by these streams, and the peoples who made this fertile and desirable area their place of residence, of refuge, of commerce and industry, of cultural development, and, ultimately, of engagement with American democracy. In doing this, the series builds upon an earlier project, "Always a River: The Ohio River and the American Experience," which was sponsored by the National Endowment for the Humanities, the humanities councils of Illinois, Indiana, Kentucky, Ohio, Pennsylvania, and West Virginia, and a mix of private and public organizations.

The Always a River project directed widespread public attention to the place of the Ohio River in the context of the larger American story. This series expands on the river's significant role in the growth of the nation by presenting the varied history and folklife of the region. Each book's story is told through men and women acting within their particular place and time. Each reveals the rich resources for the history of the Ohio River and of the nation afforded by records, papers, and oral stories preserved by families and institutions. Each traces the impact the river and the land have had on the individuals and cultures and, conversely, the changes these individuals and cultures have wrought on the valley with the passage of years. As a force of nature and as a waterway into the American heartland, the Ohio and its tributaries have touched us individually and collectively. This series celebrates the story of that river and its valley through multiple voices and visions.

The Ohio, originally published in 1949 by Rinehart & Company as number 39 in the Rivers of America Series, retains its verve, charm, and authenticity on its fiftieth anniversary. R.E. Banta's prose and grasp of the Ohio lend an epic quality to this most singular waterway. His opening chapter provides an unparalleled synthesis of *La Belle Rivière's* story. Through the succeeding nineteen chapters and extensive bibliography we come to know equally the twists and turns of the riverbed and the fortunes and fates of individuals, towns and villages, nations, enterprises, religions, and "pleasurin'"activities.

Still the river's most comprehensive history, Banta's book should be read as the best of scholarship at mid-twentieth century. His words have spurred others to investigate and bring fresh insight to specific topics. Other authors of present and forthcoming titles in the Ohio River Valley Series are indebted to Banta for having paved the way to make loving the Ohio River a fit and proper enterprise. Here, freshly presented for a new generation, is Banta's *Ohio*, in which he serves up savory delights at every bend of the river's amazing 981-mile journey through the heartland and history of our nation.

Rita Kohn
William Lynwood Montell

Foreword

AMERICA in 1934 was trapped in the morass of the Great Depression. In that era there was a distinctive stirring of nationalism on the one hand and a political restlessness on the other. The provocative influence of the New Deal with its legislative mandates and institutionalism marked a distinct historical dividing line between the old American folkways and the emergence of a new and somewhat blasé era. There lingered among traditionalists a latent fear that many traces of the heroic past would be obliterated. To date, historians had done too little to record the past in earthy human and imaginative writings. To many the historian lacked the imagination to mine the rich lode of everyday human experiences.

Nevertheless, there was a rising interest in depicting the national experience in several levels of writing. The Federal Writers' Projects, a brainchild of the Works Progress Administration, was a catalyst of sorts. It stimulated local interest by promoting the gathering of tons of raw materials preparatory to writing the state guides and local histories. This agency had a seminal influence in the stimulation of writers and publishers to take a second look at the local American scene. Capitalizing on this attention Constance Lindsay Skinner, herself a special interest-regional historian, conceived the idea of the Rivers of America Series. With fervor she expressed the concept that a much tighter lacing of American folkways and history could be achieved by writing of the rivers, of their bearings upon the mores of folk settled upon their shores and exploiting their channels as transportation and commercial intercourse arteries. To Constance Lindsay Skinner, America's rivers chuckled and roiled with the rhythm of man and nature.

Miss Skinner wrote a romantically conceived literary formula embodying her conception of the past and lamenting the lapses of current historians. She was convinced they lacked ear and soul to sense the rhythms of life at the humanistic levels of the regions of the nation. The craft of the historians she felt was served by dull fact beaglers. Only creative writers possessed the inner latitudes to view and comprehend the past combining a sense of reality with a rich poetic touch. On the strength of her essay, and no doubt persistent persuasion, she convinced the recently formed publishing house of Farrar and Rinehart to negotiate contracts with writers to produce a series of books about America's past, using the rivers as the coagulating element of folk history, but at once giving the individual books a distinct quality of diversity.

Whether Constance Lindsay Skinner advanced the individualistic nature of the books is somewhat vague. The first book to appear in the new Rivers of America Series was Robert P. Tristam Coffin's *Kennebec* (1937). There followed five other titles written by the Skinner formula. Walter Havighurst, author of *Upper Mississippi*, quickly revised his text, and Blair Niles' *James River* received such adverse criticism that it was republished in 1947 in a revised edition.

From the initial planning for the Rivers series, the Ohio was considered a major subject. Farrar and Rinehart entered into a contract with Constance Rourke, author of a biography of James J. Audubon, to write *The Ohio*. However, she became almost passionately engrossed in researching and writing another book, on American folk culture. Miss Rourke came to the University of Kentucky to talk with young women about the career of writing. She combined the lecture assignment with a search for data relating to folk culture, and I drove her around central Kentucky in search of materials. On one of those journeys she told me she was surrendering the contract to do *The Ohio*. No doubt she made a wise decision.

There might have been a creative author somewhere with the heart and competence to tackle the challenge of writing about the great heartland river, but none came forward. There perhaps could not have been a happier choice of an author to write about the Ohio River and its valley than Richard E. Banta of Crawfordsville, Indiana. He was born and spent his entire life on the bank of the Wabash River. Few people had a greater attachment to place and a more precise sense of the Ohio than Dick Banta. He was a farmer and a highly successful rare book dealer. He had an uncanny ability to locate fugitive manuscripts, pamphlets, and books relating to the Middle West, and especially materials relating to the Ohio River and its drainage basin.

Banta neither wrote nor thought in the vein of a poet or creative writer; nevertheless, he had an extraordinary talent for sounding the historical rhythm of the many phases of the Ohio River's past. Farrar and Rinehart might have had an interesting clash of personalities and ideas had Miss Skinner lived long enough to meet Dick Banta—both were strong willed, but Banta held the whip hand of intimate historical knowledge.

Few streams on the face of American geography cradled a greater lode of American history than the Ohio River. The main stream wound and cut a deep channel across half of the mid-continent, creating a significant geographical landmark and a social, political, and economic dividing line. From the dawn of human history on the American continent, the Ohio River exerted a strong environmental and economic influence on peoples of different races and conditions of life. By the very path of the river and the flow of its current, in time it was to be a mighty force in concentrating human beings along its banks. In the long draw of both Indian and Euro-American history the stream was to form a clear division of sections, and of human approaches to the land.

Paradoxically the Ohio was a river of war and of peace. Its waters at times were literally stained with the blood of conflict. North of the stream a native race had warred and competed for possession and occupancy of the land, creating both a folk history and a history of crossed destinies and conditions of culture.

Banta's treatment of the Indian presence and acts of resistance was within the context of his times, and within the bounds of the voluminous contemporary historical documentation. Bedded within the context of published history of the Valley of the Ohio, there was seldom a moralistic or judicial weighing of the great contest for possession of the land.

The in-rush of Euro-American settlers inevitably involved bitter rivalries not only between races and cultures, but among nations. The current of the Ohio and the lands of its valley were fabulously rich prizes ready for the taking. In *The Ohio,* Dick Banta clearly perceived the roily undercurrents of conquest and change to be more than romantic folk phenomena. It was within the environs of this great drainage complex that the processes of exploitation and expansion of the American landed western littoral was tempered and seasoned. It was here within this great geographical enclave of river and land that the western movement itself was hardened by Indian resistance, international conquest, and even the eternal flow of the Ohio River.

Inevitably in thrusting open new country there would come a great variety of human beings. There would in time drift down the Ohio greedy land speculators, scheming rascals, peddlers of civilization, and naive and feckless settlers who pursued the infectious westward Edenic dream without ever defining it. During the latter half of the eighteenth and first quarter of the nineteenth centuries the Ohio Valley was a rich cherry ready for the plucking. More central, however, was the all but endless drift of floating "arks" which bore within their holds that intangible but irrepressible cargo of nationalism.

Ever aware of intrigue, divisive sectional rivalries, and military adventurism, Banta writes of the Ohio's human pageantry as though he manned a flatboat or marched in lockstep to seek glory in a raid against a British-Indian stronghold. South of the river, and out of complex frontier problems, international treaty-making, and the force of population movement, sprang the Virginia-bred Commonwealth of Kentucky, and from federal committees there emerged plans for the sprawling Northwest Territory. Political planners and manipulators early, and perhaps unctuously, drew a historical borderline that in perpetuity would emphasize the realistic denominator, sectionalism. To the south aggressive pioneering westerners debated and ushered into the Union the first of the new empire states. To the north the same breed of politicians promoted and created new states, but under different land, slave, and education mandates. Almost from the moment the first settlement was made at Marietta on the Ohio there was made a comparison between the slave and free territories, a fact documented so often in the writings of Anglo-European traveler-observers.

The Ohio was from the outset of invasion of the western country a channel feeding downstream a great galaxy of humanity. The challenges of the river generated men and women who in legend became characters of dramatically enlarged human proportions, and the roll of prominent westerners who came this way is all but endless. There came James Harrod and his men to settle in Kentucky, George Rogers Clark, eventually to help wrest the western territory from Indian and British hands, Anthony Wayne to thrust the line of empire deeper into the hinterland, and later the Lewis and Clark Expedition to penetrate the breadth of the continent itself. None who floated upon the Ohio's current, however, was more colorful than the rivermen who propelled keel and flatboats and imprinted in American western literature forever the saga of the "ring tail roarers," who were men enough to challenge the river and its wayward current, maelstroms, and sucks.

That domestic vessel of the Ohio, the "ark," delivered the great procession of settlers and their possessions downstream, sometimes drifting too near the northern shore for safety from lurking warriors out to thwart the invasion of their homeland. Mixed in with the settlers, the Bradfords of Lexington, Kentucky, brought aboard a flatboat their tiny newspaper press and bag of type to establish the *Kentucky Gazette* in 1787. Lawyers came in search of clients, and preachers in search of lost souls. Companies of actors drifted downstream along with authors, artists, merchants, and skilled craftsmen. George Rapp came this way to establish his community of perfect economic and religious harmony on the banks of the Wabash. Later Robert

Owen of New Lanark purchased the community and dispatched down the Ohio his famous "Beetload of Knowledge" to establish an oasis of intellectualism and social harmony in the raw Ohio Valley backwoods. In the early decades of American penetration of the West, the Ohio River became almost as much an artery of culture and utopian idealism as one of emigration, commerce, and the spread of empire.

The introduction of the steamboat in 1811 rang down the curtain on one age and lifted it on another. This vessel within a miraculously short interval wrought a revolution in the way of life along the Ohio, a revolution best revealed in the rise of the stream's three capital cities, Pittsburgh, Cincinnati, and Louisville. The steamboats, floating bits of Victorian America, in time surrendered to the plain workhorse tow boats shoving millions of tons of coal, oil, and goods up and downstream.

From primitive paddle to sophisticated channel type propellers, the waters of the Ohio have long been churned by the restless swirl of human and commercial movement. Man has undertaken to rein in the river by building dams across its channel, erecting flood walls around the towns, and bypassing its wall with a locked and dammed canal. The river frequently strikes back with devastating floods, reminders that nature is more fickle even than man.

In her vision of America's rivers as ganglia of folk ways and culture, Constance Lindsay Skinner could hardly have conceived of the great transformations which occurred on them. Shrewdly, Dick Banta made the early assumption that the Ohio became "the great high road of American commerce." His text, however, portrays it as the great high road of history and culture, and as the tangible and imaginary border between the old East and the new West. Just as the stream physically has been a stubborn and powerful force plowing a circuitous channel through a heartland, so it has been in shaping the contours of the great and expansive American dream. In many respects *The Ohio* comes nearest to achieving in down-to-earth writing the basic concept of the Rivers of America Series.

Constance Lindsay Skinner might not have found in the text of *The Ohio* the traces of the "creative imagination" which she stressed so passionately, but surely she would have admired Dick Banta's writing as provocative to imagination and the plumbing the rich vein of folk speaking to one another, and to future occupants of the valley. The text of *The Ohio* is as vibrant and varied as the echoes of a giant calliope aboard a belle of the river steamboat.

Thomas D. Clark

THE OHIO

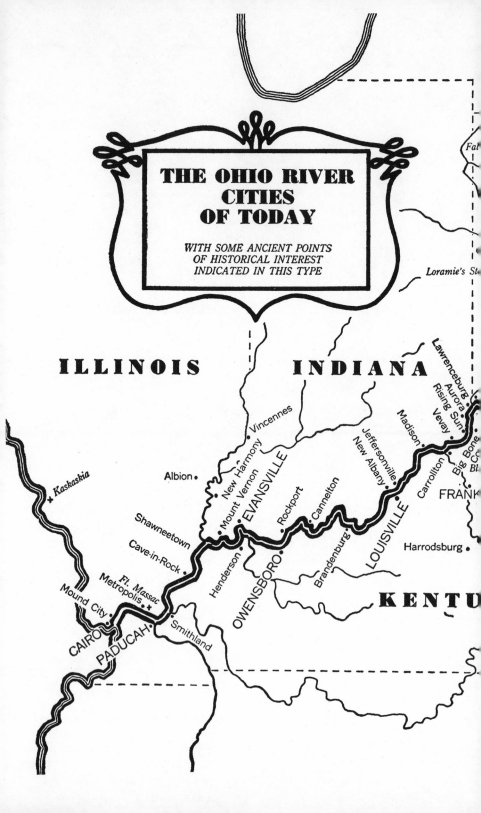

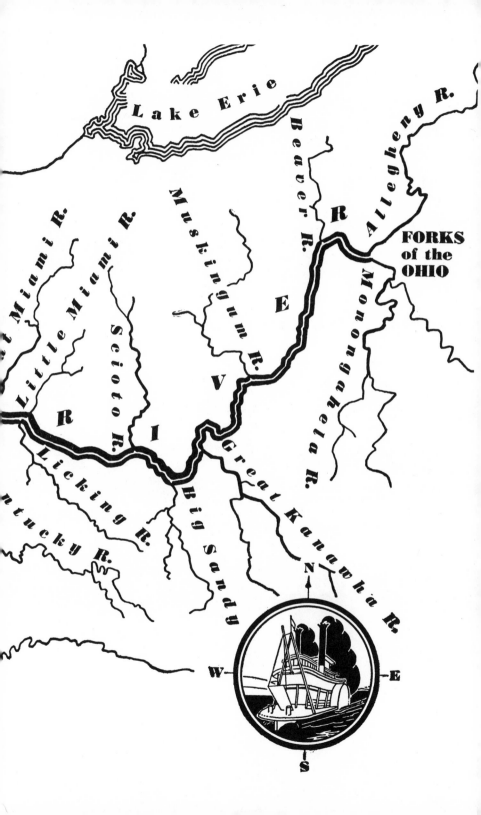

LA BELLE RIVIÈRE

WHEN the French first met the Iroquois in the seventeenth century those truculent redskin inhabitants of the eastern Great Lakes basin spoke of a great river which rose to the south of their land and flowed west.

News of a river flowing in that direction always aroused the interest of Europeans, still obsessed with their search for a short route to China, and the Indians gave a particularly glowing account of the wonders of this one. They had long coveted control of this river and when, about the middle of the century, Dutch traders sold them guns, the Iroquois put the new weapons to immediate use in a successful campaign to impose their will upon the inhabitants of its valley.

The Iroquois called the river "Ohio" and the name was later translated by the French as "The Beautiful"—*La Belle Rivière*. Not *a* beautiful river, mind you, but *the* beautiful, with no fear that it might be confused with any other.

The French may have been mistaken as to the exact meaning of the Iroquois word; linguists—possibly no better qualified as translators than were the early French—have preferred "The Great" or "The White" or "The Sparkling." But it *was* a beautiful river, of a certainty, and the first translation has seemed appropriate enough to those who have succeeded the Iroquois—be they colonial French or British, American emigrants or European travelers.

Maybe the linguists had better ear for the niceties of Indian language than they had eye for color or line. The last moments of a sunset on a westering reach of the upper Ohio, smooth-flowing water the color of burnished copper between almost perpendicular stone banks already pitch black in shadow; or the jade green of low winter water on the middle river accented by a cold sun on bordering snowbanks; or a summer morning above its mouth where the broad stream makes a sweeping curve through rich black bottom lands, bound by yellow willows and gnarled white-limbed sycamores—those sights should convince even scholars of linguistics that "Ohio" *must* mean "The Beautiful," no matter what the apparent connotation of the Iroquois noun.

But appreciation of the river's beauty was not limited to French and Iroquois. Few travelers, ancient or modern, failed to add to its fame by exclaiming in some such flamboyant vein as Charles Fenno Hoffman, urbane editor of the *Knickerbocker Magazine,* who got his first view of the Ohio at Wheeling in 1833:

> The Ohio is beneath your feet. . . . The clear majestic tide, the fertile islands on its bosom, the bold and towering heights opposite, with the green esplanade of alluvion in front, and the forest-crowned headlands above and below, round which the river sweeps away, to bless and gladden the fruitful regions that drink its limpid waters,— these, with the recollections of deeds done upon its banks—contrasted with all the luxuries of civilization that now float securely upon that

peaceful current,—these make a moral picture whose colours are laid in the heart, never to be effaced: no man will ever forget his first view of the Ohio.

In fact so much of this inflated rhetoric had been expended upon the subject by 1838 that one prosaic Englishman (who bore the appropriately prosaic name, Abner D. Jones) was prepared to find the Ohio wanting. But after he had viewed its length even Abner added his mite of commendation:

I had often heard the praises of this majestic river sung, and had curbed my expectations lest I should be disappointed. The Ohio *is* a beautiful river. There are points on the Hudson and Connecticut, and other rivers of the East, which equal any thing I saw on the Ohio; but its peculiarity is that it is *all* beautiful. There are no points bare of beauty; but every mile is as rich in scenery as it was in verdure at the time of my passage down its "winding way."

Modern dwellers on the Ohio River are not ones to buttonhole the stranger and bare their souls but even the casual sojourner is soon aware that the devotion of its people to their river is among the Ohio's most interesting attributes.

Obviously it seems to the people of the Ohio to be pretty certain that *their* river flows through the most beautiful country and has the handsomest water—jade green to cream to coffee-with-cream to milk chocolate—that it makes the sharpest bends, has the longest straightaways, the highest hills, the richest bottoms, the smoothest flow and the damnedest floods on this earth or any other. They know that along its banks lie almost all sights worth seeing, from hell-for-leather industrial progress to a stretch of bank preserved as it was before men of any race set eyes upon it.

And perhaps, one feels having seen their river, the heart is a better guide to values than the mind—perhaps the dwellers on the Ohio may not be too far from right at that!

The Ohio, besides its beauty, has another significance—a geographical and historic significance apparent even to those who know it only by its appearance on maps.

Visualize the great interior valley of North America. The continent, at the latitude of the United States, is some 2,500 miles in

width. Slightly more than 200 miles from the eastern edge is the
Appalachian crest; 600 or 800 miles from the Pacific coast rise the
Rocky Mountains. Between these heights lies a great wedge, 1,500
miles wide and almost as deep, with its point reaching the Gulf of
Mexico. It is a great segment of the continent which stretches from
western Pennsylvania to western Montana, from North Dakota to
Texas, from the edges of the Great Lakes to the delta of the
Mississippi. This region, comprising one of the most important
single areas of productive land in the world, is drained by the Ohio-
Mississippi-Missouri River system.

Earliest section of this fabulous valley to be developed—and
hence the steppingstone to the even farther West—was the region
between the Appalachians and the Mississippi. Explorers, traders,
and soldiers had thrilled to its appeal, even in colonial days; an
appeal which gave "a prospect into unlimited empires," to which, as
George Washington said, the poor, the needy, and the oppressed of
the earth might repair and abound "as in the Land of promise, with
milk and honey." After Independence, as the common run of
Americans followed destiny to the west, they found nature on a
grander scale than they had before been able to conceive—measure-
less forests, prairies extending farther than the eye could see, and
waters "whose sweep is over uncounted leagues." The more artic-
ulate described it as a "happy region! large and fertile enough for
the abode of many millions," a region where "yesterday all was
silent, save the beast and the bird" but which "becomes today the
home of the woodsman—the center of human affections—the
nucleus perhaps of an intelligent, social, virtuous community—the
focus, where, it may be, light shall emanate to other parts of the
world."

To all this the Ohio was the highroad from the early settlements on the seaboard: no aid toward the development of the great
transmountain West was more important.

*   *   *   *   *

Perhaps the surest way to locate the Ohio River for the modern
reader is to state that it heads at latitude 40 degrees, 25 minutes
North, longitude 80 degrees West (that being the city of Pitts-

burgh), which is about 320 miles west of the Statue of Liberty, a shade less than 200 miles northwest of Washington, 520 miles due north of Charleston, South Carolina, and 225 miles south of Toronto, Ontario. Flowing in a generally south-of-west course, the Ohio empties into the Mississippi at Cairo, Illinois, latitude 37 degrees, 1 minute North, longitude 89 degrees, 10 minutes West. That point is some 345 miles south of Chicago, 500 miles north of New Orleans, and 1,825 miles (as the crow is reputed to fly) east of San Francisco, California.

Debate sometimes rages for generations upon the subject of the exact source of a river: whether it is So-and-so's Pond, which is dry six months out of twelve, or Such-and-such Branch, which in an ordinary year is dry only from June to October. There is no such doubt about the Ohio.

The Ohio's point of beginning cannot be mistaken. The Monongahela River flows up from the south through its lateral valley along the western foot of the Allegheny Mountains: the Allegheny River rises in north-central Pennsylvania, flows in an arc into New York and back again into Pennsylvania, where, in its now southering course, it joins the Monongahela. Where these waters meet, at what is now the western point of Pittsburgh's Golden Triangle, *there* begins the Ohio.

Each of these two rivers which combine to form the Ohio has more than three hundred miles of course and many tributaries drawing upon the plentiful rainfall of the eastern watershed; the Ohio is off to a thriving start. When it empties into the Mississippi, after 981 miles of windings, it has flowed in every direction on the compass—though mostly south of west—to drain a watershed 203,900 square miles in extent which spreads into fourteen of the United States; it has passed through a valley as rich, as beautiful, as varied in production and as important in history as any in the Americas—North, South, or Central.

Before its windings are explored, however, one point must be made clear: in the language of old rivermen there are no north, south, east, or west banks on a river—only a *right* and a *left* bank, as those appear when one faces downstream. Since old rivermen, these days, make up only a small percentage of the population, and since the uninitiated river voyager is likely to forget whether he should be facing with or against the current while calling his direction, it seems best to consider here that the Ohio flows west, as it mainly does, and thus has banks called north and south. At times they must be designated by additional terms. While Indians were still a menace the north bank was referred to as the *Indian* shore and the opposite bank as the *Virginia* shore, for instance; after that early day they took the names of new states formed along them— Kentucky, Ohio, Indiana, Illinois, and West Virginia.

From its beginning at the Forks the Ohio starts northwest as if bound for Lake Erie but bends west again after receiving Beaver River. A few miles after crossing the Pennsylvania-Ohio line the river makes a long sweep to the south and west which holds all the way to Marietta: it flows between rocky hills indented by the gullies of numerous tumbling streams which rise in the high back country of West Virginia and the only slightly less rugged unglaciated areas of southeastern Ohio. At Marietta enters the Muskingum, beloved by the Indians—where a handwritten code of laws nailed to a tree in the spring of 1788 marked the active beginning of the first American civil government of the Northwest Territory. Scarcely a dozen miles farther downstream the Little Kanawha comes in from the south. Just below its mouth lies an island, once the home of one Blennerhassett, who had a friend and frequent visitor named Aaron Burr; that island was the place where certain conferences were held which might have resulted in a United States bounded on the west by the Mississippi River.

As the Ohio swings west, then south, it receives the Hocking from the Ohio side, continues its southward course, then makes one of its large northward bulges and starts another long reach to the south. Halfway down this southern stretch enters the Kanawha from the West Virginia side. The Kanawha, valued in pioneer times for its salt springs, has a principal tributary which cuts its way through various ridges and reaches back to the west side of the Blue Ridge in North Carolina for its headwaters. Just below the mouth of the Kanawha, but on the Ohio side, is Gallipolis—the city of the Gauls, the "semitropical" site of which the poet-preacher-salesman Joel Barlow sold to innocent French emigrants during the days of France's Revolution, thus perpetrating the most harrowing American real-estate swindle between John Law's day and the Florida boom of the nineteen-twenties. The Ohio ends this southern sweep north of Huntington, West Virginia, and again turns westward. About 311 miles from its beginning as the river flows (but less than 200 by air) the Ohio receives the Big Sandy. That stream now forms the West Virginia-Kentucky boundary line. At this point, where the Ohio is within a hundred miles of being as far south as at any place along its course, it makes another long northwestward sweep. Shortly after it again turns west it is joined by the Scioto, another favorite stream of the Indians, flowing from the north. For almost a hundred miles thereafter the Ohio's general direction is west and northwest to the mouth of the Little Miami, also entering from the north. Below that enters the river called the Licking, and a few miles farther the Great Miami flows from the north. Opposite the mouth of the Licking and between the two Miamis have arisen in the past century and a half the prosperous city of Cincinnati and its Ohio and Kentucky satellites.

After the Ohio passes the Great Miami, its major indecision as to direction is definitely over; though it indulges in momentary vagaries in which it dashes off briefly to the east, north, and south as the lay of the land and the contours of the hills make temporarily expedient, its general course is along a gradual south-of-west progression.

At one time the fifty miles above and below the mouth of the' Licking was the favorite crossing of the Indians for snaking captives out of Kentucky's "dark and bloody ground"; some famous captives were led over here—among them Daniel Boone. Later this stretch would serve as well for running slaves across into free territory to catch the underground railroad trail, operated largely by the Quakers, straight north to Canada. A Mrs. Stowe, militant wife of an always impecunious professor, got some ideas from the tales of the goings-on here and upriver, which resulted in a book that sold widely and helped to stir up a good deal of excitement one way and another along the Ohio. Today there are no sizable cities on the river for fifty miles below Cincinnati, although early real estate promoters and town planners were no doubt optimistic of the prospects when they christened Aurora and Rising Sun on the north bank. (Whoever laid out the Kentucky town opposite Rising Sun was perhaps not so hopeful of its prosperity and growth or he would not have named it Rabbit Hash.)

Hills still border the Ohio as it receives the historic Kentucky River from the south, but the heights more often retreat to leave broad and fertile fields well above high water; these areas continue to increase in both frequency and size for the next sixty miles or so, until the geological fault known as the Falls of the Ohio is reached. That rough stretch of water was always more properly a chute, with a fall of about twenty-four feet in three miles, but Falls it was and Falls it shall be. On the south side of the river, at the Falls, the gradual sloping banks were originally covered with a thick growth of beargrass which gave the region its name; they are now occupied by Louisville and its environs. Behind Louisville begin the "Knobs"—sharp hills of ancient sandstone which swing east in a broad circle to encompass the Kentucky Bluegrass Country—and across the river the Indiana Knobs approach to within a few hundred yards of the bank—from which they appear as earth and stone haystacks, sugar loaves, and hat crowns covered still with enough trees to give the appearance of being forested.

The Falls could be passed in the early days without difficulty by canoes and pirogues, and later on with the water at normal stage they could be negotiated by keelboats, flatboats, arks, and even at high water by steamboats. The last, however, since the channel wound through descending rock shelves, required the services of an experienced pilot for safety. Below the Falls is the site of Corn Island (no longer there) where George Rogers Clark rallied his harum-scarum troops to start in 1778 the expedition by which he captured Kaskaskia, Cahokia, Vincennes, and Colonel Henry Hamilton, the "Hair-buyer" British commander in the Northwest. Beyond that point the greatest dangers to navigation on the old Ohio came to an end. From there on were to be met only the normal hazards of sand bars, snags, stumps, rocks, and Indians and river pirates.

Continuing below the Falls for fifty or sixty miles the bluffs still approach close to both shores. Once past this stretch the river is edged on both sides by gently rolling prairies and woodland broken only occasionally by folds and ridges of exposed bedrock. Free of its retaining hills it begins to play some of those tricks of channel shifting at which the Missouri is the acknowledged past master. Cutoffs and abandoned channels are frequently to be seen— obvious examples of a process which has gone on for ages. In a short century and a half, during which the opposite banks have been under the jurisdiction of various territorial and state governments, these shifts of the river have caused boundary and jurisdictional perplexities—including one which resulted when the Evansville, Indiana, waterworks found itself transferred to Kentucky territory.

In the lower fourth of its course the Ohio passes through the southern end of the central lowland plain that extends by way of the Indiana-Illinois prairies north to the Great Lakes and westward to the Great Bend of the Missouri. It is in this section with three-fourths of its length already behind it that the Ohio receives, within an easy twenty-four hours' voyage, its greatest tributaries—four of them in all.

First, from the south comes the Green River. It rises far back to the east within fifty miles or so of the Great Bend in the Kentucky River and its unforgettable tinted waters pass the famous Kentucky cave country. The Green is thought to have been an underground river in its earlier history but its limestone roof eventually wore away to bring its waters to the surface. Deep and narrow, its lower two hundred miles are passable by small steamboats.

Sixty-five miles farther downstream the Wabash, after flowing some 550 lazy miles, enters from the north. The Wabash rises in northwestern Ohio, crosses into Indiana, comes within short portage distance of the head of the northeastward-flowing Maumee, swings west and south across north central Indiana, and becomes the boundary line between that state and Illinois in the lower third of its valley. Down the Maumee and the Wabash came the French from Quebec and Montreal to establish in the first three decades of the eighteenth century the military and fur-trading posts of Ouiatenon and Vincennes. By way of these streams, probably before 1720, the first regular commerce of the white man entered the lower Ohio. The French also used the Wabash-Ohio route to keep in touch with their settlements of Kaskaskia and Cahokia on the Mississippi.

The early familiarity of the French with the lower Ohio was probably responsible for an error which plagued European geographers for years. For the voyageurs rechristened the Indian-named Wabash the "River St. Jerome" and included the Ohio, below the mouth of the Wabash, under that title. Perhaps the explanation for the French mistaking the tributary for the main stream was observation of the Wabash when it was in flood from northern rains at a time when the Ohio was at a low-water stage. Whatever the reason for this error, it long added to the mystery of where the Ohio rose and where it had been before it appeared as a great river just east of the Mississippi.

Next, and also from the north, comes the Saline River, most unimpressive as to size but important because of the salt springs that lured game in early days and yielded a commodity necessary to the primitive economy of both Indians and white pioneers.

Below the Saline there is a big, right-angle turn as the river changes direction from south to west, and on the north shore, bored in one of those folds of stone which still occasionally crop out, lies Cave-in-Rock. This historic hole is not much as a cave—there are hundreds deeper and more commodious in southern Indiana and western Kentucky—but it was notorious in the early nineteenth century as a hideout for bandits, murderers, pirates, and desperadoes.

Safely past Cave-in-Rock (and there was a time when passing it safely was an achievement) the traveler finds the river's course serene for almost fifty miles before it comes to the mouth of the Cumberland; ten miles farther and also on the south side is the mouth of the Tennessee. Many a person, sufficiently informed in geography to know that there are two such rivers as the Cumberland and the Tennessee, would incline to believe that both flow into the Mississippi. To the glory of the Ohio this is not the fact.

The Cumberland rises in the eastern edge of Kentucky along the west slopes of the Cumberland plateau, flows west and south into the state of Tennessee and as it turns north back into Kentucky flows for a hundred miles parallel to and in places only a dozen miles or so east of the Tennessee. The Tennessee stems from the Powell, Clinch, and Holston rivers, which rise among the Appalachian ridges of western Virginia, and from the French Broad and Nolichucky, which begin in the Great Smoky Mountains. For 650-odd miles it flows through mountain valleys and alluvial plains, through eastern Tennessee south into Alabama, across a corner of Mississippi and back north again through Kentucky to the Ohio; in its course it drains an area of more than 40,000 square miles.

The Ohio, now a great river indeed, is away on its last long reach: sweeping in a long flat arc, first northwest and then southwest, it flows for its last few miles through a great low and often inundated bottom land to play its part in one of America's most impressive natural scenes—the juncture of the Ohio and the Mississippi.

This brief sketch of the river naturally does not do it justice; to a fuller description dozens of editions of river pilot's guidebooks have been dedicated in past years, and still many of its interesting features have not been fully catalogued. In this sketch most of the delightfully named small streams—Rolling River, the Tradewater, Pigeon Creek and Little Pigeon Creek, Indian Kentuck' Creek, Big Grave Creek, Sinking Creek, Bull Skin Creek, and all the rest— have been passed unrecognized, as have Dead Man's Island, Three Brothers Island, Little Hurricane Island, and Green Bottom Ripple. But some of these will be examined later.

Chapter Two

THE RIVER, OLD AND NEW

THE present Ohio, except for its lower third and short segments of its upper reaches, is not old, as rivers go; perhaps it achieved its present form only forty or fifty thousand years ago!

Long before that, millions of years before, during the Paleozoic era, the sea covered much of the continent of North America. During this era, which was the first to show abundant recognizable life in the water, most of the bedrock under the Ohio Valley was laid. In the Ordovician period—an early part of the Paleozoic era—the sea teemed with a strange, nightmarish life: cephalopods or octopuslike creatures, sluggish giant snails or gastropods fifteen feet long, hard-shelled tribolites somewhat like our crabs and crayfish, and fish with rudimentary spinal columns which would eventually so develop as to make the water no longer a necessary habitat for some of them. In the region surrounding Cincinnati is a huge pushed-up dome—called the Cincinnati Arch—which shows Ordovician rock exposed to view by erosion.

Upon the flanks of the Cincinnati Arch, to the east and west of the Ordovician, appear outcroppings of rock formed in the Silurian and Devonian periods. By the latter the tribolites were dying out and some kinds of fish had developed protective armor. Through most of the stretch past Indiana and a part of that bordered by Illinois, the Ohio runs through rock formed in the Mississippian period.

Near both its headwaters and its mouth the river passes through rock of the Pennsylvania period—known popularly today as the Coal Measures—the product of a time in which the most spectacular of living things must have been the giant plants that were later decomposed, heated, and compressed to be mined—after the passing of some millions of years—as the coal that now enriches and begrimes the valley.

Between the towns of Marietta and Pomeroy, Ohio, is exposed rock of the Permian period—the next after the Pennsylvanian and the last of the Paleozoic era. By the time this rock was deposited as silt at the bottom of the still warm sea, some primitive forms of animal life had developed rudimentary lungs, became amphibious, and spent part of their time on land. With the end of the Paleozoic era the region emerged from the sea and, so far as geological evidence shows, was never again submerged.

Only near the mouth of the Ohio is there an exposure of rock more recent than that of the Paleozoic era. There, where once the Gulf of Mexico encroached during the Cretaceous period when dinosaurs of different kinds wandered over the continent, are rocks laid only a few hundreds of thousands of years ago.

Though the present course of the river was shaped largely by the successive invasions of the great continental glaciers that came down from the north in inexorable creeping advances to cover most of the northeastern part of the United States, a glance at the preglacial drainage pattern of the region might be of interest.

The picture is clear only in broad outline; some of its features still lie in the realm of theory and conjecture. Briefly it is as follows:

A partial recession of the waters that covered most of the continent took place in the later part of the Paleozoic era—during the Mississippian period—and left the Cincinnati Arch the first land exposed. The coal areas developed in the adjoining swamps of the Pennsylvanian period.

At the end of the Paleozoic era there was a general upheaval of the land and the Appalachian Mountains, which had risen, been leveled, and been raised again in the earlier days of the earth's existence, rose now once more and the whole Ohio area was uplifted with them.

In the upper part of the Ohio Valley, the Monongahela River even then flowed north in approximately its present course and received a smaller stream flowing through the northern part of the Allegheny River's present valley. Joined, they flowed along the present Ohio's course, but only to the present Beaver River, near the Pennsylvania-Ohio line. There the stream—called for purposes of identification the *Ancient* Monongahela—turned north through something like the modern Beaver River valley and continued into the Grand River valley in the Erie basin. One of the small tributaries of this Ancient Monongahela arose to the east of a great ridge in the neighborhood of New Martinsville, West Virginia, and flowed north through a short stretch of what is the Ohio's present channel to join the Ancient Monongahela near the present city of Beaver, Pennsylvania.

The main stream of the next great preglacial drainage system to the west in the present Ohio Valley was a now-extinct river which geologists have named the Teays. It was a large stream which arose far to the east in the Piedmont district of North Carolina and Virginia and flowed in the valley of the present Kanawha River to the modern town of St. Albans, West Virginia, where it left that valley to go westward. It entered the present Mud River valley at Milton, continuing west, and it began to occupy the present Ohio Valley at modern Huntington, West Virginia. The ancient and modern valleys largely coincide until they reach Wheelingsburg. Here the valleys diverge, the modern Ohio turns west but the ancient Teays River continued north through the valleys of the present Scioto and its tributaries to Chillicothe, Ohio, where evidence of the Teays now disappears and becomes untraceable under the glacial drift. Recent surveys indicated that the Teays River probably crossed Indiana and Illinois in a large valley, now also buried beneath the glacial drift, to eventually enter the ancient Mississippi.

The preglacial Teays had important tributaries, some of which account for other parts of the present Ohio Valley. The extinct Marietta River had its headwaters on the south side of the divide at New Martinsville, West Virginia. It flowed southwest, following the line of the present Ohio to Little Hocking. There the courses of the ancient and modern rivers diverged, the Marietta eventually turned westward near Gallipolis and entered the Teays near present Beaver, Ohio.

The largest western tributary of the Teays flowed north in the present Great Miami valley. This river, called the Ancient Cincinnati, had three main tributaries: the Ancient Manchester, Licking, and Kentucky rivers. The preglacial Manchester headed on the west side of a divide in the present Ohio Valley near Manchester, Ohio, and along its course it joined the Ancient Licking, which then flowed through its present course but continued north through the valley of Mill Creek, at present Cincinnati. Well above their junction entered the preglacial Kentucky River, which, flowing north, had arrived at the present Ohio through approximately the bed it still occupies and continued northeast *up* the Ohio and Great Miami channels in reverse of the flow of those streams today. North of the future site of the city of Cincinnati the Manchester and Kentucky met to form the Ancient Cincinnati River, which then flowed due north to enter the Teays somewhere in the great basin near the Ohio-Indiana line which now gives rise to the Great Miami and to tributaries of the modern Wabash and Maumee rivers.

Farther west in the Ohio Valley, the preglacial rivers followed much more nearly the pattern seen today. Just where the preglacial Ohio had its headwaters is a matter of dispute but there was an ancient divide at Madison, Indiana, and this is a likely point at which rainfall may have begun to drain west rather than north; the preglacial Ohio ran west from the neighborhood of Madison, with only a few deviations, in a valley which approximates the present one.

The ancient Ohio continued west, receiving its southern tributary, the Green River, and a few miles beyond this junction it met the ancient Wabash, which still occupies its approximate preglacial valley.

After joining the Wabash, the ancient Ohio ran southwest in its present valley to a point near the modern town of Golconda, Illinois. There it received the Cumberland River, but, instead of following its present course, it turned due west and occupied a great depression across southern Illinois which, still visible, is called the Cache River Sag. Roughly parallel, five to fifteen miles south, and flowing toward the southwest, the preglacial Tennessee River flowed in what is approximately the course of the *present* Ohio west to the site of the city of Cairo, Illinois, where it joined the ancient Ohio. This junction was at about the place where the Mississippi and the Ohio now meet, but in preglacial times those two joined far to the south, for the Mississippi then occupied a channel west of its present course. Geologists believe that the ancient Ohio and Mississippi united at some point below the modern town of Helena, Arkansas, some two hundred miles south of the present junction.

Thus, the approximate line of the modern Ohio was sketched in through part of its upper course, was generally well defined below the present Falls, and was even indicated at a few points in its middle valley before the glaciers approached.

Sometime in comparatively recent geological history—possibly a couple of hundred thousand years ago, probably much later— winters began to lengthen, average annual temperatures fell, and the icecap came creeping southward from the polar region. The glacial age lasted thousands of years. As temperatures fluctuated the ice advanced and retreated; there was not one invasion but many. At their maximum the glaciers drove deepest into the United States at the lower end of the Ohio Valley: they fell short of the river by twenty or thirty miles in southern Illinois, missed it farther in southwestern Indiana and southeastern Ohio, but actually reached and occasionally crossed the present line of the Ohio into Kentucky through a 50-mile stretch on either side of the present Indiana-Ohio state boundary.

Obviously the great masses of ice that ground down to, across, or very near the line of the present Ohio were bound to have a cataclysmic effect. They filled and blocked all rivers which ran north toward their margin; they pushed along incalculable quantities of gravel, rock, and soil to fill completely the valley of the ancient Teays River and they blocked the valleys of the Marietta and Cincinnati rivers. Since the farthest-south line of the glaciers approximated the line of the Ohio, which, as has been noted, had been already sketched in by preglacial drainage, there is no difficulty in visualizing the beginnings of the present river. The vast amount of water still draining north through the old northbound tributaries joined the rain water falling constantly and the water melting from the edge of the glacier to the north. This vast flood broke through old divides where necessary, followed established drainage where available, and the modern Ohio was the result.

By the time the cycle of cold weather was over and the icecap finally retreated forever—at least we hope it has—the present drainage was established. The Kanawha, Sandy, Kentucky, Green, Cumberland, and Tennessee flowed north into the new Ohio through their old courses and new, or partly new, rivers came down from the north to add their waters. The Ohio system as we know it was established.

Even after the final retreat of the glaciers the climate of the Ohio Valley must have left much to be desired. Thousands of years would pass before the icecap shrank to even approximately its present area and for centuries after it was established the Ohio must have been frozen over during a large part of the year. Long after this was no longer true, the northern tributaries certainly continued to contribute ice water even during the warmest summer months. The refrigerating effect of the slowly retreating ice chilled and fogged the air the year round—even though the climate was growing progressively warmer.

Warmest places near the Ohio were the spring-dotted valleys

back from the main channel. They were first to be frequented by
the reptiles and that comparative innovation upon the face of the
earth, the mammals, which wandered in from the warmer lands to
the south where they had spent the glacial age.

Such springs, their waters gathering underground, partook of
a bit of the heat beneath the earth's surface. When they happened
to emerge in high-walled valleys their warmth was nurtured somewhat by protection from the frigid north winds, and an unusual
accession of grasses and edible leaves resulted. Sometimes, too,
along the Ohio, the underground seepage that supplied the springs
flowed over salt-bearing rock or sea-water silt which had been
deposited by the Paleozoic seas. Thus the water acquired a salt
flavor which gave it a further attraction—especially to the mammals, which required additional minerals to supplant those easily
gained by their sea-reptile ancestors but which they, no longer
constantly immersed in salt water, had to seek out.

Gradually, and first in these warm-spring valleys, the Ohio
region became the haunt of a varied and exotic animal life, examples of which were preserved for the amazement of modern man
in that natural pickling vat, the Big Bone Lick, which lay a short
distance from the Ohio River in present north central Kentucky.
Mastodon, arctic elephant, peccary, giant elk—the bones of all those
and probably also of tapir and giant sloth were still strewn about
the lick in historic times. Animals and reptiles even earlier extinct
had also visited the place occasionally.

Primitive man followed fairly close after the receding glaciers
in this region—at least no one has as yet succeeded in proving that
he arrived earlier. Finding what shelter his necessarily rugged
physique required under rock shelves projecting from water-worn
cliffs he existed in a state possibly so benighted that he had not yet
even developed a fear of the possible return of the spirits of his
departed associates and hence did not hesitate to eat their lifeless
bodies when appetite and necessity suggested such action.

Unfortunately the evidence of first tenancy of more primitive
man in the Ohio Valley is much less substantial than is that even
of the long-extinct primitive vertebrates of hundreds of thousands of
years before. As far as is known, man arrived in America too late

for his frail remains to receive immortality through preservation in rock and his bones were too frail long to survive the elements around the salt licks. Not until he developed some hope of a life after death and began to try to secure his own remains against the ravages of time for possible use in another world does his way of life begin to make itself discernible.

*　　*　　*　　*　　*

There remains little doubt but that mankind first reached the Americas by way of passage from the extreme eastern tip of Asia across the frozen Bering Strait to Alaska. Animals had certainly crossed long before and perhaps men simply followed the herds of wild game that fled before them as they hunted with their increasingly efficient weapons.

At first (somewhere between ten and twenty thousand years ago) their crossings may have been only expeditions; they returned when they could do so safely, or they paused where they found life temporarily supportable along the route. Presently daring or necessity drove them farther on and they began to move down the valley of the Mackenzie River to the American plains region, through the more protected and hospitable corridor that lay between the Rocky Mountains and the coastal range or even along the Pacific coast, warmed then as now by friendly currents from the tropics. The early men were followed by others—perhaps in friendship; as likely in enmity—and there was an additional natural increase and a higher percentage of survival among the immigrant peoples as their progress southward brought them to unexploited lands and herds of game in a climate of greater felicity. The settlement of America by modern men was under way.

That is only a surmise as to what may have been; actually, of course, the populating of the wondrously vast Americas was no such simple matter. Crossings from Asia continued at intervals during thousands of years by peoples already widely divergent— who continued to change for better or worse in the new land.

Artistic and mechanical improvements may have been developed and lost or forgotten among scattered bands, leaving no trace of their passing. The products of a highly developed skill in woodworking, for instance, could leave no trace after ten thousand years, should their makers be wiped out by some more numerous but inferior people who used only the crudest of stone implements —while the stone artifacts of the conquerors would survive to give evidence of their lesser culture.

Certainly the people of different waves of emigration from Asia varied greatly—there is no other way to account for the great variety in size, skull measurements, and facial characteristics of ancient burials discovered at different levels and in different local- ities, and in the great diversity in appearance, character, and in- tellectual capacity among the tribes of American Indians that continued into historic times.

The more primitive, less competent, peoples were probably earliest to come—or be driven—across to America. Some of the first of them may have been of longheaded stock; some, even more primitive, may have been short in stature with something like the prominent cheekbones, broad face, heavy brows, and flat noses that survive in the Australian Bushmen and the "Hairy Ainu" aborigines of Japan; others had round heads, still others low fore- heads but rather long skulls. Of course there were mixtures and permutations of these strains before, and certainly to a great extent after, the migration to America.

Combinations of these stocks could easily produce all types of American Indians of historic times; especially those with narrow heads, highbridged noses, black hair, and eyes usually brown in color but slightly Mongoloid in shape, which seem to have been typical of the more advanced tribes at the time of their discovery by Europeans. Those primitive Mongolians who, as Eskimos, stayed in the arctic to build a peculiar economy of their own, probably filtered in after the greater migrations had moved on south in America leaving a cold, vacant, and inhospitable wasteland behind them.

Since these various people who crossed the Bering Strait made up the ancestral stock of the original Americans, they must have carried within themselves alike the seeds of the artistic genius of Aztecs, Mayas, and Incas and the depravity of the grub-eating

Digger Indians of the West. Dreamy speculation notwithstanding, there could have been few contributions to the race from other sources before the visits of white Norsemen after A.D. 1000—and these few only occasional canoeloads of unfortunates blown to the west coast from the islands of the Pacific.

Whatever happened in those time-obscured days when men were moving into the Americas from the extremity of upper Asia, however the succeeding waves may have varied in appearance, whatever their customs, their most pressing motives, their hopes may have been, their advance guard certainly spread rapidly through the New World. By 5000 B.C. men were probably living on the Atlantic and Pacific shores of North and South America and scattered throughout the interior of both continents in much the same average state of cultural advancement as were their fellow beings in Europe and Asia.

Those men already mentioned as probably the first to reach the Ohio Valley were doubtless either representatives of one of the earlier emigrations from Asia or members of a later wave upon whom circumstance in the New World had brought degeneration. Although they may have depended upon caves or natural rock shelters for what protection they had from the elements, they must not be confused with those cave-dwelling American Indians who left their cultural evidence in the Southwest. These new arrivals on the Ohio scarcely "lived" at all by comparison—rather, they huddled indiscriminately together under any shelter which deflected rain and existed from day to day as fortune saw fit to preserve them. So scanty were their achievements in the working of tools or weapons in any material which could survive to tell their story that their presence is manifest mainly by occasional bits of charred bones embedded in layers of the broken shells and fire-blackened earth which mark their gathering places.

They must have been few at best; straggling family groups, probably, attached to each other only by common necessity, well able to transport their worldly possessions on their backs, and eking out an animal existence motivated mainly by hunger and regulated by a common fear of destruction. By exactly what route they arrived will probably never be known but some of them probably survived to develop an identifiable culture in the valley. Called the Indian Knoll Culture, it is the earliest upon the Ohio River to be dignified by a name of its own and to have a probable time of existence assigned to it.

The people who followed its practices in the Ohio River region seem to have lived in greatest numbers upon the Kentucky side of the river with outposts in southern Illinois, Indiana, and Ohio. They were quite short in stature—the men averaged 5 feet 4 inches, the women 5 feet ½ inch—with medium long heads. Particularly fond of shellfish (or so unproficient at hunting livelier quarry that they had to depend upon them for their chief supply of food), they strewed shells around their crude pole and brush shelters until hummocks were formed upon which they continued to live, as the surface rose, until their villages occupied well-defined knolls. They, too, sometimes lived in rock shelters or caves—whether only in time of danger or from choice cannot be determined—and their existence must have been only a little less precarious than had been that of their predecessors.

By the time of their passing another influx of settlers had arrived on the Ohio—this time of a culture apparently motivated by aims almost directly opposite to those of the primitive rock-shelter people and considerably different from those of the Indian Knoll Culture. Where the cave dwellers had sought to exist by stealth, by concealing themselves and their possessions, by a protean blending with the natural background, these new people made every effort to decorate their persons, to call attention to their communal sites, and to preserve their own remains after death. They succeeded to an extent which has intrigued and amazed all those who have visited the sites of their activities, from the first ignorant backwoodsmen to the most eminent men of science. Few today are not familiar with at least some of the monuments of the so-called mound builders of the Ohio. The most spectacular of the thousands of mounds in the valley (built at different periods, possibly hundreds of years apart) are probably the Great Serpent Mound in Adams County, Fort Ancient in Warren County, Ohio, and the Grave Creek Mound in Marshall County, West Virginia.

The Serpent Mound (an egg and serpent, actually; that combination of symbols which appears in the primitive religions of most races) is of the effigy type, examples of which were also constructed in the Midwest in the image of birds, men, bear, foxes, turtles, and lizards. The Serpent Mound has been thus described:

The Serpent, beginning with its tip end, starts in a triple coil of the tail on the most marked elevation of the ridge and extends along down the lowering crest . . . curving gracefully to left and right and swerving deftly over a depression in the center of his path and winding in easy and natural convolutions down in the narrowing ledge with head and neck stretched serpentlike and pointed to the west; the head is apparently turned upon its right side with the great mouth wide open, the extremities of the jaws, the upper or northern one being the longer, united by a concave bank immediately in front of which is a large oval or egg-shaped hollow eighty-six feet long and thirty feet wide at its greatest inside transverse, formed by the artificial embankment from two to three feet high, and about twenty feet wide at its base. The head of the serpent across the point of union of the jaws is thirty feet wide, the jaws and connecting crescent are five feet high. The entire length of the serpent, following the convolutions is thirteen hundred and thirty-five feet. Its width at the largest portion of the body is twenty feet. Here the height is from three to four feet, which increases towards the center of the body to a height of five or six feet . . .

Fort Ancient, located on the Little Miami River west of Wilmington, Ohio, and evidently intended by its builders as a defensive work, is an area of about 100 acres enclosed in earthen walls still, after centuries of erosion, from 6 to 20 feet high. The walls supplement the natural defensive qualities of the irregular plateau on which they are built and were themselves topped by palisades of stakes during the time the fort was occupied.

The Grave Creek Mound was constructed to contain the burials of either several people of importance or of an important individual and his connections. A cone seventy feet high, it is typical of hundreds throughout the valley. Many are smaller, and the Miamisburg Mound, south of Dayton, Ohio, is of similar size.

Works of these monumental proportions would be sufficient, alone, to distinguish the mound-building peoples but they had far more interesting accomplishments as well.

Modern archaeologists place the appearance of the first of those who constructed burial mounds, the Adena Culture, on the Ohio before A.D. 900. The Adena people were the first to build mounds according to preconceived design as a place in which to bury their dead; they were also first to supply those burials with an extensive stock of tools, ornaments and eating utensils for use in a future life—and thus all unwittingly to furnish posterity a record of their own achievements as craftsmen.

Most skeletal remains show the Adena people as of medium stature with large heads which were frequently artificially flattened from front to back after the fashion of the Aztecs of historic times. Their houses were round, with walls framed up of poles and closed with a basketwork of woven twigs. The walls were made to slope outward from base to eaves and the houses were probably roofed with bark. The Adena people grew corn (remember that corn was an American innovation—along with tobacco, long-fiber cotton, tomatoes, potatoes, and rubber) and they supplemented their gardening by hunting and fishing with efficient weapons made of stone, shell, bone, and wood.

They used raw copper to make ornaments and they carved shapely pipes and decorative objects from stone. Since the conical mounds they constructed were built up by dirt carried in baskets which held only a bushel or so at most, the vast labor involved indicates that their builders must have achieved a high degree of civil organization.

Remarkable as was the Adena Culture, however, it would presently be surpassed. About the time the Norse were exploring the fogbound North Atlantic around Greenland, a new people penetrated the Ohio region bringing a new way of life and introduced it by either force or persuasion in the valley. Theirs is called the Hopewell Culture.

Hopewellian mounds, for defense, for burials, for ceremonial purposes, with traces of the foundations of their wooden houses, public buildings and stockades and the mortal remains of the people themselves who partook of the culture abound upon the northern tributaries of the Ohio—with weapons, utensils, jewelry, pipes, statuettes, and even parts of their clothing preserved. Individuals differed greatly—as do the citizens of the same country today—but they appear to have been average in height. They may not have all employed the same language and they certainly differed from place to place in minor points of custom and ceremony, but they made use of the same materials, they exchanged innovations and improvements, they borrowed designs for decoration of utensils, apparel, and mounds, and they probably worshiped the same gods, spirits, or fetishes. In their highest development they achieved a semicivilization superior to any north of the Pueblos; they were, at their best, the most artistic, the best workmen, had the highest degree of specialization in manufacturing, the widest trade, and probably the most complete social organization seen on the river until the third quarter of the eighteenth century.

The people of the Hopewell Culture apparently devoted much time to the construction of mounds. They built Fort Ancient and also, at dozens of points in the Ohio Valley, those remarkable mounds in the form of perfect circles, squares, rectangles, rhomboids, and other geometric figures even more difficult to lay out and construct.

Hopewell villages were usually located in the "second bottom" fields along streams tributary to the larger rivers. The most remarkable of them were along the northern tributaries of the Ohio River and almost invariably these same spots were chosen for townsites by the whites. The Hopewell people were farmers, hunters, and fishermen and manufacturers of weapons, ornaments, and household implements. Their houses were probably pole frames with bark, wickerwork, or skin sides and roofs, but few of their village sites have been explored and much less is known of their building than of their other skills.

Hopewell mounds were certainly spectacular, but it was in the genius of their artists and artisans that they excelled all others of the mound-building peoples and equaled any race in a similar stage of development. Their probable appearance, dress, and manner of living are worth closer inspection and, thanks to more than a century of tireless work on the part of archaeologists, amateur and professional, such an inspection is possible.

Now, among men of science, scholarly archaeologists are most

reticent, most fearful of misapprehension, most given to understatement and to qualifying adjectives.

Their caution cannot be attributed to lack of imagination, for,
persuaded to speak on subjects outside their scholarly field, they
show an ample supply of that commodity: most likely it springs
from some of the glaring and embarrassing errors made by their
fellows of two or three generations ago. At any rate it is at present
their most obvious characteristic, particularly demonstrated in their
cautious treatment of the mound builders of the Ohio Valley.

Listening to the archaeologist's carefully restrained discussion
of these interesting peoples—especially the Hopewellian—one is
tempted to interrupt: "But this culture is remarkable! A talented
people, a people with a vast interest in mechanics and art, with
preparation for a life hereafter always first in their thoughts; a
trading and traveling, peaceable and well-governed people! How
could they have lost so much of this culture, these arts, this way of
life and degenerated to become only the Indians of history within
three centuries or so? You know they were different; your very
anxiety to gather every shred of information about them proves
you think them to have been something very special!"

The archaeologist? He purses his mouth, restrains his burning
enthusiasm, and replies, "There is as yet no proof that they were
other than ancestors of historic Indians; they may have been building mounds of sorts even down to the Iroquois invasion of the Ohio
Valley around the year 1650."

And there you are!

But we are not archaeologists; we may draw what conclusions
we please. (And the obvious conclusion is one which very probably
the archaeologist himself will applaud, in his innermost heart,
though he must frown upon it publicly in order to maintain his
scholarly escutcheon unsmirched.)

Let us look at the Hopewell people:

If a man, dressed and accoutered as the archaeologists of the
Ohio Valley have proved indisputably the Hopewellian mound
builder might have been at the time of his highest development,
should appear today he might be thought queer but his costume
could not be considered either cheap or slovenly. He might be sus-

pected as possibly a fugitive from a lunatic asylum, or believed to be on his way to a lodge meeting—but he would be immediately recognized by all as a prosperous lunatic or the high potentate of his strange lodge.

For, assuming that he was a leader among his people, he could well appear thus, as evidence on display in several great museums proves beyond a doubt:

He is almost 5 feet 6 inches in height, broad-shouldered, rather long of leg, and large boned. His forehead has been flattened slightly and caused to slant in a flat plane back from brows to above the hairline. (This characteristic is artificial, having been carefully induced by his mother while he was a baby.) His eyebrows themselves are upon a ridge, strong, but not anthropoidal, his nose is high bridged and narrow, and his jaw is fairly prominent. His eyes are slightly epicanthic, although far less so than even the highest classes of modern Orientals. His hair is straight and black but through the haze of intervening centuries we cannot quite identify the colors of his skin and eyes; possibly the former is a yellowish tan and the latter a very dark brown.

His hair is gathered upon the back of his neck in a sort of cord snood, with one plait, laid upon a strip of colored leather or painted cloth, falling down his back. A regular and gracefully curved swirl is tattooed upon each cheek and there are designs upon the backs of his hands and his shins. The swirls are reminiscent, though less elaborate, of those employed by primitive New Zealanders, and similar ornamental designs on various parts of his costume have a suggestion of the Oriental symbol "yang and yin."

It is in his costume and jewelry—and "jewelry" even in its strictest connotation is the correct word—that the gentleman is most remarkable.

Upon the crown of his head rests a shallow wooden skullcap, custom shaped to fit exactly. To it is fastened a complicated crest which consists of a strip of beaten copper projecting over his forehead as a cap bill and continued back to follow the line of his skull to its termination at the nape of his neck. The edges of this helmet crest are finished in a cutout scalloped design with regular projecting lobes. From the top of the crest a pair of double copper wings, shaped like those of a dragonfly and each about six inches long, extend to right and left. The body of the crest is studded with regular triangles of mica and upon the wings are mounted fourteen large fresh-water pearls, each more than one-third of an inch in diameter. From the back of the wooden skullcap and beneath the copper crest, a netlike veil falls over his shoulders and a short way down his back. The veil is decorated with a passementerie of pearls, the claws of young bears, and brilliant feathers of small birds, with the preserved head of a hawk sewed on as centerpiece of the design.

The rims of his ear lobes have been slit and he wears inserted therein a pair of copper ear spools covered with beaten silver foil of almost the exact size and shape of that modern toy, the yo-yo.

Highest around his neck he wears, choker fashion, half a dozen strings of matched fresh-water pearls. Below them is suspended a string of bangles, made from the canine teeth of young bears, each of which has been sawed into sections and restrung so that they do not quite meet at the places of sawing and make a slight rattling sound as the wearer moves. In the largest section of each tooth is set an enormous pearl.

His single garment is not of tanned leather—as would be that of an Indian of historic times—but of a linenlike woven cloth which is painted in an allover design of interlocking whorls in tan and green, suggesting the convolutions of serpents and reminiscent also of his tattooing. The garment is cut in the style of the Mexican huipil, or poncho—a 6-foot length of cloth, two feet wide, with a transverse slit midway of its length to admit his head and throat. Half the length of the cloth falls before and half behind, the edges pinned together below his armpits by polished copper skewers with decorative heads.

Across his breast, suspended from thongs around his neck, hangs a rectangular copper plate, shaped in decorative scallops at the ends. Each corner has a cutout design—the shape of a comma or the half of the yang and yin symbol—the plate itself being expertly worked and polished. Copper medallions are sewed to the shirt in pairs; one pair in the comma pattern but elongated this time, so that each resembles a modern curved-stem, round-bowled tobacco pipe; a pair of swastikas; a pair of wheellike medallions containing a cutout design marked by four smaller circles, a central cross, and four partial spokes.

The huipil-like shirt is gathered in at the waist by a leather belt. To this belt are attached a copper-headed hatchet, a copper-bladed knife, and a large salt-water conch shell, its inner whorl ground away to make it a serviceable cup. What may be a leather bag also is attached—containing a small terra-cotta figurine, a shell comb, a set of tubular copper whistles fastened together pipes-of-Pan fashion, a cake of red clay, tobacco, some pearls and copper buttons for use as currency, and perhaps a piece of rare meteoric iron, sharpened to a chisel edge.

The shirt ends above his knees, but below, front and back, fall the ends of a strip of rougher, burlaplike cloth which he has pulled between his legs and through a cord around his waist next the skin to form a breechclout, his only undergarment.

His feet are encased in moccasins similar to those worn by some historic Indians. Around his ankles and his upper arms are neat bracelets of coiled copper rods and on his fingers are copper rings.

Over one shoulder and down his back falls a brilliantly colored feather cloak; whether it is actually made of feathered skins of birds, sewn together, or only an animal skin or woven fabric embroidered with feathers cannot be distinguished; at any rate it presents the appearance of being covered entirely with curled feathers (again that sworl-comma motif!) each about two inches in length.

In one hand our mound builder holds a slender ceremonial staff, decorated with copper bands and tufts of feathers and surmounted by a figure cut from sheet mica. The figure is about sixteen inches long and eight inches high, a dragonfly with the head of a man represented as wearing a headdress and executed in the pictorial style familiar in Aztec and Mayan art.

In his other hand he holds what is probably his most prized possession, a beautifully polished green steatite pipe, the bowl of which is a hollow drilled in the back of a perfect representation of a frog, carved in the round, seated upon a platform which is slightly curved upward longitudinally: its front (that is, the end below the frog's head) is pierced to form the mouthpiece.

If the mound-building gentleman's wife is present she is probably attired in a knee-length wrap-around skirt of cloth painted in curvilinear designs in tan, black, and a rust-red tone and pinned with decorative copper skewers. Around her shoulders and upper body she clasps a fur cloak—or one of featherwork if the family is sufficiently opulent to afford two such garments—and her moccasins are similar to her husband's. She wears no headdress, but locks of her hair, pulled through polished copper tubes, fall in front of her ears and her back hair is gathered under a carved tortoise-shell comb and tied with strings of pearls.

Her ornaments are even more lavish than her husband's—similar ear spools, armlets, and rings are augmented by anklets of strung copper beads and her neck and bosom are almost concealed by tier upon tier of strings of graduated pearls, three or four hundred of them in all, the largest of which appear to be almost a half inch in diameter. Her chin, forehead, and hands are tattooed and her cheeks decorated with spots of vermilion paint.

The lady's taste is a bit on the exotic side, perhaps, but she is by no means unprepossessing and her costume and jewels would attract their share of attention even among the most expensively and fashionably dressed ladies of today.

Taking a last look at our mound builders returned, we note some interesting similarities—the sworl-pattern tattooing and feather cape to those of the Polynesian; the commalike yang and yin symbol with the modern Oriental. Both of these are scientifically permissible—after all, Polynesians and Indians are both generally agreed to be of Asiatic origin. But if one is to maintain the scholarly approach, one must not be reminded of the Norsemen by the winged metal helmet or the crested ceremonial staff; for to suggest that some Norse fashions might have penetrated to the Hopewell people before the end of their glory would be a scientific heresy of the most damning.

And what happened to the proponents of this remarkable Hopewell culture? Were they exterminated by some plague which preceded those lavish ailments brought by the white men from

Europe? Were they conquered by some race of lesser intellectual stature but more virility? Did they move off into some as yet undiscovered Shangri-la in the arctic?

No, they are thought to have come to a much less romantic end. Archaeologists believe they lived on, their culture degenerating, and that they were the ancestors of the Shawnees, Cherokees, and perhaps some related tribes of comparatively high intellectual attainment among American Indians at the time of the invasion by whites!

True it is that there is a similarity in stature and skeletal measurements. It is conceivable that the race, degenerating, might have turned from its ancient gods and adopted a burial ritual less laborious and expensive; that the incursion of enemies, or a temporary change in climate, might have made a more nomadic life necessary; might have brought about a greater dependence upon hunting, a lesser on agriculture—but could *all* those Hopewell arts have been forgotten between the supposed end of their era, 1300-1400, and the first contact of the Cherokees and Shawnees with whites three hundred years or so later? Could their secrets possibly have been forgotten so completely?

The Shawnees and Cherokees of history loved personal adornment as well as any Indian; the leisure, especially of their men, was largely occupied in self-adornment and the toilet. But *they* made beads only of mussel *shells,* not of fresh-water pearls (they possessed no drills small enough to pierce pearls for stringing anyway). They wove nothing finer than the coarsest of burlaplike cloth—for they had nothing resembling the thread from which the Hopewell fabrics were woven—and they made little cloth of any quality, preferring deerskin. They used no copper for utensils and what few copper gorgets or medallions they shaped were executed by pounding raw ore with stones into a flat slab, undignified by much plan, design, or finish. The working of gold or silver into foil was far beyond their comprehension, ability, or imagination. Apparently they made no use of iron as a tip for spear or arrow, knife, chisel, or awl. Feathers they used, but only quills or tufts of down as headdress or trimming on pipes and weapons.

Shawnees and Cherokees could only chip their stone—of *carv-*

ing it into gorgets or effigies they knew nothing; in execution of balanced and repeated designs upon ivory, bone, or skins they were most inept; their mounting and attachment of minerals or other decorative materials upon garments or weapons was of the crudest, and their pottery was sorry stuff. In only one art, the working of leather, did the modern Indians apparently equal their predecessors.

Could a people, so loving adornment as did these, have forgotten those arts, those materials, those glories, so easily?

* * * * *

After the culture had disappeared, some Hopewell mounds seem by rather sketchy evidence to have been used by a more primitive people. This trespassing evidently took place around 1400 and is called the Intrusive Mound Culture. It was in every way the manifestation of an inferior people who made use of existing mounds only as a place in which to inter their own dead in shallow graves. But its proponents could have been more warlike if less intellectually and socially advanced than the Hopewellians and may, indeed, have pushed down from the northern wilderness to occupy some Hopewell moundsites by force. Perhaps these intruders were forebears of Indians known at the head of the Ohio and around the Great Lakes two or three centuries later and simply moved on north and east after a short stay in the Hopewell country. In any case, by the beginning of the fifteenth century the central Ohio Valley (notably a large area extending north from the Ohio on both sides of the Great Miami valley) was dominated by a culture called the Fort Ancient, and the lower Ohio, through southwestern Indiana, southern Illinois, and western Kentucky, was host to the people of the Middle Mississippi Culture. Both these cultures survived into historic times, though the survival of the latter presents much the clearer picture.

The Fort Ancient people did not build the great work in Ohio that bears that name, although one of their chief characteristics was the building of similar hilltop fortifications surrounding their villages, mounds, and burial vaults. One of their principal sites is at Charleston, Indiana, overlooking the Ohio, another is at

Merom, Indiana, overlooking the Wabash, and there are hundreds besides, usually on hilltops, the approaches to which are guarded by rivers and gullies as well as the artificial defenses the people reared. The number of these sites, and the fact that many give evidence of extensive occupation, indicates that the Fort Ancient people were numerous and powerful.

Perhaps they came up from the region in Tennessee and Kentucky surrounding the headwaters of the Cumberland River, where similar works are found; very likely they were related to the ancient Iroquois, who fathered the Iroquois of historic times as well as the Cherokee, their relatives, and other peoples familiar to whites.

Not so much is known of the lives or the arts and crafts of the Fort Ancient people as of many of their predecessors. Their implements seem to have been chipped and ground stone; as in other mound cultures, they sometimes buried their dead underground in vaults lined neatly with flat slabs of stone and they employed loose stone as well as earth in building their defensive walls—sometimes facing these walls with the same sort of stone they used in lining graves. Curious depressions in the surface of the level ground within their earthworks have led some investigators to surmise that their houses may have been constructed partially belowground—perhaps after the fashion of those of the Mandans and other later Indians.

The Fort Ancient folk were probably once powerful; from the evidence of their fortifications one can guess that they were either a warlike people or were living in constant fear of aggression, but no evidence except fragments of their well-tempered pottery has yet been found which indicates a degree of artistic or mechanical development comparable to that of either the preceding Hopewell or their contemporary neighbors of the Middle Mississippi Culture.

The Middle Mississippians seem to have stayed upon the Ohio almost—but not quite—to the time when the French explorers first penetrated to the river at and below the mouth of the Wabash. They must have been gone when the French arrived because there is no mention of them by French missionaries and explorers, who seemed much interested in people of the same culture whom they visited on the lower Mississippi. The same explorers would certainly have mentioned similar communities had they still existed upon the Ohio.

People of the Middle Mississippi Culture, alone of the moundbuilders, survived almost unchanged to be examined (and to be finally exterminated) by Europeans. Undisturbed at the gigantic Cahokia Mound and at the Angel site, near present Evansville, Indiana, they were carrying on their arts and crafts, building their palisades of posts interwoven with vines, with bastions every 125 feet or so apart (a good bow shot), perfecting their truncated pyramidal mounds, so like the stone pyramids of the Mayas and Aztecs, even as the Spaniards were destroying the latter people.

The results of archaeological surveys upon their Ohio River village sites, and the accounts of them by the Spanish and French who actually visited them in the South, give a particularly clear and interesting picture of them.

The people of the Middle Mississippi Culture apparently excelled all their predecessors north of Mexico in the perfection of design and building to which they brought their palisades, houses, temples, and mounds. The artistry of design and execution that they exhibited in etching portraits, designs, and religious symbols on slate, wood, stone, and mother-of-pearl was comparable to that achieved in Central America. On the Mississippi River their religious worship of the sun and the serpent was the guiding factor in their daily lives—probably the villages on the Ohio had the same or similar gods—but after the main body of the proponents of this culture had made contact with de Soto's brutal followers and the other Spaniards who soon penetrated from the Gulf Coast the religion developed, with tragic and well-justified foreboding, a cult which included the skull as a principal symbol and the worship and welcoming of death. By this religious temporization one hopes they were able to soften the blow of their almost complete extermination within a few decades of the permanent settlement of the lower Mississippi by the French.

One of those tribes whom the early French—mostly traveling missionaries—describe as practicing the typical Middle Mississippi

Culture at the end of the seventeenth century were the Taensas: "These people are very mild, give a warm welcome and have a great esteem for the French; they are sedentary and cultivate the earth. . . . They have rather fine temples [this from a Christian priest!] the walls of which are of mats. That of the Taensas has walls seven or eight feet thick on account of the great number of mats one on another. They regard the Serpent as one of their divinities . . . when they receive anything it is with a kind of veneration which they turn towards this temple . . ."

The Tonicas were a similar and related tribe who were "dispersed in little villages. . . . Their houses are made of palisades and earth, and are very large; they make fire in them only twice a day, and do their cooking outside in earthern pots . . . very peaceable people, well disposed . . . [and, in 1699] They are dying in great numbers. . . . They have a temple on a little hill; we went there to see it: there are earthen figures which are their manitous . . ."

Both these tribes were akin, though but poor relations, to the mighty Natchez whose civilization was likewise built upon the worship of the sun, serpents, and other typical objects; who practiced human sacrifice (voluntary victims killed upon the death of a chief) and who appear by the evidence of contemporary description to have brought the Middle Mississippi Culture, which we have noted on the Ohio, to its highest development. Powerful as they were they feared the Spanish after their first meeting; intelligent as they were they made the error of welcoming the French.

There was to be no question as to how the arts and crafts of *these* people happened to be lost quickly and irrevocably! By the hundreds their artists and craftsmen went to meet the death their new religion taught them to welcome—burned in their own stockades, dismembered in their temples, shot as they fled or, if preserved, sold into slavery in the Spanish West Indies for a more lingering and only slightly less painful death. Those few who escaped to join the nomad tribes of Indians had little heart for continuing their way of life—although their influence may, at that, have had a bearing upon the rapid progress toward civilization made within the next hundred and fifty years by the Cherokees, Chickasaws, and other tribes who took them in!

The Middle Mississippi Culture upon the Ohio (recall that it is believed to have flourished there through the fifteenth, sixteenth, and probably into the seventeenth century) was marked most notably in its village sites by the truncated pyramid mounds, the platforms of which were approached by long ramps of an easier pitch; by the wall surrounding the houses and public buildings, which was constructed of stakes interwoven with saplings; by rectangular houses and temples with plastered side walls, often painted in red and black and topped with steep-pitched thatched roofs (quite identical in shape with a form of building typical of the Mayan civilization of Central America), their temples (also like those of the Mayas) occupying the top of the pyramidal mounds.

The Middle Mississippian folk raised corn, squash, tobacco, and possibly beans, hemp, and flax, in fields outside their towns— much in the fashion of the feudal villages of Europe—and they used spears, bows and arrows, hand axes, nets, and traps in hunting in the neighboring woods and fishing in the streams. They made excellent pottery, superior to any of their predecessors including the Hopewell people, which they sometimes modeled in the shape of animals and men and usually decorated with artistic designs in color. They made baskets, wove cloth, and worked leather—employing all these techniques in manufacturing their sandals and clothing—and they, also, made robes of featherwork. They used shell extensively, as well as copper, bone, wood, and stone, for making jewelry and decorative objects, and their masks and amulets employing engraved designs were distinguishing marks of their culture.

The bones in their burials sometimes showed evidence of syphilis but there is no reason for speculation as to where they could have acquired the infection—the men of de Soto's expedition had been in Europe after the return of Columbus and had themselves visited the West Indies before they came to the country of the Middle Mississippi Culture.

But, as has been said, these mound-building people had left the comparative safety of the lower Ohio Valley before 1700, apparently of their own volition in the absence of any evidence to the contrary, and returned south to rejoin their relatives and to meet extermination. The Ohio was thereafter only the home, where it was occupied at all, of typical forest-dwelling, seminomadic tribes of Indians who existed on the results of their hunting and fishing supplemented by corn and squash which they produced in their rather haphazard summer gardening.

Even before the last of the mound builders had been exterminated upon the lower Mississippi, friction began between these tribes of Indians who had moved down from the north into the Ohio Valley. The Iroquois had driven westward from their home in the northern part of present New York State and had cleared the upper Ohio of other tribes; the present Kentucky shore, except for its western end, was already established as a neutral hunting ground. The north bank of the river, through present Indiana and Illinois, was sometimes home to a few Piankishaws (or other related tribes later to become parts of the Miami Confederacy) and possibly some of the Illini and Kaskaskias, but the river was, in the main, almost devoid of residents strong enough to defend themselves. The way was open to exploitation of the Ohio by whites.

Chapter Three

SOMEONE FINDS THE OHIO—
AND FRANCE AND ENGLAND
CLAIM IT

The Ohio River is unusual in that its beginning can be exactly defined; identifying its discoverer is no such simple matter.

The most famous character for whom the distinction was claimed—there is no positive proof that he ever claimed it for himself—was the tragically unfortunate René Robert Cavelier, Sieur de La Salle; if he was actually the discoverer, the distinction was clouded by one of the evil dispensations that plagued so many of his ventures.

By 1650 the French to the north, the Dutch to the east, and British to the southeast had all heard Indian-borne rumors of the great southwest-flowing river which eventually reached salt water. Being still blinded to the wealth of America by their dreams of the Indies and the Orient, they naturally hoped that the salt water would be that of the Pacific Ocean.

Certainly the French Jesuits knew such a river existed, for Father Lalemant, writing from Montreal in 1662, reported that a band of four hundred Iroquois from his neighborhood were "proceeding rather westerly than southerly . . . from here in pursuit of a nation whose only offence consists in its not being Iroquois. . . . Furthermore if we believe our Iroquois who have returned thence, and the slaves whom they have brought thence, it is a country which enjoys none of the severity of our winter, but enjoys a climate which is always temperate. . . . Their villages are situated along a Beautiful River, which serves to carry people down to the Great Lake (for so they call the sea), where they trade with Europeans who pray as we do. From their account we suppose these Europeans to be Spaniards." Father Lalemant, being a bit shy of geographical knowledge, like his contemporaries, did not stop to think that a westward-flowing river might turn south and join the waters of the well-known Atlantic; he also probably thought the "sea" was the Pacific.

In 1663 that gaudy monarch, Louis XIV, assumed direct control of the hitherto more or less independent French settlements along the St. Lawrence, and thus gave some official substance to Samuel de Champlain's dreams of a French empire in the New World. The system of control over the fur trade was strengthened, encouragement was offered to gentlemen-adventurers in the St. Lawrence valley and French interest in the New World was renewed. The era of the exploring French gentleman and his ever-present companion, the missionary father of the Roman Catholic Church, had begun.

They made a competent team, the worldly gentleman-adventurer and the reverend father, and aided by Indian guides and canoeists, or the no less able native voyageurs of New France, they mapped and made first settlement on the Great Lakes, the Mississippi, upper and lower, and the lakes and rivers of the eastern half of present Canada.

Unlike the representatives of the same professions who had gone out from Spain to South and Central America a century or so before, these early French seemed willing to show the Indian the cross without crucifying him; to give some boot of reasonable value when exchanging the pox, large or small, for his furs, and to take his lands without taking his life—except in isolated cases like that of the Natchez. The result of this singular behavior on the part of the French was that they gained, and kept, the Indian confidence for a half century or more.

One of the most famous of these Frenchmen was young René Robert Cavelier who gave up study for the Jesuit priesthood to come to New France in 1667. Notwithstanding centuries of memorial-occasion orators and Sunday newspaper feature writers "Cavelier" was his family name, not his title. His brother, the Abbé Jean Cavelier, was a priest of St. Sulpice already stationed in Montreal and his influence and the vast abundance of unoccupied lands available resulted in a concession, or grant of land, on the St. Lawence River for Robert. Thus the new arrival became René Robert Cavelier, Sieur de La Salle, and named his plantation Côte de Saint-Sulpice in honor of the saint who, through the agency of his brother, had thus favored him—though shortly the concession was being called "La Chine" by witty neighbors because of its owner's avid interest in the possibility of finding, somewhere, a short route to that principal country of the Orient.

Perhaps La Salle knew the story that Lalement had reported six or seven years before and probably, in his conversations with visiting Indians, he heard many others to the same purpose. In the winter of 1668-69 he got Governor Frontenac's permission to explore in the west, sold or mortgaged his concession, and with the proceeds bought four large canoes and hired fourteen boatmen.

The Sulpician order had planned a missionary tour to the west at the same time and Governor Frontenac suggested that the missionaries and La Salle join forces for mutual aid and protection. La Salle agreed—with perhaps no great enthusiasm—and on July 6, 1669, he set out with the two priests, Galinée and Dollier, and the boatmen—a total of twenty-one men in seven canoes. Preceded by two canoes filled with Senecas who were returning south to their home, after having spent the winter near La Salle's plantation, the party moved up the St. Lawrence from Montreal.

An account of 'the first stages of this voyage may be drawn from a document purporting to be the journal kept by Galinée. This journal was discovered and first published in the eighteen-thirties in *Découvertes et Etablissements des Français dans L'Améri-*

que Septentrionale, compiled and edited by that distinguished though sometimes slightly biased historian of France in America, Pierre Margry. The original manuscript was later reported lost. This lost document, some maps which had obviously been tampered with (to be described later), and Margry's strong partisanship for La Salle as discoverer and first explorer of the upper Ohio combined to produce one of those fierce scholarly vendettas more common, happily, to the preceding century than to this. One party of competent historians supported La Salle; another, as competent, denied him. La Salle himself, as far as can be known from proved writing extant in his hand, made no statement. He, in his day, made not the smallest documentary contribution to the controversy that would begin two centuries later. Taking his course to be the wise one, neither shall we.

This, however, is supposed by various historians, to have been what happened:

La Salle, the Sulpicians, and their rather extensive retinue set out from the neighborhood of Montreal. In due time they passed into Lake Ontario. Galinée understood something of surveying and planned to make a map of the lakes and kept (or did not keep, depending upon whose opinion one takes) his journal. Supper was the one big meal of the voyageur; an occasion for songs, stories, and conversation. Naturally the conversation sometimes lingered on the subject of the southwest-flowing river that might lead to China, and Father Dollier conceived an ambition to go and preach along its banks, rather than upon the shores of the Great Lakes, as had been intended. If La Salle did anything to encourage this ambition it was probably with tongue in cheek; his strongest motive, then and later, was to play the discoverer, and to build up a great inland empire based upon trade rather than soul saving.

After thirty-six days (according to that ephemeral journal of Father Galinée) the travelers reached a Seneca town located a short distance south of Lake Ontario: they estimated it to be 300 miles from Montreal. They had expected to use the Jesuit missionary stationed at the Seneca town as an interpreter, but he had left the day before their arrival and the Senecas took advantage of his absence to torture an Indian prisoner they had brought in. La Salle and his associates were both disgusted and fearful for their own safety.

The Senecas told them—or so they gathered from the sketchy interpreting available—that the river that so interested La Salle lay to the south, about 360 miles overland, or about three days' portage from Lake Erie, should they wish to continue that far by water.

They soon decided on the latter course. Hugging the south shore of Lake Ontario, they passed the mouth of the Niagara River and reached the western point of the lake. Here they camped at an Indian village where La Salle suffered an attack of fever. They bought two Indian slaves, one for the priests, one for La Salle—and the slaves also claimed to know of the river for which La Salle inquired and said they could guide him to it in a matter of six weeks or so. On September 22, as soon as La Salle had recovered sufficiently, the party started overland across the Niagara peninsula.

Two days later, while stopping at an Ottawa village, they met Louis Jolliet, born and educated in New France, who was returning from a search for a copper mine on Lake Superior. He had found copper but, more important to the development of the west, he had made the acquaintance of a priest at the mission of Sault Ste. Marie who shared his interest in exploring the new country. The names of Louis Jolliet and Father Jacques Marquette will both be remembered long after Lake Superior copper deposits have been exhausted.

La Salle and the Sulpicians must have spent many fireside hours in questioning Jolliet; must have gained greatly in confidence as he described to them the routes and the best manner of travel and living. Happily for La Salle's supposed aims Jolliet had a special word for the priests; he had met a large band of Potawatomi on Lake Superior who had no spiritual shepherd and he commended these Indians strongly as a worthy congregation. Galinée and Dollier decided to go to them but La Salle said (still according to that journal) that the state of his health would not permit him to make such a voyage. On October 1, 1669, the priests embarked for the northwest.

There ends the part of Father Galinée's journal that has to do with the storied river to the south; probably it is truthful as far as it goes but the questioned activities of René Robert Cavelier, Sieur de La Salle, follow his parting from the priests. Did he and his men follow the Indian slave across Lake Erie, over the three days' portage, and down the Ohio? Even down the Ohio below the Falls—described by some of his partisans among historians as having been seen by him as a roaring cataract, presumably comparable to Niagara—and on to a great swamp, of interminable extent?

Perhaps La Salle started for the river, perhaps not; perhaps he found it. Or did his associates in France eight years later so report in order to bolster his reputation and further his hopes of completing greater projects in the west? Evidence that he made such an exploration stems mainly from the writings of Margry and consists of those two documents which do not bear close scrutiny and a reproduction of Jolliet's map which had certainly been tampered with—and in the time of Margry, rather than La Salle, at that! The last piece of proof submitted by Margry in support of La Salle's discovery is a document purporting to have been written after conversation with La Salle by the Abbé Arnand, presumably in Paris in 1678. The authenticity of this evidence is vouched for by La Salle's niece, Madeleine Cavelier, Dame Le Forestier, *in a letter dated January 21, 1746,* seventy years after Arnand had supposedly taken it down, and at a time when the heirs of the long-dead La Salle were trying hard to establish claim to the lands he had discovered in America. That document, at very least, should be questioned.

Shall we say tnat La Salle *could* have spent the winter of 1669-70 on the Ohio and there let the matter rest?

The British had begun to talk seriously of a search for the storied river as early as 1648 and between 1654 and 1664 one Abraham Woods, an English fur trader, claimed to have visited rivers later identified as Ohio tributaries. Woods's claims, like La Salle's, were delayed of publication and did not extend in any event to a view of the river itself. It is likely that at least by 1669 other British had actually seen the Ohio. Any ordinary season's trapping or trading trip from the upper Virginia settlements or even from New York might well have led to its banks, but unfortunately

British adventurers did not carry literate priests, obligated to report regularly to superiors, on their expeditions.

Many Europeans may have seen, even traveled, the Ohio before the date claimed for La Salle. Usually for sufficient reasons of one sort or another white colonists had certainly begun to wander inland from the settlements as soon as settlements were made. Those with the most pressing reasons for wandering far usually had equally good ones for not returning to tell of their experiences, and there was little to encourage the trapper or trader to report a waterway leading easily to profitable fur country for the benefit of his competitors.

Certainly the Ohio River was known, although not described, when the French laid formal claim to the American northwest in 1671. The Ohio's lower reaches began to appear on French maps soon after 1674, when Jolliet returned to Montreal to report the trip he and Father Marquette had made down the Mississippi. Very shortly these maps were showing the lower Ohio with fair accuracy (although the French were some years in deciding whether the river should be called La Belle or Rivière St. Jérome below its junction with the Wabash) but the upper Ohio remained a blank for decades to come. No less an authority than Henry Popple, in his vast "Map of the British Empire" of 1733 showed the river with considerable accuracy from its mouth to a point which would be thirty or forty miles from its beginning and there broke it off sharply in full flow, with a single chop of his engraving tool!

Whether or not La Salle explored the Ohio above the Falls in 1669, plenty of Frenchmen, including La Salle himself, were to see its mouth during the next twenty-five years.

Louis Jolliet and Father Jacques Marquette paddled their canoe past it between June 21 and July 10 on their trip down the Mississippi in 1673 and noted it not only then but upon their laborious upstream return. In 1682 La Salle passed the mouth between February 13 and 20 as he was paddled down the Mississippi to the Gulf, where, on April 9, he took possession of the country drained by the Father of Waters for France and named it Louisiana. In 1686 La Salle's lieutenant, Tonti, passed the mouth of the Ohio as he floated down the Mississippi to keep a rendezvous with his superior, who was then returning to the Gulf from a voyage to France. Even the guilty hurry of the seven of La Salle's followers, who murdered him in what is now Texas in 1687, could scarcely have been sufficient to keep them from noticing the mouth as they fled up the Mississippi to the Illinois country. There is no doubt that La Salle and his associates knew the *lower* Ohio well before La Salle's death, for the next year, 1688, the Marquis de Denonville, writing upon the subject of the French limits in North America in a memorial to the King, said "Sieur de la Salle having afterwards employed canoes for his trade . . . as he had already done for several years in the rivers Oyo, Ouabache, and others. . . ."

Even though the *upper* Ohio was unknown to the French in the last decade of the seventeenth century and a description of it had not reached the scholarly ear of Henry Popple in England forty years later, the river had certainly by then been traversed, end to end, by one rather prominent Dutchman of Albany, New York.

Arnout Viele (his name suffers those variations of spelling with which any delver into New York Dutch genealogy is familiar) had been employed by an Albany trader for duty as an interpreter among some of the tribes of the Iroquois League. Having had a misunderstanding with his employer, he set up in the Indian trade for himself and, taking a fancy to some Shawnees he met in his travels, decided to go to their country, which was at that time the land between the Cumberland, the Mississippi, and the south bank of the Ohio in present Kentucky.

With eleven white men and a few Mohawks and Shawnees, Viele left Albany in the fall of 1692. He crossed Pennsylvania by way of the Wyoming valley and the Susquehanna and the Allegheny rivers to the Ohio. The men then passed down the Ohio to, or very near to, its mouth, where they stayed until the summer of 1694.

The trading operations were presumably successful. They were at least satisfactory to the Shawnees, a large party of whom followed Viele back up the Ohio to settle near their old home above the Delaware Water Gap. In 1682 La Salle, then at Fort Crèvecœur on the Illinois River, had promoted an alliance between the Shawnees, Miamis, and Mascoutens which somewhat alleviated the fear

these Indians had of the Iroquois and eventually gave them courage to return to the upper Ohio in force.

Those Shawnees were an interesting people. Their complexion was yellowish, like that of their cousins the Cherokees, rather than the typical reddish brown of most Indians of historic times; a fact which hinted at some rather special racial strain. They, again like the Cherokees, seemed to possess rather unusual intellectual capacity. This enabled them to develop some outstanding leaders who eventually caused them to play an important though foredoomed part in the final effort to keep the Ohio for the Indian. They had wandered over much of America during the seventeenth and eighteenth centuries and in these wanderings (the course and sequence of which is disputed by historians and ethnologists) they had left their tribal name all over the eastern half of the United States.

Just where the Shawnees lived and when is the subject of a great body of disputatious scholarly writing which has been published—and an immensely greater body which never found itself immortalized in type. There is no need to cast a gantlet in these lists at present; let us only note that, apparently, these are some of the possibilities:

In 1614 all or part of the Shawnees seem to have been located on the banks of the Delaware River, near its mouth. They were, or appeared to the English colonists to be, an unwarlike race, pacific to a fault; the colonists therefore soon bent their attentions to driving them from their homeland.

Either this group was not *all* of the Shawnee people or else they divided upon leaving the beautiful Delaware, for the *Jesuit Relations* for 1647-1648 mentioned some of them as being then on the south shore of Lake Huron; Father Marquette remarked, as of 1673, that the Ohio "comes from the country to the east, inhabited by the people called Chauanons, in such numbers that they reckon twenty-three villages in one district, and fifteen in another . . ." and Nicholas Perrot, who was living with the Miamis and Foxes in 1664, stated that when the Iroquois were driven south of Montreal by the French they dispossessed the Shawnees of their country and the latter moved to the Carolinas, where they lived in his time.

The Shawnees may have divided after leaving the Delaware;

certainly they did divide after the Iroquois had driven them from the Ohio, for no entire tribe could have lived in all the localities in which the Shawnees were reliably reported during the next half century. Robert Morden's map of 1687 showed Shawnees in what is now North Carolina; part of them, at least, were reported on Mobile Bay before 1713; the Savannah River is called the "Sawano" River as late as 1739 and de l'Isle's map in his *Atlas Noveau* calls it the "Rivière des Chaonanons" and locates a "Chaonanon Village" upon it. (Could "Savannah" itself be an accidental and coincidental corruption of the French version of the tribal name?) And of course that deep-south Suwanee River is another of their locations.

In any event a sizeable body of Shawnees were south of the lower Ohio in 1692-1694, with a memory of their former homeland and in close enough contact with their southern relatives to invite them to come along after the leadership of Viele and La Salle gave them the opportunity to return.

Although no capital was made by the British of the voyage and sojourn of Viele's party it was well and widely known. In a letter dated August 30, 1699, Pierre Le Moyne d'Iberville, founder of the colony of Louisiana, who, of all people, had least reason to support any claim of exploration of the western country by British subjects, said:

"I am well aware that some men, twelve in number, and some Maheingans—started seven years ago from New York, in order to ascend the River Andaste Susquehanna, in the Province of Pennsylvania, as far as the River Ohio, which is said to join the Oubache, emptying together into the Mississippi. . . ."

Le Moyne's report of the correct number of white men is entirely too pat to refer to any other expedition. If the knowledge of this exploit had come even to him, the British in the eastern colonies were more marvelously obtuse than usual if they were not also aware of it.

In 1700 another party of English traders traversed a part of the Ohio—and approached it in such a way as to indicate an extensive knowledge of its southern as well as its northern tributaries. One Jean Coutoure had served La Salle and Tonti as commander of their post on the Arkansas River in the 1680's. After La Salle was killed and French trade became stagnant on the Mississippi, Coutoure made his way to the Carolina seaboard and succeeded in interesting a group of Charleston traders in undertaking a venture in the west. In 1700 he guided a party through the Cherokee country, down the Tennessee River to the Ohio, thence to the Mississippi and Arkansas, where they set up trade. This group must have continued to use the same route to and from their base at Charleston until the French government granted the lower Mississippi to Crozat in 1712 and trade by British subjects presumably was stopped in the area.

By 1700, then, the Ohio—*La Belle Rivière* of the French—was fairly well known in its lower reaches, say below the mouth of the Great Miami. Arnout Viele, the Albany Dutchman, had traded upon it; La Salle's canoes had traversed it, from the Wabash down, for "several years" before 1688; Coutoure had led a party to and down a part of it. The upper half of the river had been navigated twice by Viele's men, perhaps once in full and once in small part by La Salle's, and possibly by others. This upper half was known, surely, but it was still unmapped and thus far largely disregarded officially by both French and British.

*   *   *   *   *

The development of North America was considerably retarded during the century following 1650 because the attention of England and France was distracted continually by wars, rumors of war, preparations for wars, and recoveries from wars in Europe. In addition England was confused by the distractions incident to the conflicts aroused by the spread of Puritanism and the struggle against Stuart absolutism.

England was interested in her growing colonies in America, but only occasionally and somewhat absent-mindedly so. Samuel Pepys, for instance, anxious as he was to discover sources of better timbers, masts, knees, pitch, and cordage for the fleet of his Majesty James II, wasted only a comparative drop or two of ink in mention of America for every pint he expended on gossip of the court and in recording his bouts with the band-settler's lass.

French North America, before 1663, when it became a crown

colony, was under a casual government distracted not only by wars
but by complicated affaires d'amour which must have required all
the statesmanship that leaders of the country could muster. Actually
the French received less practical good from their interest in the
continent than did Britain: the latter had at least, besides the
colonies of which it was rather dimly aware, a useful haven in
North America from which her privateers and pirates could prey
profitably upon the Spanish commerce passing across the Atlantic
to the south.

The attitude of the French crown toward its American lands
at this time was that of the proprietor of a vast preserve dedicated
almost entirely to the production and export of furs. So long as
revenue from this commerce maintained a reasonably satisfactory
level there was no complaint; and because of this complacent attitude the French monarchs were, of course, mulcted of a sufficient
part of the income to ensure that it never arose above the "reasonably satisfactory" level.

The commerce and development of New France had been
chartered as a monopoly to various companies and individuals by
the crown. Licensed trading houses were established at Montreal,
Quebec, and Three Rivers through which all transactions in fur
trading were supposed to pass, but most officers at the scattered
military posts conducted a thriving illicit side line in the trade.
Bourgeoisie and voyageurs took to the woods and became coureurs
de bois or unlicensed traders and enterprising civilians grew suddenly rich and purchased seigneuries back on the St. Lawrence
around Montreal. The fur traders' rough-and-ready woods- and
watermen left a tradition as romantic as did the frontier hunters
and the cowboys of the United States.

A colony of the crown in which the crown took little interest,
improved only occasionally by competent individuals who found
themselves in power more or less by accident, it is not surprising
that France in Europe more or less ignored even such an impressive
stretch of country as that between the Great Lakes and the Ohio.
Had there been more men available of the caliber of La Salle and
Frontenac the story might have been different. But La Salle, whose
own personality—detached, dignified, rather given to dreaming in

the grand manner—rendered him perfectly understandable to the headmen of the Indians, was killed in 1687 by lesser men among his followers who could not appreciate his genius and therefore mistrusted him; Frontenac, who supported La Salle, and who had the wisdom to make peace with the Iroquois and the audacity to defy both the powerful Quebec traders and the Jesuits, died in 1698.

There were also other reasons. The French were in constant fear of the tribes of the Iroquois League to the south, and then there existed practical, protected waterways to the great fur country on the upper lakes either by way of the lakes themselves or by the Ottawa River and Lake Huron. After the Iroquois had driven the Shawnees and other Indians from the Ohio Valley that country became a magnificent game preserve, overflowing with valuable fur-bearing animals but with no Indians to hunt them. It was only after the Shawnees and the kindred of the Miamis began to return to it that most of it became a·profitable trading territory.

At the end of the seventeenth century the French finally began to garrison a line of posts extending from the Great Lakes to the junction of the Ohio and the Mississippi. They included Cahokia, 1699, on the Mississippi and, seven miles from it, Kaskaskia, founded ·about 1703; Detroit in 1701, Post Vincennes, on the Wabash, actually commissioned in 1735 or 1736 but possibly a trading post for forty years or so before; and Post Ouiatenon, on the Wabash, near the mouth of the Tippecanoe River. There were other posts which completed the chain from the St. Lawrence to the Gulf—Biloxi in 1698, New Orleans about 1718, Michilimackinac shortly after 1670, and Fort Chartres in 1719—but none of them was located upon the Ohio. That river still had no settlement, no post—not even a properly expressed claim to sovereignty over it.

During the first two decades that saw the establishment of most of this chain of forts (perhaps posts would be a more accurate description) Europeans at last began to show a genuine and effective interest in North America in general—especially French and British. Even in competition with the wars, intrigues, plots, and counter-plots that blossomed in the Old World, many people learned more of America—some to their infinite sorrow.

The people who gained this expensive knowledge were mainly

of the merchant class or of the outer fringes of aristocracy—the bourgeoisie and the petite noblesse in France and the tradesmen, remittance men, half-pay officers, and second cousins of the minor barons in England. Since most of the people were bankrupted in the process of their lesson in geography and economics, and since their reduction to penury left France, which had had a strong middle class, with only the very rich and the very poor in their previous status and with the bourgeois badly incapacitated, in so far as the sous in the family sock were concerned, the episode under discussion had a considerable part in bringing about the French Revolution a half century later.

The mania that resulted in this glorious bust began with the incorporation of England's South Sea Company in 1711 and culminated in John Law's Mississippi Bubble.

The South Sea Company has been wrongfully suspected to have been the brain child of none other than Daniel Defoe, man of many interests. It was incorporated for the purpose of receiving from the government of England a monopoly upon the trade of South America and the Pacific Isles—in return for which the company offered to assume the national debt (a little matter of £51,300,000) and to pay £3,500,000 for the privilege! (The debt consisted mainly of terminable annuities held by thousands of individuals—for which the company expected to exchange its highly watered stock.) The original proposition was not accepted, but when the company doubled its cash offer Parliament took it up. Even while John Law's similar bubble was bursting in France in 1720, stock of the South Sea Company in England rose from 128½ to 1000. Stock in anything could be sold in those days—just as in 1929: utopian schemes were welcomed then, just as they were a century later when Robert Owen appeared upon the scene—and even in the more recent history of America and the British Isles!

It was John Law's project, however, that had the most direct effect upon North America:

When, in 1717, Antoine Crozat relinquished the charter he had been granted for the government and commercial operation of French Louisiana, a financier of great reputation was ready to take it over. He was John Law, a Scotch banker who had enjoyed a highly successful career in France. Law organized the Compagnie de l'Occident, which, chartered like Crozat's, was to govern Louisiana, to pay a percentage of its profits to the crown and to keep the balance for its trouble. Law's reputation for success attracted wise investors to the company's stock; prices rose and investors less wise bought also—and prices rose still higher. Then followed one of those wild scrambles for investment which occasionally sweep a population, as the dancing mania swept Europe in the Dark Ages. Shares were divided into the minutest fractions and almost everyone who had money or the means of raising it bought in.

Law acquired other companies holding grants of one sort or another in the Western Hemisphere—none of which had been doing business at any attractive profit—and renamed his organization Compagnie de l'Indies: the stock continued to increase in price. Thus far there had been no profits and, in the light of the past experience of Crozat and the original holders of the other grants, there was no reason to anticipate any: only the magic of John Law's name and the mass hysteria of which the people were victims backed the boom. The hypnotic spell dissolved itself suddenly and spectacularly in 1720, Law left the company, and the only profit accruing was that greatly increased but somewhat embittered knowledge of America.

In the second and third decades of the eighteenth century not much of importance took place upon the Ohio. The Shawnees and others were moving back to its waters and a Frenchman, Chaussegros de Lery, conducted an expedition of French troops down the Ohio in 1729. M. de Lery, an engineer, reported that he surveyed the river with a compass and his good work shows itself in N. Bellin's map, "Carte de La Louisiane," published in 1744. Bellin gives credit to de Lery and shows, in his map, an easily recognizable representation of the head, as well as the lower waters, of the river.

A second little-known French military expedition made use of the Ohio ten years after de Lery's: in 1739 Charles Le Moyne, 3rd, second Baron de Longueuil, led an expedition of four hundred and forty-two men, including Indians, from Montreal to assist in a campaign against the Chickasaws on the lower Mississippi. Little information has survived regarding the journey to the south, except that it was made; that M. Céloron also took troops from Michili-mackinac to the Mississippi by way of Lake Michigan and the Kankakee and Illinois rivers, and that the British, under whose "protection" the Chickasaws then existed, were considerably irritated by the French movements.

The Lower Shawnee Town was built before the seventeen-forties near the junction of the Scioto and Ohio (present Portsmouth, Ohio) where the great Warriors Path crossed the latter river en route from the Great Lakes to the southern country. Judging by the prehistoric mounds around it, that townsite had been important for ages past, and this new settlement there would continue so for fifty years following.

In 1744 delegates from the British colonies of New York, Pennsylvania, Maryland, and Virginia met at Lancaster, Pennsylvania, with representatives of their old friends the Iroquois League. After deliberations well lubricated with rum the Iroquois sold to the colonies their conqueror's interest in a loosely defined tract of land to the west of their then borders. This sale would later form a basis of British colonial and United States claims to the upper Ohio River.

In 1748 the Peace of Aix-la-Chapelle terminated the European wars for a spell, but it gave little consideration to the Ohio Valley and made no decision as to its ownership.

Then, on March 18, 1749, the British took a step which left no doubt of their intention as to the future of the Ohio. The Ohio Company of Virginia, organized two years before, was chartered by the crown and was granted something over 500,000 acres of land to be located on both banks of the river by the governor of Virginia at the instruction of his Majesty's government. One of the avowed purposes of the company was "to anticipate the French by taking possession of that country southward of the Lakes to which the French have no right." Word of this piece of impudence soon reached France and shortly New France in America. In the spring of the same year the Marquis de La Galissonnière, governor of New France, decided that the time for action had come.

The action took the form of a military expedition, led by Pierre Joseph Céloron de Bienville, which was directed by the governor to plant the arms of France upon the Ohio. Carried out with as much of the gorgeous pomp customary to the French of that day as the small force and the difficulty of transportation permitted, some of the proceedings of the expedition must have rivaled the court scenes from Gilbert and Sullivan, but it was all deadly serious business to the commander, and of the greatest importance to the future of the river and of the continent. Once Céloron had recorded the claim of Louis XV and France to the Ohio under the Treaties of Ryswick, Utrecht, and Aix-la-Chapelle, the stage was set for a fight to the finish for the rich region that had heretofore remained neutral land by default.

The expedition left Lake Erie at the Rivière aux Pommes, a small creek near present Westfield, New York, passed overland to Lake Chautauqua, crossed it to Conewango Creek and thence to the Allegheny River. Shortly it reached the waters of the Ohio.

In the party, besides Céloron, were the Jesuit priest and mathematician, R. P. Bonnecamps, twenty soldiers of France, including one captain, eight subalterns, and six cadets, one hundred and eighty civilian boatmen, guides, hunters, and servants, and a detachment of Indians. Céloron, his officers, and his men carried parade uniforms and banners; most certainly his Indians had received an extra issue of vermilion paint for the occasion and the luggage of the expedition contained copies of the arms of the king of France "engraved on a sheet of white iron" and arms, accouterments, supplies and gifts for the Indians in proportion to the importance of the occasion.

The ceremonies observed from time to time on the voyage were designed to exceed the most lavish pageants the Indians could imagine. They were at worst inconceivably more elaborate than the poor simple gesture by which La Salle had claimed the Mississippi for France.

Céloron, upon reaching the valley, had nailed the engraved arms of France to a tree and at intervals upon the Ohio he proclaimed, in a speech delivered under waving banners, the French ownership of this country. The proclamation was accompanied by prayers from the priest and a salute of firearms. At each point where the ceremony was repeated he buried a lead plate by the

Ohio's bank, engraved with the manifesto by which his government stated its claim:

> Year 1749, in the Reign of Louis Fifteenth, King of France, We, Céloron, commanding the detachment sent by the Marquis de la Galissonniére, commanding general of New France, to restore tranquillity in certain villages of these cantons, have buried this plate . . . as a token of renewal of possession heretofore taken of the aforesaid River Ohio, of all streams that fall into it, and all lands on both sides of the source of the aforesaid streams, as the preceding Kings of France have enjoyed or ought to have enjoyed it, and which they have upheld by force of arms and by treaties, notably those of Ryswick, Utrecht and Aix-la-Chapelle.

By Céloron's frequent verbal pronouncements British traders and their agents were ordered to withdraw and councils were held at the various Indian towns when Céloron warned the chiefs formally against carrying on commerce with the British. Some of the Indians appeared none too gracious, when these commands were given, and their manner became even less cordial as the expedition moved west.

Perhaps Céloron's approach was not the most appropriate to win the Indians' support, much as they admired pomp and ceremony. Unlike La Salle, who had possessed a native dignity which had appealed to the chiefs and headmen, and unlike the French woodsmen, who understood the redmen so well, Céloron was not particularly notable for either tolerance or diplomacy. Of his appearance at Logstown, then and for some time to come an important Indian and traders' town located about eighteen miles below the Forks of the Ohio, he wrote, for instance:

> I observed three French flags and one English. As soon as I was observed salutes of musketry were sent from the village. . . . This salute is made by all nations of the South; often accidents happen from it. . . . I told them, . . . to cease firing in that manner, or I should fire on them. I told them at the same time, to lower the English flag, or I should pull it down myself. This was done instantly; a woman cut the staff, and the flag has not reappeared. . . . Near the village I established my camp, which I made to appear as extensive as possible. . . .

The village is of fifty cabins, composed of Iroquois, Chaounons and of Loups. . . . About five o'clock in the evening, the chiefs, accompanied by thirty or forty warriors, came to salute me, to compliment me on my arrival to their home. . . .

Céloron discussed trade on the Ohio with these men and, even thus early in the voyage, discovered that

Their interest engages them to look with favor on the English, who give them their merchandise at so low a price that we have reason to believe that the King of England, or the country, bears the loss which the Traders make in the sale of their merchandise to attract the nations. It is true that the expenses of the English are not nearly so great as those which our Traders will be obliged to make, on account of the difficulties of the route. . . .

The Ohio certainly should have been more easily accessible from Philadelphia than from Montreal but there was another factor unrecognized by Céloron; England was already developing as a manufacturing and exporting nation and the British were able to offer a larger variety of goods in exchange for furs. They might have received full French prices but being eager to establish themselves and to draw the fur trade from the Detroit, Montreal, and Wabash and Mississippi River posts of the French, they gave better values as an attraction even without the royal subsidy which Céloron suspected. Besides there was the attraction of the Irish employees of the British, who seem to have been even more singularly en rapport with the Indians than were the low-caste Frenchmen.

Céloron proceeded to the Great Miami (just above present Lawrenceburg, Indiana) and turned up that swift, shallow stream's narrow valley to the locality in which were settled some Miamis under the chief La Demoiselle, who had formerly lived near French posts in the Wabash country. Whether this Indian leader was possessed of marked feminine characteristics (early travelers reported that many Illini and Miami men were so afflicted) or his name was only a translation of the Indian for "dragonfly" is not recorded; in either case he had a mind of his own, as he later demonstrated. Céloron tried to persuade the people of this band to return to their old country but received scant encouragement.

Now probably as far up the Great Miami as it could be traveled at the low summer stage of the water, Céloron burned his canoes and marched north to Kiskakon (present Fort Wayne, Indiana) at the head of the Maumee. Here he held a conference with the local Miamis still friendly to the French. Another singularly named chieftain was in command at this town but Céloron, a literal-minded man of serious purpose, probably did not think to inquire whether Le Pied Froid was so called because of a poor circulatory system or because of an absence of ardor for warlike action. Le Pied Froid assured the French of his own friendship but could only confirm Céloron's fears that La Demoiselle's band was not likely to return.

Beyond question Pierre Joseph Céloron had done his best to secure the interests of France by pageantry, show of arms, and solemn council but he had no illusions as to his success. He had traveled more than twelve hundred leagues but he wrote, "All I can say is, that the tribes of these localities are very badly disposed toward the French and entirely devoted to the English. . . ."

The Ohio Company of Virginia, which so threatened the interests of France, was a partnership which included John Hanbury, of London, and a number of such prominent Virginians as Thomas Lee, Daniel Cresap, George Mason, George Fairfax, and Lawrence and Augustine Washington. The last two had a young half brother, George by name, who would eventually take an interest in western lands in connections and capacities other than those of speculation.

The grant that King George had been pleased to make to this company on March 18, 1749, included first 200,000 acres to be located on the south bank of the Ohio between the Monongahela and Great Kanawha rivers in present West Virginia. Later, if its first project succeeded, the company was to be privileged to locate 300,000 acres farther downstream on either the north or the south bank. The company was relieved of the payment of quitrents for ten years and, in return for this favor, it guaranteed to colonize these lands with at least three hundred families within a limited time, and to build two forts to defend it—one at or near the head of the Ohio (thus Pittsburgh was conceived) and the other at the mouth of the Kanawha.

Chartering this company and granting this land would serve two purposes for the British: it would encourage colonization and resulting trade after the fashion of grants in the British development of America from its beginning; and it would serve as an experiment to test the spirit of the French—to discover whether or not they were sufficiently interested in the Ohio to fight for it. The immediate dispatch of Céloron's expedition from Montreal partially answered this question; the French were now undoubtedly interested in the river.

The Ohio Company of Virginia was enormously important in settling the ultimate fate of the Ohio Valley, but like so many similar pioneer ventures, even down to reclamation and irrigation projects in the twentieth century, the investors in it presently lost their figurative shirts. Its grants, its privileges, and its exemptions were extended from time to time but still it was necessary to merge it in 1773 with the Grand Ohio Company, the leading stockholders in which were Thomas Walpole, brother of Horace, Governor Thomas Pownall of Pennsylvania, Samuel Wharton, and one Benjamin Franklin, printer, of Philadelphia. This merger, however, was a gesture as futile as may be imagined, for within two years of its consummation the crown of Great Britain ceased to have authority for granting Ohio River lands and its successor in governmental power looked with disfavor upon the free granting of public lands to private persons for no other reason than that they happened to be in favorable standing.

But to return to the first days of the great new concession on the Ohio.

The American partners in the Ohio Company of Virginia were quite well aware of the mode of procedure most likely to ensure success, if success was possible. They delegated a competent individual named Thomas Cresap to open a road from Cumberland, Maryland, to the Monongahela River. Though Cresap had no inkling of it, that road, after it had been considerably widened, would eventually be called Braddock's rather than Cresap's and would lead to the scene of a defeat bloody enough as a catastrophe to set a mark for persistently unadaptable British arms to shoot at for generations to come. The company also ordered a warehouse constructed on the site of Cumberland, Maryland, then on the far frontier. It also selected surveyor Christopher Gist as its agent to explore the upper Ohio and to locate the lands most eligible for its purposes.

By his agreement Gist was to receive a fee of £150 "and such further handsome allowance as his service should deserve." He was instructed to explore the Ohio as far as the Falls with particular instructions for securing the best possible lands for the company:

> You are to go as soon as possible to the Westward of the Great Mountains . . . to Search out and discover Lands upon the river Ohio. . . . When you find a large quantity of good, level Land, such as you think will suit the Company, You are to measure the Breadth of it, in three or four different Places, & take the Courses of the River & Mountains on which it binds in Order to judge the Quantity: You are to fix the Beginning & Bounds in such a Manner that they may be easily found again . . . the nearer in the Land lies, the better, provided it be good & level, but we had rather go quite down the Mississippi than take mean broken Land. . . .

In addition to these positive instructions from the partners, Governor Robert Dinwiddie of Virginia privately commissioned Gist to talk with any Indians he might encounter and to invite them to a treaty meeting to be held at Logstown in 1752. Christopher Gist's substance was vouched for by this confidential mission as well as by the £150 fee—no small sum in those days, regardless of what "further handsome allowance" he might merit. The total force that made up Gist's "expedition" consisted of himself and one young Negro slave, in contrast to Céloron's two hundred and twenty-two men. This ratio was typical of the frequently neglectful attitude toward the rich American hinterland that Englishmen exhibited: it resulted, twenty-five years later, in the loss of the lands by Britain, not to France but to her own neglected people.

Christopher Gist was the first officially accredited agent of Virginia known to have carefully explored and reported upon the upper Ohio. (He was indeed the first to report, but as for exploring there is in truth no telling but that some eighteenth-century Virginia prototype of Lawrence of Arabia may not have poked about the country in Shawnee garb long before the Ohio Company was organized.)

Born in Maryland, Gist had, judging by his journal, received a reasonably good English education. He had studied surveying, had acquired some property, and at the time of his selection as agent for the Ohio Company of Virginia was living near the home of Daniel Boone on the Yadkin River in upper North Carolina.

Gist left the company's warehouse at Cumberland and followed the road Cresap had blazed to the Forks of the Ohio. His journal began in October, 1750, and the early portion gave an excellent technical description of this trace. He found the land "mean, Stony and broken" to the Forks—as, indeed, it appears to the eye of the casual traveler today.

From the Forks he passed down the Ohio to Logstown. At that Shawnee-Mingo town, where a trading post had been established two years before and where Céloron had been offended by the British flag, he found a "Parcel of reprobate Traders." Probably there was nothing incontinent about the expletive; traders, certainly traders' employees such as these, were frequently of an order to make the term "reprobate" smack rather of flattery. Except for some of the owners of the posts and stocks, and an occasional young man who took employment temporarily for the purpose of learning the business and acquiring a foothold in the West, most traders were refugees from former haunts grown too hot for comfort or were men whose distaste for the restraints of civilization exceeded even that of the Indians. They were mainly a loose, dissolute, depraved lot and nothing is to be gained at this late date by gilding their sinful memories.

The traders at Logstown, despite their laxity of moral fiber, had a news item of importance both to Christopher Gist and to the cause of Virginian colonial sovereignty of the upper river; they reported that "George Croghan & Andrew Montour who were sent upon an Embassy from Pennsylvania to the Indians, were passed about a week before. . . ." Therein lies a story.

Both Pennsylvania and Virginia made fairly reasonable claims to the headwaters of the Ohio under the terms of their charters.

Under certain stipulations of the original description, the western border of Pennsylvania was supposed to take the same curves as the Delaware River—obviously a difficult surveying assignment to be carried out in the trackless wilds. Hitherto there had been no call for a showdown on the matter; the country was not settled, and, since Virginians had engaged mainly in agriculture and had left Pennsylvanians to their comfortable Quaker mercantile pursuits, there was little competition for trade. Now, however, with the activities of the Ohio Company of Virginia obviously aggressive, the governor of Pennsylvania considered action expedient.

The very difference in the economies of the two colonies made Pennsylvania's interests more in accord with those of the Indians than were Virginia's. The Ohio Company of Virginia planned to colonize, to develop lands, and to farm, with trade only a secondary aim—the policy that the Indians had seen as fatal to themselves since settlement began—while Pennsylvania hoped only to retain control, to protect her licensed trade, and to maintain the Indians in their hunting grounds. Certainly the fur trade could not flourish in a land of cornfields and tobacco patches.

Even aside from considerations of this practical nature, however, Pennsylvania appeared to wish to continue the policy of keeping faith with the Indians which William Penn had inaugurated in the preceding century. Only the year before, in 1749, at the very time the British government was granting Indian lands to the Ohio Company of Virginia, Pennsylvania had taken stern action against a group of her own citizens who had encroached upon Indian rights. Several families had settled on Sherman's Creek (present Perry County) and on complaint of the Indians the government had burned their cabins and moved the squatters back to the Susquehanna, above present Harrisburg.

One of the returned offenders celebrated the building of his new home on the Susquehanna with a housewarming festival which achieved such a degree of merriment that the host was accidentally shot by an Indian guest called "The Fish"; and The Fish was in turn killed by John Turner, a lodger with the family who soon after married the bereaved widow. The host, though mourned but briefly, left four sons behind to carry on his name. Two of them,

Simon and George Girty, achieved an eminence of infamy as cruel renegades that has conferred on them an evil immortality.

Christopher Gist hoped that the Pennsylvania traders at Logstown would not realize his true mission for fear they would stop him. He was allowed to pass on, however, and at Beaver Creek he met one of the few traders as yet employed by the Ohio Company and together they went to the mouth of Muskingum River (present Marietta, Ohio). En route Gist commented upon the scarcity of good farm lands.

At the Wyandot town located near the junction of the Muskingum with the Ohio a certain George Croghan, recently appointed agent for the colony of Pennsylvania, had located one of his principal trading posts and Gist, noting the "English Colours hoisted on the [Wyandot] King's House, and at George Croghan's," decided to wait there for the arrival of the owner.

Now, Pennsylvania had chosen its agent as wisely, for its purpose, as had Virginia. Few educated men in America knew the Indians better and few were more respected by them. George Croghan, a literate Irishman, had come to America in 1741 as a lieutenant to Sir William Johnson, Indian agent of New York. He had left the service to set himself up as a trader, had been licensed by Pennsylvania in 1744, and had married a Mohawk woman, daughter of Nichos, a man prominent in the affairs of the Iroquois League.

Croghan's talents were soon recognized by both the Iroquois tribes and the colony of Pennsylvania; in 1746 he was made a counselor of the League and in 1749 he was appointed a justice of Cumberland County. He retained the confidence of the Indians on the Ohio and its tributaries throughout his life.

Waiting at George Croghan's post at the Wyandot town, Christopher Gist was told by Croghan's five men that Croghan himself had missed a meeting with Céloron's expedition of the preceding year by only a fortunate few days, and that Céloron had specifically ordered them to withdraw. Only recently, they said, the French had captured several English traders on the Ohio and had carried them and their "seven horseloads of Skins to a new Fort

that the French were building on one of the branches of Lake
Erie." Presently Croghan and Andrew Montour arrived.

This Andrew Montour was a character of stature equal to that
of either Gist or Croghan. Son of Louis Couc Montour and his
Indian wife, Madeleine—called always simply Madame Montour—
and related to those other fabulous French-Indian Montours,
Queen Esther, Queen Catharine, French Margaret, he and his
family deserve a volume of their own; failing that, it must suffice
to say that Andrew Montour was as important to the early opening of the Ohio Valley as any man.

Gist told Croghan of "my Business with the Indians [al-
though not, one assumes, of the part of it relating to the aspirations
of the Ohio Company of Virginia to the ownership of lands "good
and level" on the north side of the Ohio River] and talked much
of a Regulation of Trade with which they were much pleased, and
treated Me very kindly."

On Christmas Day Gist "intended to read Prayers, but after
inviting some of the White Men, they informed each other of my
Intentions, and being of several different Persuasions, and few of
them inclined to hear any Good, they refused to come." Sectarianism was already rearing its head on the Ohio! However, "one
Thomas Burney a Black Smith who is settled there went about and
talked to them, & several of them came, and Andrew Montour
invited some well disposed Indians, who came freely. . . ."

Gist read "The Doctrine of the Salvation Faith . . . from the
Homilies of the Church of England . . . in the best manner" he
could summon. He had, unwittingly, performed what must have
been the first Protestant service on the Ohio or, indeed, in any part
of what would become the Territory North-West of the River
Ohio!

Apparently impressed, the Indians asked him to settle among
them and to teach and baptize them—one trusts that Gist suffered
a twinge of conscience at this—and even told him that he might
build a fort to defend himself and them from the French. Gist
declined, saying that his duties were to his governor.

Gist stayed at the Muskingum until January 15. Almost every
day reports came in of the troubles between the British traders

and their partisans with the French and the Indians attached to French interests. Leaving the Muskingum he traveled down the north bank of the Ohio, describing the country and the small Indian towns—two, five, ten, twenty families—he encountered along that way. He was now accompanied by Croghan and Montour.

At Windaughalah's Delaware town "of about twenty families" on the "S. E. Side of Sciodoe Creek" [Scioto River] a council was called; Montour repeated a message from the Governor of Pennsylvania warning the Indians against the French, and invited them to the council at Logstown. "This," reported Gist, "is the last Town of the Delawares to the Westward—The Delaware Indians by the best Accounts I could gather Consist of about 500 fighting Men all firmly attached to the English Interest, they are not properly a Part of the six Nations [the Iroquois League] but are scattered about amongst most of the Indians of the Ohio. . . ."

Gist, Croghan, and Montour then moved down the Scioto to the Lower Shawnee Town, Chillicothe, and Gist recorded their experiences in his journal. It should be noted that this document, already quoted, is of particular interest because Gist was making an effort to give a detailed and lucid account of Indian life for the benefit of the generality of partners in the Ohio Company of Virginia. The British investors certainly, and many of those of Tidewater Virginia probably, were as ignorant of the minutiae of Indian life as most of us are today. Gist tried to give these men the whole picture of this country they soon hoped to claim: in doing so he did an invaluable service to history.

The party arrived opposite the town and

Here we fired our Guns to alarm the Traders, who soon answered, and came and ferryed Us over to the Town—The Land about the Mouth of Sciodoe Creek is rich but broken, fine Bottoms upon the River & Creek—The Shannoah [Shawnee] Town is situate upon both Sides the River Ohio, just below the Mouth of Sciodoe Creek and contains about 300 Men; there are about 40 Houses on the S. Side of the River and about 100 on the N. Side, with a kind of State-House of about 90 Feet long, with a light Cover of Bark, in which they hold their Councils—The Shanaws are not a Part of the Six Nations but were formerly at Veriance with them, tho now reconciled; they are great Friends of the English, who once protected them from the Fury of the Six Nations, which they gratefully remember.

The party remained at this town, one of the most important Indian towns located immediately on the Ohio River in historic times, from January 31 to February 11, 1751. Gist took interesting notes:

While I was there the Indians had a very extraordinary kind of a Festival, at which I was present. . . .

In the Evening a proper Officer made a public Proclamation that all the Indians' marriages were dissolved, and a Public Feast was to be held for three succeeding days after, in which the women, as their custom was, were again to choose husbands.

The next Morning, early, the Indians breakfasted, and after spent the Day in dancing till the Evening, when a plentiful Feast was prepared; after feasting, they spent the Night in dancing. The same way they spent the next two days till Evening, the Men dancing by themselves, and then the women in turns around the Fires, and dancing in their Manner in the Form of the Figure 8, about 60 or 70 at a time. The women, the whole Time they danced, sung a Song in their Language, the Chorus of which was,

"I am not afraid of my Husband,
I will choose what Man I please."

singing these lines alternately.

The third Day in the Evening, the Men being about 100 in Number, [danced] some time at length, at other Times in a Figure 8, quite round the Fort and in and out of the long House, where they held their Councils, the Women standing together as the Men danced by them; And as any of the Women liked a Man passing by, she stepped in and joined in the Dance, taking hold of the Man's Stroud whom she chose, and then continued in the Dance till the rest of the women stepped in and made their choice in the same manner; after which the dance ended, and they all retired to consummate.

(Note that this took place at Chillicothe, Ohio, in 1751; not, as might be imagined, at Reno, Nevada, two centuries later.)

Now Gist and Croghan made a sharp detour overland to the north, crossing the headwaters of those Ohio tributaries the Little and Great Miami and even approached the source of the Wabash, 150 miles or so north of the Ohio. Here they were away from Iroquois influence and in the land of the Miamis (or Twightwees, as Gist and other Englishmen called them). The Miamis had once suffered a terrific defeat at the hands of the Iroquois on the bluffs of the Wabash above Terre Haute, in present Indiana, but had now again grown strong and independent.

They had, since La Salle's day, been a loose association or confederacy which consisted of the Miami proper, possibly an eastern branch called Pickawillanies or Pickwaylinese (so called, perhaps, only by the British, who had a talent for misinterpreting Indian names) and certainly the Piankishaws, Kickapoos and Weas or Ouiatenons, according to the accepted French version of the name. These tribes appear to have been by no means definitely established as to division. It is not unlikely that a band called the "Pickwaylinese" when on the Pickaway Plains or in the neighborhood of Piqua (now a prosperous little Ohio city) might become known to whites as the "Eel River Miamis" when it chose to move to that tributary of the Wabash.

At any rate the Miami Confederacy was of considerable importance and Gist and Croghan gave its representatives the same softening words from their respective governors: "fear the French; trust the British and come to the council at Logstown for talk and gifts." The Miamis, even though they had been long-time friends of the French, appeared agreeable to the suggestion; meanwhile Gist noted the great eligibility of their lands to white settlement.

After a few days Gist, Croghan, and their party left, and presently Gist and his slave parted from the other two and followed the Little Miami back toward the Ohio. He reached the "Shannoah town" which he and Croghan had passed before; revealed to the inhabitants the friendly spirit of the Miamis and was tendered an "Entertainment in Honour of the late Peace with the Western Indians." Here he met one of the Mingo chiefs, who had "been down to the Falls of the Ohio" and "informed him of the King's Present, and the Invitation down to Virginia" (not, it will be noted, to *Logstown,* on the north shore of the Ohio, as had been the case when George Croghan was within hearing!). Three days later he crossed the Ohio and started downstream along the rough south bank.

Eighteen miles on the way "I Met two Men belonging to Robert Smith at whose House I lodged on this Side the Miamee River, and one Hugh Crawford, the said Robert Smith had given Me an Order upon these Men, for Two of the Teeth of a large Beast, which they were bringing from the Falls of the Ohio, one of which I brought in and delivered to the Ohio Company—Robert Smith informed Me that about seven Years ago these Teeth and Bones of three large Beasts (one of which was somewhat smaller than the other two) were found in a Salt Lick or Spring upon a small Creek which runs into the S. Side of the Ohio about 15 M. below the Mouth of the great Miamee River, and 20 above the Falls—He assured me that the Rib Bones of the largest of these Beasts were eleven Feet long, and the Skull Bone six Feet wide, across the Forehead, & the other Bones in Proportion; and that there were several Teeth there, some of which he called Horns, and said they were upwards of five Feet long, and as much as a Man could well carry; that he had hid one in a Branch at some Distance from the Place, lest the French Indians should carry it away— The Tooth which I brought in for the Ohio Company was a Jaw Tooth of better than four Pounds Weight. . . ." This is the first British-American report on the Big Bone Lick of Kentucky.

For four days Gist continued along the south bank of the Ohio—seeing little, as he would today, of eligible farm land. He reached the mouth of the Licking River, but here he began to see the fresh tracks and newly set traps and to hear the gunfire of Indians; naturally he feared they were "French Indians," who, he had been warned, were hunting near by. With some misgivings as to the Ohio Company's satisfaction with his course, Gist decided to rely upon the reports he had had as to the configuration of the Falls—that they were "not very steep, on the S E Side there is a Bar of Land at some Distance from the Shore, the Water between the Bar and the Shore is not above 3 feet deep, and the stream moderately strong"—and to go south across country to a pass which would soon take him safely out of reach of "French Indians" and back to Virginia.

Shortly after their tours, word of Gist and Croghan's movements reached New France and within a year the French began to assemble stores at Presque Isle, at present Erie, Pennsylvania, and to open a line of communication—with some rudimentary fortifications—on the path to and down the Allegheny River toward the Forks of the Ohio. Their next aim was the construction of a fort at that point; a project to which the Ohio Company of Virginia was already committed. Those two conflicting plans were bound to cause trouble soon.

Chapter Four

THE FRENCH LOSE A RIVER— AND SHORTLY A CONTINENT

THE treaty meeting of which Christopher Gist and George Croghan had given such wide notice was held at Logstown between June 9 and 13, 1752. While there survives no comprehensive eyewitness account, it is safe to assume that the red brethren were welcomed with open arms; that there was a generous distribution of blue stroud (a coarse cloth woven of reclaimed wool, which was considered an ideal gift because of its negligible cost), axes of uncertain temper, pipes, Virginia tobacco well diluted with willow bark, and most generous of all, rum.

From the first settlement of North America until the final capitulation of the natives almost four hundred years later the

Indian could depend upon but one occasion when he came into his own—at a treaty meeting during which he received gracious hospitality, mainly in liquid form, before being requested to cede his lands or some of his rights of tenure. At treaty meetings the Indian was saluted as a friend and a brother. At other times he was a barbarian—in his domestic arrangements as primitive as had been the run-of-the-mine ancestors of the English, Scots, and Irish a couple of thousand years before; his warriors as merciless and blood-thirsty as Crusaders; his children as wild as those of the London slums; his sachems as sedulous at burning at the stake as the earlier Christian clergy. Of course there was no reason to give him decent treatment—except when he was about to be separated from his property.

At the Logstown Treaty the western Indians whom Gist and Croghan had visited—Shawnees, Delawares, and the tribes of the Miami Confederacy—ratified the dubious grants which the Iroquois League had made at Lancaster eight years before.

Two years back the British Parliament had passed a law limiting the erection of iron mills and forges in America in order to prevent exportation of wrought iron from the colonies. His Majesty George II, less incompetent than his father, was now on the British throne; the cabinet system was firmly established; the old colonial policy based upon trade and mercantilism was beginning to change into one of imperialism; Great Britain was becoming imperial in scope. The mother country was attempting to gain a firmer grip on the American colonies and to extend British authority westward to the Mississippi. Control of the upper Ohio was at last recognized by the British as a factor vital to this end; the river itself began to be regarded in its true magnitude—as the highroad to the west.

Only one obstacle intervened before the exploitation of the Ohio by the British—the French, who persisted in their claim to the river and all the country north of it. Since the formal claim had been laid by Céloron, French forces scattered throughout the land between it and the Great Lakes had seized those British traders who came within their reach—or within reach of those Indians who were attached to the French interest. George Croghan had already lost considerable goods as well as several men who ventured too close to the French posts on the Wabash and the Maumee. The Iroquois sale of Ohio lands at Lancaster in 1744 had not been taken too seriously—those Indians had always been friends of the British—but the ratification of that sale by the western tribes at Logstown appeared to the French to be a serious matter; many of these people had been friends of New France since La Salle had aided them in combining to resist the Iroquois. Now the French immediately took drastic action.

They began a grand roundup of those British traders who still remained in the disputed area north of the Ohio River, and Indians friendly to the French were encouraged to range even below the Ohio, especially in the west, and to secure and bring in any British they found—keeping the captured stocks of merchandise in payment for their trouble.

John Finley, an Irish-born trader, was a member of one such party which French Indians took while far afield. He and some companions from Pennsylvania had come down the Ohio with a cargo of trade goods sometime after 1750. Near the Big Bone Lick, Finley met a party of Shawnees who were on their way to hunt in the Kentucky interior. The traders decided to join them. With the Shawnees they crossed the Bluegrass region to the point where the Warriors Path, which led from the mouth of the Scioto River, on the Ohio, to the Cherokee towns in modern Georgia, crossed Lulbegrud Creek. There, in a sharply rolling country, some geologic quirk had left a small area of level, fertile field which the Indians had appropriated as a townsite. Called Es-kip-pa-ki-thi-ki the place came to be known as Indian Old Fields. Both the tiny creek and the town were interesting, the first for the name later given it, the second for its unusual character.

The creek arose only a few miles away from the town in an area of sulphur springs whose waters also contained a trace of oil. The vicinity of such springs always provided a likely place to look for game and hence made a good campsite for hunters.

Almost two decades after this first visit Finley was back again, this time with Daniel Boone and others, and with the purpose of founding homes in the Kentucky country. An interesting yarn tells of how they named the stream, on this occasion, and why:

A member of the party had brought a book across the mountains—one of those little volumes poorly printed in small type which were produced to be sold cheaply by peddlers at home and in the colonies—but it must have been highly prized indeed, since its owner was willing to allow it space in his pack which might otherwise have accommodated an equal bulk of powder to stand between him and starvation in the wilderness. The book was *Gulliver's Travels* and Boone and the others had been entertained by readings from it by the fire of evenings. The doings of Gulliver had made an impression upon the frontiersmen, for one young man is said to have returned from a day's hunting and reported that he, too, had been to "Lorbrulgrad and killed a couple of Brobdingnagians." After he led his companions to the little stream and found the two buffalo he had shot, the name stuck, though corrupted by mapmakers of lesser literary experience to "Lulbegrud Creek."

The town of Es-kip-pa-ki-thi-ki had considerable interest even beyond the fact that it was probably the largest of the few Indian towns located in Kentucky east of the Cumberland River.

It is not reported as having been a settlement of any particular tribe: Cherokees may have made up the bulk of the inhabitants, but Shawnees, Delawares, Miamis, and others were evidently welcomed there and it is supposed to have been a sort of miniature "open city" upon the great north and south trail; a cosmopolitan community of well-built houses where northern and southern ideas and goods were probably exchanged to the cultural and economical benefit of all Indian visitors.

Finley built a shelter to display his merchandise and was soon joined by other traders. He took a fancy to the place and he must have lived a comfortable life there, where he believed himself to be far from the troubles that beset the upper Ohio until, on January 26, 1753, he learned the error of his complacence.

On that day those distant troubles found him, even on the bank of Lulbegrud Creek: they were embodied in the form of a party of French-attached Indians under the command of a renegade

Dutchman whose name was probably Philip Philipse although history preserves it as simple "Philips." Whatever the spelling of his patronymic, the Dutchman and his Indians seized the trade goods, burned the thriving town, killed some of the traders, took six of them back to Montreal as captives and permitted the rest—Finley among them—to escape to the southeastern settlements.

The western posts—Ouiatenon, Vincennes, Kaskaskia—were being strengthened by the French and steps were taken toward completing the forts already begun between the head of the Ohio and the Great Lakes, for Governor Duquesne, of New France, was definitely a man of action. Fort Presque Isle (at Erie, Pennsylvania) was soon built, as was a station at Le Bœuf (now Waterford, Pennsylvania), and Fort Machault (on the site of Franklin, Pennsylvania) near the post of Venango, which had itself been seized by the French in 1753. These, with Fort Niagara, already established for a quarter of a century, constituted a line of fortification from the eastern end of the Great Lakes to the head of the Ohio. The French had a long-established hold on the lakes proper, and the old western forts on the Wabash, Illinois, and Mississippi formed a north and south line from the mouth of the Ohio to Detroit and Michilimackinac. New France had now accomplished the fortification of a gigantic quadrangle of which the Ohio River was the southern base line. To the eventual chagrin of the French, the river itself was held only by the ephemeral claim of Céloron, his buried leaden plates, and the painted arms of France he had nailed upon riverside trees: that was not enough.

If the managing partners of the Ohio Company of Virginia saw the French threat to their hopes of settlement in its entirety they would have done well to find means to follow Governor Duquesne's example of activity. The company probably did its best, in view of its lack of capital, but progress was slow even after it finally began to build the fort at the Forks of the Ohio—Fort Prince George, it was to be called. As events transpired, haste in its building, had there been much, would certainly have been waste, for the fort was soon to be out of the Ohio Company's hands.

As for the colonial governments, Pennsylvania and Virginia believed themselves to have most at stake, but a quarter of a century was to pass before any considerable community of interest would exist between any two American colonies; both proceeded to extend their governments independently to the Ohio—even in conflict, as far as the neighborhood of the Forks was concerned—and in a leisurely manner.

In May the Pennsylvania Assembly was formally notified of possible French encroachment upon the borders it claimed. Word of increasing French activity also reached the Virginia capital and in June commissioners were sent by that colony to warn the French against further aggression.

But only two practical moves were initiated by either colony at the time; guns and ammunition were sent to what the governments hoped would be permanently friendly Indians along the Ohio and two attempts were made to secure additional—and more enthusiastic—ratification of the Treaty at Logstown by the Indians who had agreed to it originally; one meeting was held at Winchester, Virginia, the other at Carlisle, Pennsylvania. One of the commissioners at the latter was the Philadelphia printer Benjamin Franklin, but his great talent for diplomacy may not have developed at that time—in any case the Indians, still feeling themselves the victims of a squeeze action between the two European powers, confirmed their former agreement reluctantly and showed no hearty friendship for either British or French.

Such was the state of affairs when, in November, 1753, Governor Robert Dinwiddie of Virginia decided to send one more warning by messenger to the French. This messenger, to be guided by Christopher Gist, was a 21-year-old planter, surveyor, and an officer in the Virginia militia, Major George Washington.

George Washington—certainly one of the most competent if not the most brilliant young men in the colony of Virginia—was already identified by those who knew him as a youth of serious mind and tenacity of spirit; one who could be depended upon to shoulder his way through formidable obstacles and to arrive at his destination, battered perhaps, but with his aim accomplished and with the full confidence of his associates, high and low.

By his appearance in early portraits—tall, slender, with his proud, even then somewhat aloof bearing, and the high-bridged nose and ruddy coloring of his English ancestors—he seems one who would be more at home riding to hounds in the Tidewater, extending the hospitality of a Virginia manor house or leading a cotillion than in cutting his way through a wilderness on a mission such as that to which he had been assigned. But appearances were deceptive; neither the wilderness nor the battlefield offered rigors which aristocratic George Washington could not face as successfully as could the most base-born and case-hardened of his followers. As his achievements later demonstrated, Washington could do almost anything, in any field of human endeavor, which was required of him—and he could do it superlatively well.

To casual history he is "The Father of His Country"; his generalship won the American Revolution (to Frederick the Great, who understood such matters, he was "the greatest military genius of his age"); he was the first president. But probably as important as either of the last two was his work in initiating the opening of the Ohio and a way to the west. Whether the fact was or even yet is appreciated by the states that succeeded the original colonies, such territorial enlargement was vital to the successful beginning of the nation, once independence was secured.

Washington and Gist reached Logstown in November, frequently a season marked by no striking felicity of climate upon the upper Ohio. They began immediately to interview Indian chiefs and headmen in residence with a view to winning them to certain British support or, if this was not possible, learning the exact state of their temper toward the two competing nations.

Presently Washington got dependable information upon a phase of the French activity in arming the Ohio Valley, though not from the Indians, when he encountered four civilian residents of Kaskaskia, faraway French post on the Mississippi above the mouth of the Ohio. These four had been part of a contingent of ten who had made a trip to New Orleans and returned in the company of a force of one hundred men with eight canoeloads of supplies for the French posts to the north. The four had deserted, for some reason, and had come on up the Ohio. Washington wrote in his journal:

They informed me that there were four small forts between New Orleans and the Black Isles [Washington, whose education in the languages had been less comprehensive than that of his fellow Virginian Thomas Jefferson, apparently mistook "Illinois" for *"Iles Noires"* and so translated it]. They also acquainted me that there was a small pallisadoed fort on the Ohio, at the mouth of the Obaish. . . .

With Gist and some friendly Indians the future Father of His Country moved on to Venango, where the good liquor and hospitable ways of the French almost won away the Indians who had accompanied him. From there he went to Le Bœuf, the fort at the head of French Creek, and delivered Governor Dinwiddie's ultimatum to the French command on December 11, 1753.

Washington returned overland to Virginia and his report of the unhappy state in which he had found matters in the Ohio Valley led to the calling out of volunteers in Virginia within the month. The militia would be useless, if there was trouble far outside Virginia's actual boundary, because colonial militia was not required to march more than five miles beyond its own territory and many a commander had learned, by experience, that the independent militiamen were fully competent to measure that five miles and were inclined to pull up sharp when it was completed. Since Virginia maintained a tenuous claim to the Ohio's Forks, militia might be used at least to that point but there was evidently a suspicion that the action was likely to take place farther afield.

The French, meanwhile, were playing no waiting game. Their activity was stimulated rather than retarded by Governor Dinwiddie's warning, and in February about eight hundred men, under the command of one Contrecœur, left Montreal bound for the Forks of the Ohio. On April 17, 1754, this force took the Ohio Company's unfinished fort without any show of resistance by the occupants.

Contrecœur destroyed this work—it is to be feared that, built as a requirement for securing the grant of lands and with no specifications as to its size or design, it may have been planned with a view rather to inexpensive construction than to strength— and the French began a larger structure on its site. Fort Duquesne, when completed, occupied the exact point of the triangle formed by the junction of the Allegheny and Monongahela rivers, with these streams protecting two of its flanks. The wall on the third side was twelve feet thick, faced with squared logs and filled with tamped earth. The approach was further protected by a moat bordered on the outside by a stockade of upright logs. The fort's various buildings were of log with convenient emplacements for cannon and portholes for rifle fire.

The Virginia troops were moving northwest during the construction of Fort Duquesne, carrying with them orders to fortify the same spot. Their progress was slow, for as they advanced they were enlarging the Ohio Company's road—it could have been little more than a marked trail—to accommodate their wagons and their ten cannon. Colonel Joshua Fry was in command, with newly promoted Lieutenant-Colonel George Washington as second in rank.

Eventually an advance party of about one hundred and fifty Virginians reached a mountain-surrounded valley called Great Meadows, west of the Youghiogheny River and some ten miles east of present Uniontown, Pennsylvania. Here they were met by Christopher Gist with news of a French detachment waiting on their route just ahead. Lieutenant-Colonel Washington moved on to attack it.

On the morning of May 28 the two detachments made contact and in the skirmish that followed ten of the French, including the commanding officer, M. Jumonville, were killed, twenty-one were taken prisoner, and the rest were routed. The occasion was important, small as were the forces involved, for it was George Washington's first skirmish and the beginning of a world war.

The French survivors claimed that their commander, Jumonville, was not killed in honorable conflict but was assassinated after his capture. It was entirely possible; enraged frontiersmen certainly had done such things before and would do them again, time after time, in the next century. What no one then suspected was that the question of Jumonville's manner of death would one day become the subject of propaganda directed against the young commander of the Virginia force which would have national and international repercussions.

Unfortunately the auspicious success in this first fight for the Virginians was far from indicative of the final outcome of the campaign.

Colonel Fry had been left behind, ill, at Wills Creek (present Cumberland, Maryland) on the line of march. On May 31 he died, leaving young George Washington in sole command.

Another company had arrived from South Carolina to augment the force, making a total of about three hundred sixty men. Except for a guard left at Great Meadows, all of them had been brought on to camp near the scene of the meeting with Jumonville's party and to continue widening the road beyond that point. Washington kept on this work until, on June 28, spies brought in word that the French at Fort Duquesne were making preparations for an attack in force. The problems of espionage that marked this and all American wars through 1865 must be kept constantly in mind: parts of the forces involved on both sides in all our wars to the latter date were of similar race and background and spies were difficult to identify. Who could tell, for instance, by the look of a Miami Indian whether he was friendly to French or British? Who could say whether an Irish trader was in French or British employ? After six months in the woods who could tell by his appearance whether the casual passer-by on the trail was a Virginia backwoodsman or a citizen of Kaskaskia? Spying was a comparatively simple business under those circumstances—but there was often considerable doubt as to who might be spying for whom!

The alarming news of the preparations of the French, known to be well supplied with rations and ammunition, found Washington's situation dangerous indeed. Supplies had not come through to him, either because of scarcity and the condition of the road or simply because of characteristic governmental neglect, and there was little hope of anything less than total defeat if there should be a fight. Washington and his staff decided to withdraw.

The retreat began and soon experienced that difficulty common to all such moves involving frontier volunteers; the men began to get out of hand. By the time Great Meadows was reached the fact became dreadfully apparent that there was no chance of continuing it in an organized manner. Washington called a halt: he had decided to make a stand.

These militiamen could be depended upon to obey when there was work to do, and to fight when fighting offered, and Washington set them to improving the sketchy fortifications they had left at Great Meadows as they passed on their way out. "Fort Necessity," he called it—first recorded of the many apt phrases a more mature Washington would leave to posterity.

At eleven o'clock on the morning of July 3, 1754, the French attacked and thus, in an obscure sheltered valley to the southeast of the head of the Ohio, continued both the French and Indian War, which would decide the imperialistic control of North America, and the Seven Years' War of Europe, which, from 1756 to 1763, not only ravaged that distant continent but extended to India, in the Orient, as well.

The force that attacked the Virginians and Carolinians consisted of about five hundred French and four hundred Indians. Fort Necessity, frail stronghold that it was, had been wisely located in order to be as far as possible on all sides from cover of trees and underbrush. The long-range fire thus necessary was not effective for either side except that, as ammunition began to be exhausted, the French had an open road to Fort Duquesne and further supplies while the colonials, surrounded, had no hope of reinforcement. Soon their provisions, grown scanty enough before the return to Great Meadows, were exhausted; there was no choice but capitulation.

Colonel Washington secured good terms, under the circumstances; his force was allowed to retain its colors and its arms (which were mostly the property of the individual volunteers) after giving hostages to ensure the return of the twenty-one French prisoners from Jumonville's party who had been sent back to Williamsburg. Carrying its wounded, the little army marched back to the settlements.

The French now held the Ohio and the Indians, actually with little choice in the matter, were their allies. The French did what they could to win Indian approbation; they rebuilt Logstown, which had been burned by an Oneida chief friendly to the British as soon as the French took the Forks of the Ohio, and brought on traders from Montreal to supply the Indians with goods.

In January, 1755, France made an offer of compromise upon the control of the Ohio country but official England, far from the scene and totally unfamiliar with it, refused to accept. France made a second proposition: that her claim should be only to the country "west of the Ohio" and that the British colonies should not attempt to extend themselves west of the Allegheny Mountains, leaving that range as neutral territory. The British eventually agreed to this, providing the French would destroy their Ohio River fortifications. This time it was the French who refused. Meanwhile the diplomats of both countries were dispatching ships, officers, and supplies to their American ports and preparing for trouble at home as well.

In April, General Edward Braddock and Governor William Shirley of Massachusetts, commander in chief of the British forces in America, laid out a plan of attack against the French at four points: the strong Fort Duquesne which Contrecœur had now completed, Crown Point on Lake Champlain, Niagara, and Nova Scotia. It was not a bad plan; if all four points could be captured by the British the French would be cut off from assistance from Europe, their line from Montreal to the west could be controlled from Niagara, Fort Duquesne might serve as a base from which to move against Vincennes and Kaskaskia, and Crown Point would stand as a threat not only to Le Bœuf and Venango but also to Montreal and the settlements on the St. Lawrence.

After a conference with the colonial authorities at Alexandria, Virginia, Braddock left on April 20 for Wills Creek, which had been selected as the rallying place for the forces that would move against Fort Duquesne.

Edward Braddock was a British officer formed in the traditional spit and polish, stiff-necked, bullheaded mold of his day but the catastrophe that overtook his expedition cannot be loaded entirely on his proud shoulders. True he was unfamiliar with the frontier; true he could not conceive of warfare in any but the European style—wide roads prepared in advance, troops in solid columns marching four abreast, battles fought in formation with volleys fired at order, headquarters at nightly encampments supplied with every luxury, and the duty of the enlisted man only to do or die as commanded. Braddock took no stock in such frontier innovations as sharpshooting from cover, individual combat with ax or hunting knife, or imitation of Indian stealth for purposes of infiltration, but the Virginia authorities, dazzled by Braddock's past reputation and current show of genuine old-country brass, had agreed to his plan at Alexandria. The rigid discipline upon which he insisted must have appeared to Washington and others to promise a method of controlling the men, where they had failed, in retreats and maneuvers other than actual fighting. Colonial enthusiasm did not, however, extend to the delivery of the financial support that had been promised the expedition; the colonial governments regularly failed to meet their quotas of supplies and money—although meanwhile, according to Benjamin Franklin, the citizens of Philadelphia raised a fund by subscription to purchase fireworks to celebrate the anticipated victory!

At the Wills Creek rendezvous Braddock gathered about 1000 professional soldiers and the same number of colonials. It had taken four weeks to move from Alexandria to Wills Creek—130 or so miles by road—and three more weeks were lost there in waiting for wagons to be brought on from the settlements. Finally, on the 8th of July, what was euphemistically called the "fast-moving" division arrived at the junction of the Youghiogheny and the Monongahela (present McKeesport, Pennsylvania), 15 or so miles from Fort Duquesne.

Colonel Washington, acting as a personal aide to Braddock, had started with the second, or "heavy," division of the army. He had now recovered from an attack of fever and joined the first division. He found the men in good condition, in view of the vast labor they had performed in cutting the military road through the primeval forest and over the mountains, and the British officers apparently in high spirits in spite of the facts that they knew the men of New France to be efficient fighters and that the very thought of Indian enemies, whose mode of warfare had already been made known in England through widely circulated penny-dreadfuls, gave them all a touch of the jitters.

The approach to Fort Duquesne was carefully scanned by General Braddock and his staff. It was decided that the army should march down the Monongahela, fording it twice in order to follow the most eligible terrain, and attack the fort from the eastern side.

The force marched on the morning of July 9: George Washington often said later that he had never seen a sight so impressive as was that of Braddock's army as it forded the Monongahela—uniforms clean, accouterments polished, flags flying, formation perfect.

Both fordings of the river were made without incident, the formation still such as North America had not seen off the parade ground before. The river bottom was crossed after the second ford (at the site of present Braddock, Pennsylvania) to the mouth of a ravine which, while narrow, offered a gentle slope rising to the highlands over which the remaining twelve miles to Fort Duquesne were to be traveled.

Indian fighters knew that a ravine was the favorite spot for an Indian ambush but either the Indian fighters, awed by the majesty of a full-dress British command, had failed to speak or more likely they had not been consulted: an advance guard had not been provided either (and *that* violated even European custom) and the column marched up the declivity.

The French adapted, as they had since the days of their first arrival, the more useful of the Indians' ways; they also continued the program of offense that had already given them control of the Ohio. French regulars, volunteers, and their Indian allies were concealed in the Indian manner among the trees and brush that outlined the summit of each side of the declivity.

Their fire began as soon as the space below was well filled with General Braddock's troops marching in close formation: they scarcely found it necessary to sight their rifles, they needed only to aim in the general direction of that wide column of human flesh and bone. It was scarcely possible to miss an effective shot.

No cowardice was exhibited by either the general or his aristocracy of British officers in these strange surroundings, against this fearsome new warfare; they did their best to rally, to charge, to re-form, and in the end they died like soldierly gentlemen—but only the colonial volunteers, characteristically ignoring the general's orders to form in platoons or in hollow squares and dropping individually behind rocks, stumps, and fallen trees to shoot at anything which showed a red skin or a French uniform, prevented the massacre from becoming unanimous. By the same token these backwoods individualists also ran like the wind when it became evident that any man who hoped to save his hide must preserve it himself. They

Ran off, leaving the enemy the artillery, ammunition, provision and baggage, to Gist's plantation . . .

Well, after all, artillery was no use on a frontier farm and a woodsman in his accustomed haunts could usually find *something* to eat.

The General had five horses shot under him, and at last received a wound through his right arm into his lungs, of which he died on the 13th . . . Secretary Shirley was shot through the head . . . Col. Washington had two horses shot under him, and his clothes shot through in many places . . . Sir Peter Halket was killed . . . Col. Burton and Sir John St. Clair were wounded . . .

Also wounded was Colonel Gage (later the general of Revolutionary War fame in Boston), and in all twenty-six of the eighty-six officers were killed and thirty-seven were wounded; 714 privates were casualties, either killed, wounded, or missing.

Of the missing many were captured, and such a large proportion of the enemy force being Indians, and the Indians requiring a proper regard for their customs as the price of their loyalty, a most unhappy fate awaited some of those captives.

One of the more fortunate of them—he had been taken before the battle—was 18-year-old James Smith; but even his adventures in the following weeks were harrowing enough.

James Smith had enlisted in Colonel Burd's party of three hundred Pennsylvanians which was opening a road designed to bring supplies to Braddock at Cumberland. (It is now a part of the Lincoln Highway near Bedford, Pennsylvania.) When the road-builders were within a few miles of Fort Bedford, Smith was sent back to hasten the approach of a provision wagon detail (oh, characteristic failure of the campaign!) and after carrying out his mission he was returning to his outfit with another young man. They were attacked by lurking Indians, Smith's companion was killed and Smith's horse bolted and threw him, making him an easy prey to the redskins.

He was taken west across the Laurel Mountain some fifty miles to an Indian village. Having been well treated on the way, he received what seemed to him to be a welcome when he was brought into the town, apparently that of some neutral band.

Next day he was carried on to Fort Duquesne, where cannon were fired in salute upon the approach of his party and French and Indians alike came out to welcome Smith's captors, who camped well outside the walls. This time the captive's reception differed from that at the Indian town; the Indians from the fort, armed with hatchets and ramrods, formed two lines leading to the main gate and brandished their weapons. At first Smith did not understand the order he received but it was soon made plain to him; he was expected to "run the gantlet." He made a good dash of it until, near the gate, he was knocked down by a war club. As he regained his feet and started on, a handful of sand was thrown in his face and, before he could wipe it from his eyes, there was a heavier blow and Smith knew no more until he regained consciousness inside the fort and under the care of a French military surgeon.

Soon one of his Indian captors visited him and inquired as to his health. He was naturally surprised, being unfamiliar with the ceremony of the gantlet, and asked why he had been so treated. He was told that the gantlet was the customary greeting for a prisoner; —"Like English how d'ye do," explained the Indian—and that his future disposition depended somewhat upon the manner in which the prisoner came through the ordeal—there being no personal grudge against him. Smith may have been proud, and should certainly have been relieved, to learn that his conduct on the occasion was considered irreproachable.

Smith asked what word there was of Braddock and the Indian told him that scouts reported the progress of the army every day;

he said Braddock's men were passing through the woods in close order—taking four red sticks from his medicine bag he laid them side by side and pressed them close together as an illustration—and added that, at the proper time, Braddock's men would be shot down "like pigeons." Recalling the fact that this was the day of the close-packed flights of the now-extinct passenger pigeons in the western country, the Indian's simile is uncomfortably apt.

Young Smith was presently able to hobble about and was allowed the freedom of the fort. He saw the French and Indians leave on the morning of the battle and—noting what he immediately recognized as their inferior numbers—could hardly wait for their defeat and his release by friends whom he expected to take the fort by evening.

His hopes withered in midafternoon when an Indian runner returned to report the successful attack: at dusk the first dozen British regulars were driven in, stripped naked and with their faces painted black as the universal Indian sign of condemnation to death—by torture and fire. They were followed by yelling Indians dressed in odds and ends of scarlet coats, belts, and epaulets from British uniforms. Braddock's pack-horses and wagons loaded with heavier plunder and still-dripping scalps brought up the rear.

There was no restraining Indians upon occasions like this, be they allies of French, British, or later, Americans, and this fact was perfectly well known to those who solicited their aid. Presently painted stakes were set up in the Indian camps outside the fort and the burning of the prisoners began.

Smith could not help but hear the affair—nor, for a time, but watch it. As an old man the scene was still clear to him:

The prisoner was tied to the stake with his hands raised above his head, stripped naked, and surrounded by Indians. They would touch him with redhot irons, and stick his body full of pine splinters and set them on fire—drowning the shrieks of the victim in the yells of delight with which they danced around him. His companions in the meantime stood in a group near the stake, and had a foretaste of what was in reserve for each of them. As fast as one prisoner died under his tortures, another filled his place. . . .

Poor lads, those British regulars (for they, who followed their officers' commands, made up most of the prisoners), many had never before enlisting for this campaign been five miles from the Thames! They had heard of burning at the stake (it had been a common punishment in Europe a few centuries before) but this wild scene must have made its horrors seem a thousand times more terrible than they—fearful as they had been of the Indians—could have imagined.

A few days after the battle, when the Indians dispersed, Smith was taken by the tribe whose members had captured him. Nothing is known of his life before he enlisted with Braddock but he had evidently not lived near the frontier, for he was totally ignorant of Indian customs. When, on reaching one of their permanent villages, his captors took his clothes, painted him and dressed him in a breechclout, he thought that he, too, was to suffer at the stake; when a group of young squaws dragged him to the river and ducked him repreatedly he decided that drowning was to be his fate instead. As a matter of fact the painting and dressing in Indian fashion signified only that he was to be adopted into an Indian family and the ducking by the buxom Indian girls was designed to wash the white blood out of his system!

After living with the Indians in what is now the state of Ohio some time Smith took his leave and went back home to Pennsylvania. In his later years he moved to Kentucky, where he became an influential early citizen.

Back south of the Forks of the Ohio Braddock's Field was no longer Braddock's, no longer Pennsylvania's, no longer Virginia's; unquestionably, it was French territory—as was also the whole of the country northwest of the Ohio. Only watchful waiting was possible for the colonies now, as far as the Ohio was concerned, until such a time as Indian unrest might rob the French of some of their allies and another army could be enlisted from the now thoroughly frightened colonials to make possible a third expedition.

In 1756 Colonel Adam Stephen, stationed at Fort Cumberland (present Cumberland, Maryland, site of the Ohio Company's first storehouse and Braddock's base) reported the state of affairs at the Lower Shawnee Town. The Indians had grown tired of the floods that covered it annually and their current friends, the French, had relocated and rebuilt it in fine style up the Scioto on what were called the Pickaway Plains. This town was known in turn, to the endless confusion of readers of midwestern history, as the "New Lower Shawnee Town," the "Shawnee Town on the Pickaway Plains," "Upper Chillicothe," and "Old Chillicothe." It was about four miles below present Circleville, Ohio.

Colonel Stephen got his information from a former female captive of the Shawnees about whose marital status before her captivity he seems to have been in some doubt. According to his report she was

A woman who once belonged to John Fraser (his wife or mistress) and has now, after being prisoner with the Shingas &c [one of the typical British interpretations of an Indian tribal name—"Shawnee" in this case] thirteen months, made her escape. . . .

The Shanoes [another variant spelling introduced by the colonel] are gathered all together there [the newest Shawnee Town] and are forced to borrow a Captain of their Cousins, the Delawares, having had all their Captains and sixteen warriors killed on our frontiers last Spring and summer, except one, who was sentencing one of our soldiers to be burnt, whom they had taken prisoner on our frontiers. The soldier took the advantage of them, and sitting close behind three Indians and the Captain, set fire to a bag of powder close by, and blew himself and the four Indians to pieces. . . .

Virginia made a gesture toward continuing the war that year by aiming an attack at the Upper Shawnee Town near the mouth of the Great Kanawha; the force did not even reach the place. This failure, with the defeat of Braddock the year before (the French saw that the Indians were made fully aware of the importance of *that* event, one may be sure) gave the Indians confidence and by May, 1756, the Virginia back country was full of Indian raiding parties, picking up what they could in the way of scalps, prisoners, and plunder from the outlying farms.

It was in that month that war was formally declared between France and England; the "Seven Years' War," it is called in Europe, where a long and complicated raison d'être is supplied, all involving the Eastern Hemisphere: actually one of its principal causes was nothing more nor less than the contest for the Ohio River country and the war had started, in fact, on May 28, 1754, with Lieutenant Colonel Washington's skirmish with Jumonville and subsequent stand at Fort Necessity in the valley called Great Meadows.

Little immediate activity on the Ohio frontier marked the formal declaration of war; in September a foray was made by Pennsylvania troops against Kittanning, on the Allegheny River, and Indian depredations against outlying settlements in that region were discouraged for a time, but at best it was a minor action.

The year 1757 marked no change except for the capture by Montcalm of Fort William Henry and the construction of Fort Massiac on the lower Ohio (at modern Metropolis, Illinois) below the mouth of the Tennessee River—the first fortification on the river below the Forks.

This Fort Massiac has been the subject of a tremendous amount of misapprehension by local historians, some of whom have ascribed the original settlements on the site to de Soto in 1542. Even visitors within a decade of its abandonment in the early nineteenth century appeared unable to learn its true history. The fort was probably never the scene of any important military action, historic or pre-historic, although its ruins made an excellent foundation upon which to build romantic conjectures. It was built by Charles Philippe Aubry by order of the commanding officer of the Illinois country. Located on a rolling plain well above the river, which may or may not have been the site of an early French trading post, it was first called Fort Ascension. Later, and for obvious reasons, it was re-christened Massiac; the Marquis de Massiac was then French minister of the marine. At its best it was a well-situated square picket fence of logs with four bastions probably surrounding barracks. It mounted eight cannon and was manned by a hundred men. Abandoned by the French in 1764 under the terms of the Treaty of Paris it was not garrisoned by the British and was soon afterward destroyed by mischievously inclined Indians.

The mouth of the small stream to the east of the fort, and called by the same name, was the point where George Rogers Clark's army debarked in 1778, and during Anthony Wayne's well-planned and successful Ohio campaign of 1794 it was rebuilt by his order as Fort "Massac," probably along the original lines. It was placed under the command of Major Thomas Doyle and later Captain Zebulon Pike—then a military man not yet of sufficient stature to warrant giving his name to mountains. General James Wilkinson made it his headquarters at one time and some of his associates entertained ex-Vice President Burr there briefly. Abandoned in 1814, it was unused until 1864, when its site became a camping place of Union troops in the Civil War.

Clark was mistaken in referring to the place as "Fort Massacre," as was Wayne in calling it "Massac." Both assumed that the French had so named it because of an Indian massacre which had once taken place on the site, and their errors seem to have been the origin of many of the myths that sprang up about it—except for that de Soto theory which must have been the inspiration of a later and more fertile imagination.

But the original Fort Massiac was of little value to the French, for the tide of their fortune turned in 1758. Pitt, the British prime minister, projected a comprehensive attack on their strongholds in North America which, with full British governmental support, seemed likely to succeed. British forces were victorious by sea and land in Europe and in New France Louisburg and Fort Frontenac were taken. In the early spring an expedition was planned in conformance with Pitt's general scheme, the purpose of which was to retake Fort Duquesne and the upper Ohio.

General John Forbes was in command, assisted by Colonels Henry Bouquet, George Washington, John Armstrong, James Burd, Hugh Mercer, James Byrd, and George Dagworthy—competent officers all. Troops included colonial militia from Pennsylvania, Virginia, and Maryland; with a body of British regulars, Highlander and Royal Americans.

Colonel Henry Bouquet was instructed to open a road toward the Forks of the Ohio, and the selection of the route to be followed, subject of somewhat acrimonious discussion at the time, was of great importance in shaping the future development of the Ohio River. If it passed mainly through Virginia, that colony would gain preponderant influence over the headwaters of the Ohio as well as its middle and lower sections; if it passed through Pennsylvania, it would furnish easy connection with New York, New Jersey, and other middle seaboard territory. Naturally, in view of the fact that both Pennsylvania and Virginia charters gave them reasonable but conflicting claims to the area around the upper Ohio, these two colonies were in direct conflict in the matter.

Colonel Washington's violent (and quite logical) contention that the old Braddock's Road should be utilized was finally overruled by General Forbes and Colonel Bouquet and a new line, beginning at the end of an already established road from eastern Pennsylvania, through Lancaster, Carlisle, and Chambersburg to Bedford, was projected from that point over the mountains via Ligonier to Fort Duquesne. Bouquet employed almost fourteen hundred men in the actual work, as guards and in his service of supply. As both the Pennsylvania and Virginia proponents had foreseen, this new road became, during the next half century, the chief line of access to the Ohio for the westbound emigrant (U.S. Highway 30 now follows the general line).

Intentionally or not, Colonel Washington was making a special plea for his native Virginia's interests in the matter. From a financial standpoint the opening of Forbes's Road was a wasteful operation. However, since it eventually assisted in populating the Ohio Valley with a people of widely various backgrounds, rather than with former citizens of the Old Dominion exclusively, and since this successful amalgamation of peoples eventually became one of the valley's most important distinctions, it is as well that Colonel Washington was overruled.

John Forbes became ill in the spring, and details of the campaign devolved largely upon his subordinates, but once the question of route was settled the plan was carried forward. One of the first steps was to dispatch an emissary with instructions to try to win back the support of those Indians who had originally been persuaded to British partisanship by Gist and Croghan but who had mostly been lost again to the French.

This emissary was Christian Frederick Post, a member of the Moravian (Church of the Brethren) sect who had lived among the Indians and who had an Indian wife. Post traveled across the headwaters of the Ohio's northern tributaries and conferred with Indians, who finally appointed a council to be held near Fort Duquesne; there Post spoke eloquently and was believed—for even then Moravians had the well-deserved confidence of the Indians. When he began the journey back to Philadelphia, on September 9, Post had assurance of peace with the Ohio Indians.

The forces still resolutely commanded by the invalid Forbes— who now had himself carried along the advance on a litter—continued to push west on the new road. In spite of the fact that the army floundered in the mud resulting from perversely early Fall rains and among the stumps left standing only a few inches less than wagon axle height by the roadbuilders, it made satisfactory progress. On September 25, 1758, the advance guard at last approached the long-feared Fort Duquesne—to find only blasted earthworks and smoking timbers. The French had withdrawn about six hours before.

Most of the great army was dispersed. General Forbes was carried back to Philadelphia, where, after surviving three months and receiving the praise of the city, he died and was buried with great pomp in Christ Church. Colonel Washington went back to Virginia to attend to his plantation and most of the militiamen were at home in good time to shock their corn and bury their root crops for winter use.

At the Forks a force assigned to garrison duty immediately constructed a temporary fortification about two hundred yards up the Monongahela from the ruins of Fort Duquesne; next year General John Stanwix personally supervised the construction of the permanent Fort Pitt, a five-sided structure which was more than two years in the building and was strong enough to protect the upper Ohio—except temporarily from a future governor of Virginia—as long as that land required protection. Its garrison and inhabitants made up the first permanent Anglo-American settlement on the Ohio River.

Immediately after the evacuation of Fort Duquesne George Croghan and the faithful and astute Andrew Montour started downstream to learn what they could of the present temper of the Indians—especially of Croghan's old friends and good customers the

Shawnees—and to see what might be salvaged of Croghan's trading business. He was bankrupted by the war but his standing was such, and he was so regarded by Pennsylvanians, that his major creditors formally resolved that since "the said George Croghan has been for some time and is now at . . . [his home] in the most melancholy and deplorable circumstances, in a condition very defenceless, destitute of all kinds of provisions but what is procured at the hazard of his life . . ." he should be protected from court action by his minor creditors. That was a rather handsome gesture, a considerable tribute to George Croghan and one of great importance to him. America still was governed by the drastic British debtors' law, which in the ordinary course of events might well have put Croghan in a colonial jail till he, or it, rotted away.

This petition of his creditors was wisely acted upon by the Pennsylvania Assembly and immunity from creditors' action was granted to Croghan for a period of ten years in order that he might start in business anew—though, in the typical bungling manner of the British government toward American colonials, this measure was overruled three years later by the ministers of his Majesty George II!

There was comparative quiet immediately upon the Ohio in 1759 and 1760. Trade was gradually being resumed and Fort Pitt, which was still in process of building, was used only as a base from which Colonel Bouquet advanced up the Allegheny, captured the French posts at Presque Isle and Erie, and made a masterfully executed dash into Canada which broke forever the eastern line of French forts from Niagara to the Ohio. Fort Pitt was visited in the latter year by a colorful character who commanded a rather unpredictable military force called Rogers' Rangers, and who was then on his way back from taking over the French posts of Detroit and Mackinac—now fallen, as had all outposts of New France south of the Great Lakes to Louisiana; as would fall, presently, all of New France in America.

Not all those who passed Fort Pitt were military men: Christian Post returned west, this time with the intention to settling on the Muskingum and establishing the first Protestant mission in the Ohio Valley (it was actually begun by Post's friend Conrad Zeisberger on the Allegheny in 1767) and many a less distinguished citizen passed on his way to settle on the rich Ohio Valley lands.

Such a move was legally impossible, however, for other than licensed traders. A quarrel between two factions of the proprietors of the Ohio Company of Virginia had resulted in a case in chancery and all hope of permanent settlement appeared to be blasted by the new British governmental policy expressed in the provisions of the Proclamation of 1763, which forbade settlement west of the line of the Appalachian Mountains.

Those independent souls who shortly undertook to enter the millions of newly acquired acres were violently ejected by Colonel Bouquet's regulars (parts of the 60th Foot and Royal American regiments, which had been enlisted in America). The colonies were, in short, being forced by the measures of the proclamation to maintain an army of their own soldiers, at their own expense, for the sole purpose of keeping their own people within the bounds set by the British crown. Freedom to trade and freedom to manufacture also had been abolished by the proclamation and, most irritating of all, though relatively unimportant, was the sudden interest which Parliament, the king, and the king's ministers began to take in even the most insignificant matters relating to his Majesty's American subjects—the length of time to which trader Croghan's moratorium from creditors was to continue, for instance.

No further significant development of America was possible under such a regimen. Soon a good many Americans were coming to see this fact plainly and to discuss means by which a change might be made. There began to be a good deal of talk about taxes and as far away as Virginia a brash young man named Patrick Henry presently made some pointed arguments in favor of the right of people to govern themselves. Under their common burdens the colonies were beginning to see a common enemy.

Chapter Five

1763-1775: RED SCALPS HANG
FROM THE LODGE POLES

THE Treaty of Paris was signed on February 10, 1763. The Seven Years' War of Europe was officially ended; its American phase, the French and Indian War, had been over for some time. New France in America was no more, officially as well as in fact. The renewed interest that England took in America as a result of those wars now began to be felt.

Vacillate as the Indians of the Ohio did between French and

British allegiance, depending upon which government was bringing most pressure to bear at the moment, the tribes around the Great Lakes had known all along where their sympathy lay; they were not at all enthusiastic over the prospect of British rule. They had always been closely attached to the French as a people, and they had far less cause for complaint of ill treatment by whites than did Indians whose contact had been with other European races elsewhere. With the passing of French rule they were naturally wary of what might lie ahead for them. They were understandably restless and ready to listen to any able leader who might offer a plan which could ensure their future welfare.

Such a leader was among them in the person of Pontiac, a minor chief of the Ottawas, and his ideas for the proper conduct of Indians were ready formulated.

Indian resentment had become evident as soon as Montreal was surrendered and Major Robert Rogers had gone out to take over the western posts. A great council had been held at Detroit in September, 1761, and Sir William Johnson, British Indian agent in the North, had tried to conciliate the Lakes tribes gathered there—the Chippewa, Ottawa, Wyandot, and Potawatomi tribes and their minor satellites—with what appeared to be a measure of success.

The council over, however, the British individually had resumed their customary arrogance toward the Indians. Traders' prices had grown much higher, since Céloron had found them remarkably low, for there was no longer competition and in truth British traders were now even farther from their sources of supply than had been the French when trading on the Ohio. Besides, many of the northwestern Lakes Indians had always preferred to be robbed by a French, rather than a British, trader, because of the agreeable manner in which the operation was carried out by the former.

The outlook for future security became even darker to the western Indians when news came that France, hoping to keep some foothold in North America for the house of Bourbon, had ceded Louisiana to Spain in 1762; every reasonably well-informed Indian knew what to expect from Spaniards! All these matters combined to increase discontent among the followers Pontiac was gathering.

George Croghan, now an assistant to Sir William Johnson in the Indian service—as well as a private trader—did what he could to combat this increasing disaffection in the ways he, of all the British, knew best. In 1762 he sent Ensign Thomas Hutchins, a competent officer and Croghan's assistant in the service, on a tour of the Great Lakes. (This was the first of the exploratory tours from which eventually resulted Hutchins's celebrated map of the country between the Ohio and the lakes.) He gave gifts liberally to those Indians who he knew could help the British cause, but he was soon brought up short in this wise course by instructions from Lord Jeffrey Amherst, British commander in chief in America, who held, in his abysmal ignorance of the matter, that such a policy was wasteful and unnecessary. Croghan knew that the giving of gifts was essential to transacting business with the Indians and he continued the practice, taking the goods out of his own trading stock until he could carry the expense no longer. Late in 1763 he resigned his position and went to England in hope of regaining at least a part of what he had lost in the French and Indian War through claims against the government of France, and of recovering the cost of his recent Indian gifts through a plea to Amherst's superiors in the British government.

By the spring of 1763 Pontiac, by the power of his earnest eloquence, had gathered a following which included not only most of his own people but also some of the Miami Confederacy, the Delawares, the Shawnees, and part of the Senecas, one of the member tribes of the Iroquois League. Everything was ready for a desperate effort to drive the British from the west. His well-planned blow against all the western posts simultaneously was struck in the middle of May.

It came very near to being successful. By June 16 the Indians had taken Posts Ouiatenon, St. Joseph, Sandusky, Miami, Mackinac, Presque Isle, Venango, and Le Bœuf; only Niagara, Fort Pitt, and Detroit remained in British hands. Dozens of unfortified outlying settlements also had been raided and the borders of the west had suffered more than at any time during the French and Indian War.

Pontiac himself directed the surprise attack on Detroit; the fort held, with its garrison of about one hundred, but it remained under siege for more than a year.

Fort Pitt was surrounded after early June, but the redoubtable Swiss, Colonel Henry Bouquet, in command of five hundred men (parts of the two regiments already mentioned) was fighting his way west to its relief through the bands of Indians who had swarmed into Pennsylvania.

Bouquet had left Carlisle, Pennsylvania, and was attacked by Indians on August 5 near Bushy Run, twenty-five miles east of Fort Pitt. The fighting raged on nearly even terms through the day and until night—when Indians ordinarily held their fire and contented themselves with keeping their opponents hemmed in— but the hot weather and the absence of water within Bouquet's lines caused suffering almost unsupportable by his men.

At daybreak Bouquet—having regulars more amenable to discipline than were usually the militia and volunteers—ordered a feigned retreat and, as the Indians pressed their apparent advantage, closed in his flanks and routed them completely. His loss was great—perhaps twenty per cent of his men—but he gained his object and passed on to Fort Pitt to raise the siege.

The Battle of Bushy Run—little known in comparison with dozens of lesser engagements—was vastly important. It was one of the greatest victories won over the North American Indians to its time and not least significant of its effects was that of regaining the prestige of British and colonial arms, which had been lost by Braddock's overwhelming and unnecessary defeat and by the recent triumphs of the western Indians. Bouquet's victory may be said to have broken the back of organized Indian resistance to the settlement of the Ohio. The river was now definitely available to whites; only the narrow and obstinate mind of the British government barred the way, and Bouquet and his regulars were on hand to implement that blockade.

By winter of 1763 Pontiac's Indians began to fall victim to the restlessness and lack of concerted purpose in any long undertaking that was always their characteristic, even in success. Discouraged by their failure to take the three principal forts and heedless of maintaining the advantage they had, they straggled back to their villages. Soon only Pontiac's own command, the now ineffectual force around Detroit, was engaged in the war and during the winter even it was inactive except as a bar to the escape of the garrison.

The following spring General Thomas Gage, now commander in chief in America, ordered Colonel John Bradstreet to Detroit and Colonel Bouquet was instructed to follow up his triumphal relief of Fort Pitt by an expedition into the country north of the Ohio River.

Bouquet realized that his force was inadequate to attack the Delawares and Shawnees on their own grounds. He applied for additional men and the Pennsylvania Assembly graciously answered his call with a levy of militiamen who deserted in such numbers that there was little net gain. Bouquet then called for Virginia volunteers to assemble at Fort Pitt. Though the Virginians came slowly enough, he had fifteen hundred men by the fall of 1765, when he moved to the Muskingum, nearest important waterway which led north to Indian country. Bouquet's newly acquired reputation and the size of the force he commanded was enough; he found the Indians ready to treat for peace.

Taking advantage of this happy circumstance, he demanded the return of all white prisoners among them and two hundred were soon brought in. The Indians were further directed to apply to Sir William Johnson for terms and to take with them the hundred white captives who were known still to be among the Shawnees. As insurance that these instructions would be carried out he held several of their principal men as hostages. The Indians lived up to his terms scrupulously; not only Colonel Bouquet's military force but also the wisdom he had displayed in selecting hostages who represented their best qualified leadership assured that.

According to tradition, not a few of the white prisoners had to be brought in to be "released" by force. It is a strange fact that white prisoners of the Indians, no matter what their backgrounds had been, often came to prefer the Indian life to that in which they had been reared, once they had avoided the stake and been adopted by a tribe.

With children—even in their teens—this was almost invariably the case, but it was frequently true also of adults—even some who had been known before their capture as Indian fighters.

No great stress was laid upon this feature, naturally, by the contemporaries who wrote the early paper-backed "Indian Captivities," which made harrowing and therefore popular reading matter from 1750 to 1850. In such accounts the captive's "longing for his loved ones" always formed an important feature of the narrative. Even those writers, however, were forced to admit that captives often waited two, three, and even more years to attempt an escape from the Indians—when it is recalled that the ingenuous Indian considered the captive literally a brother and beyond all reason for wishing to escape once he had been adopted, the question arises whether these long captivities were not often due to the captive's own enjoyment of his status.

The features that made Indian life attractive must have varied with the individual captive. Perhaps in some cases the wild life offered an escape from responsibility and conflicts at home; in others there may often have been some much more concrete reason for avoiding the haunts of law and order. Too, the fact must be considered that the domestic arrangements of many a frontier family were far less removed in comfort from those of the Indian than idealized pioneer reminiscence indicates. Indian cooking was usually reported by the more fastidious as leaving much to be desired—"They put fire under their kettle—threw the two cubs in it, feathers and all, without even ablution, and made their choicest dish of soup; and this delicate mess was all the food they appropriated. . . The flesh of the bear, which to the uninitiated white man was so much more desirable, was left to be consumed by the wild beasts." Do not condemn the Indian too hastily for this breach of taste; a modern traveler in the United States back country can testify that comparable aberrations are still conceivable among individuals of the lower order even of the white race—whose ancestors of two centuries back may easily be imagined as having relished bear cubs stewed in the hide as well as did their Indian neighbors.

Whatever the reason, there is no question but that many a former captive returned to "his" tribe after years spent back in the settlements, and sometimes in high places in the settlements at that! Sam Houston certainly went back to the Indians when the exigencies of his strenuous life as a politician, revolutionist, lover, and empire builder drove him to a change of scene and there are records of dozens of others, of reasonably gentle upbringing, who either refused to leave the Indians or spent their old age recalling their happy days in the forest.

To return to Pontiac, however: he was pardoned by the British and, beaten but still hopeful of the ultimate success of his plan, he moved on west, where he lived for a time with a remnant of the ill-fated Illini in Illinois. Still dreaming of an Indian empire, he was assassinated at Cahokia in 1769 by a drunken Indian, supposed to have been hired for the purpose by a trader in the neighborhood.

So ended the first great Indian plan for saving the Ohio Valley and the Great Lakes from white incursion; the second and last, in 1811, had more effective leadership but by then even less hope of success.

George Croghan had returned from England and on January 5, 1765, Colonel Bouquet, recognizing Croghan's great talent for dealing with the Indians, recommended to General Gage that he be sent to take over the old French posts on the Wabash for the British. General Gage agreed and on May 15 Croghan, Major Thomas Smallman, a few other whites, and some representatives of the Senecas, Shawnees, and Delawares moved down the river from Fort Pitt.

They stopped to hold friendly talks at the various small Indian villages along the Ohio and reached the mouth of the Wabash on June 6. Here, Croghan reported, they found a "breastwork" erected. He must have seen some works of the mound builders, for there is no evidence of any fortification by whites upon the lower river of that day except the briefly occupied Fort Massiac, many miles below. The party now turned its boats up the Wabash and

Croghan's published journal—most of his erratic but graphic spelling evidently edited out by some latter-day busybody—gives the incidents involved in this final securing of the chief line of the French to the Ohio:

At day-break we were attacked by a party of Indians, consisting of eighty warriors of the Kiccapoos and Musquattimes [Mascoutens or Muscatins], who killed two of my men and three Indians, wounded myself and all the rest of my party, except two white men and one Indian; then made myself and all the white men prisoners, plundering us of everything we had. A deputy of the Shawanees who was shot through the thigh, having concealed himself in the woods for a few minutes after he was wounded—not knowing but they were Southern Indians, who are always at war with the northward Indians—after discovering what nation they were, came up to them and made a very bold speech, telling them that the whole northward Indians would join in taking revenge for the insult and murder of their people. This alarmed those savages very much, who began to excuse themselves, saying their Fathers, the French, had spirited them up, telling them that the English were coming with a body of Southern Indians to take their countrie from them . . .

In spite of the speech of the Shawnee, the Kickapoo captors divided the portable property that pleased them and started up the Wabash with Croghan and his party toward Post Ouiatenon. They soon reached Post Vincennes, consisting then of "about eighty or ninety French families settled on the east side of the [Wabash] River. . . . The French inhabitants hereabout are an idle, lazy people, a parcel of renegadoes from Canada, and are much worse than the Indians. . . ."

On the 23rd they arrived at Ouiatenon (present Lafayette, Indiana): "here I met several chiefs of the Kickapos and Musquattimes, who spoke to their young men who had taken us, and reprimanded them severely for what they had done to me. . . ." This finally accomplished the desired effect and the captives were released.

Croghan delivered his peace talks to the Kickapoos, Weas (Ouiatenons), Miamis, and Piankishaws and made a trip into the Illinois country where he met and talked to Pontiac and to visiting representatives of the Iroquois League, Delawares and Mascoutens. He visited the great Miami village located on the site of present Fort Wayne, spent some time at Detroit, and returned to New York by way of Niagara. He reported success in quieting the western Indians and made a further claim for goods advanced by himself and lost on the mission.

The following winter, that of 1765–66, the British took formal possession of Fort Chartres and the Illinois country. In April, 1766, General Gage dispatched Croghan, now again acting as deputy to Indian agent Johnson, to distribute gifts to the neighboring Indians. Both Thomas Hutchins, the future geographer of his Majesty's colonies in America, and Captain Harry Gordon accompanied him. It was upon this trip that Gordon made the first accurate detailed map of the Ohio.

Little is known of Harry Gordon. He disappeared from the public record within a few years of this assignment, and of his parentage and station in life nothing has been discovered. From his wide-eyed wonder at the beauties that he saw, and from his lively interest in what went on around him, one inclines to think of him as a recent arrival in America: perhaps a younger son of a family of means; doubtless a more attractive person than his more materially fortunate eldest brother, guileless, and perhaps a bit ineffectual—a sort of better-off Thomas Traddles of a fellow. Perhaps it may not even be too much to picture him as a rosy-cheeked 20-year-old, with no particular military aptitude but with a cheerful willingness to carry out the duties of the commission which his family had purchased to solve the embarrassing question of just what to do with him.

It is from Gordon's journal of this tour that the following account is taken, the first entry being of May 18, 1766:

. . . having engaged the sufficient Number of Battoeman, we embarked on the Ohio at 1 P.M. By Rains that fell this and the preceeding Day, the River Ohio had risen between 2 and three Feet the largest Battoes of the Merchants [Baynton, Wharton & Morgan, of Philadelphia, who would continue to trade at Fort Chartres and throughout the Ohio Valley] that were sent under our Escort, which consisted of Indians, never touched, altho 7 Tons Burthen.

The 19th we arrived at the Mingo Town [near present Steuben-ville] which by our Reckoning is 71 Miles below Fort Pitt. The Country between broken with very high Ridges; the Valleys narrow, and the Course of the River plunged from many high Grounds which compose its Banks . . . Indian Business detained Us A Day . . . the River rose for several Days; and run so rapid as to carry us with moderate rowing from 6 to 7 Miles P. Hour.

Proceeding downstream:

Our Indians killed several Buffaloe between the Mingo Town and the Muskingum. We first met with a herd of this kind of Animal about 100 miles below Fort Pitt. . .

The River Ohio, from 50 M. above Muskingum to Sioto is most beautifull, a number of Islands are to be seen of different Sizes, but all covered with the tallest of Timber. The long Reaches, among which is one of 16 Miles and a ½, inclosed with the finest Trees of different kinds, of various Verdures, and Leaves of the largest Sorts, afford a noble and enchanting Prospect. The Stillness of the Current and a calm Sunshine put a Face on the Water from which was reflected the most beautifull Objects of simple Nature that I ever beheld. The glorious Vista was terminated by two small sugar Loaf Hills, of an easy Ascent, from which can be discovered all this magnificent Variety. The Rivers Hockhoking and Canawha fall in to the Ohio in this Space, besides others . . . The Country is every where pleasant. In the Bends of the river's Course, are large, level Spots of the richest Land; and on the whole is remarkably healthy, by the Accounts of Traders . . . in those parts. One Remark of this Nature may serve for the whole Tract of the Globe comprehended between the western Skirts of the Allegheny Mountains, beginning at the Post of Ligonier, thence bearing S. westerly to the Distance of 500 Miles opposite the Ohio Falls, then crossing them Northerly to the Heads of the Rivers that empty into the Ohio—thence East along the Ridg that separates the Lakes and Ohio Streams . . . It may be from proper knowledge affirmed, that it is the healthiest (as no sort of chronicle Disorder ever prevails on it), most pleasant, and most commodious Spot on Earth known to European People (supposing a State of Nature).

—and it must be recalled that Captain Gordon had no land to sell, no ax to grind, no discovery of his own to glorify! Captain Gordon simply admired the Ohio Valley, and said so.

Stopping at the principal towns to give assurance—reinforced by gifts—of British interest in the Indian inhabitants, the expedition reached that point on the south side of the Ohio nearest the Big Bone Lick of present Kentucky on June 16:

. . . we encamped opposite the great Lick, and next Day I went with a Party of Indians and Battoemen to view this much talk'd of place. The beaten Roads from all Quarters to it easily conducted us; they resemble those to an inland Village where Cattle go to and fro a large common. The Pasturage near it seems of the finest kind, mix'd with Grass and Herbage, and well watered. On our Arrival at the Lick, which is 5 Miles distance S. of the River, we discovered laying abt many large Bones, some of which the exact Patterns of Elephants Tusks, and others of different Parts of a large Animal. The Extent of the mudy Part of the Lick is ¾ of an Acre. This mud, being of a Salt Quality is greedily lick'd by Buffaloe, Elk, and Deer, who come from distant Parts, in great Numbers, for this Purpose. We pick'd up several of the Bones, some out of the Mud, others off the firm Ground, and returned . . . next Day arrived at the Falls.

As have most other visitors, Captain Gordon remarked that the Falls should not be so called, even though they were placed there by design of Divine Providence:

the Stream on the North Side has no Sudden Pitch, but only runs rapid over the Ledge of a flat Limeston Rock, which the Author of Nature has put there to keep up the Waters of the higher Ohio; and to be the Cause of that beautifull Stillness of the River's Course above it. That this bed or Dam should not wear, it is made almost flat and Smooth, to resist less the current . . . Mr. Morgan unloaded one third, and with the assistance of the Indians who knew the Channel best and were usefull and willing, got his Boats safe down and raised on the N. side. The carrying Place is 3 Qrtrs. of a Mile on this Side, and half as much on the S.E. This last is safer for those that are unacquainted, but more Tedious . . .

They reached the large island opposite the mouth of the Wabash on the 31st, reconnoitered and found only the "Tracts of some small hunting or War Parties, but none of any Number together," halted near "the Saline (present Shawneetown, Illinois), where Croghan dispatched the Indian deputies to the northern tribes, and moved on down the Ohio to the abandoned French Fort Massiac.

The Reason of the French's sending a Garrison to this Place was, to be a check on the Cherokee Parties that came down the River of that Name [the Tennessee was then called the Cherokee River], which is navigable for Canoes From their upper Towns, and who harassed extremely the French Traders intending to go among the Wabash and Shawanese Nations. The situation of this Fort is a good one, jetting from a Point a little into the River, the Reach of which up and down it discovers to a considerable Distance. A Garrison here will protect the Traders that come down the Ohio, untill they have Accounts from the Illinois. It will prevent those of the French going up the Ohio or among the Wabash Indians. Hunters from this Post may be sent amongst the Buffaloe, any Quantity of whose Beef they can procure in the proper Season and the Salt may be got from the above-mentioned Saline, at an easy Rate, to cure it for the Use of the Troops. . . . The Situation is a good one, no where commanded from, nor can the Retreat of the Garrison (a Consideration in the Indian Countries) ever be cut off. . . . It will, in a political Light, hold the Ballance between the Cherokee and Wabash Indians, as it favors the Entrance of the former, across the Ohio, into the later's Country, and covers their Retreat from it. . . .

About forty miles below Fort Massiac the party reached the mouth of the Ohio. "The gentle Ohio is push'd back by the impetuous Stream of the Mississippi, whose muddy, white Water is to be seen above 200 yds. up the former. . . ." (This happens, it should be noted, when the Ohio is low and the Mississippi high.)

Croghan, Gordon, and their associates moved up the Mississippi to Fort Chartres. The garrison was sickly, but Croghan was able to report that the western Indians were in a happy state of mind. In October the party passed down the Mississippi, Gordon continuing his journal meanwhile. Croghan returned to New York in the following winter, was back at Fort Pitt in June, 1767, and made a trip to Detroit in the fall of that year.

The Treaty of Fort Stanwix was concluded on November 6, 1768. The Proclamation of 1763 had anticipated that a line between the Indians and the colonies would be established and, as already remarked, Colonel Bouquet had moved to enforce it. In 1768, at the insistence of the colonies, the Indian agents were instructed to call a meeting of the Iroquois League and their allies to further define—and liberalize—the boundaries. In September, Sir William Johnson arrived at Fort Stanwix with twenty boatloads of goods especially selected to catch the admiring eyes of the Indians. This property, the Iroquois were told, was the price of a satisfactory grant of land.

The goods were tempting enough and the Iroquois relinquished their claim to the territory lying east and south of a line beginning near Fort Stanwix (present Rome, New York) and running south to the Delaware River, southwest along the Susquehanna, on to the Allegheny River, down it to the Ohio, and thence along the Ohio's course nearly a thousand miles to the mouth of the Tennessee River. This was a sizable tract indeed to exchange for even *twenty* boatloads of trade goods! George Croghan was present—perhaps it was his good judgment that provided such attractive merchandise. The Iroquois also here ratified a purchase of lands near the Forks of the Ohio which Croghan said he had made at Logstown on August 2, 1749, but which, in the turbulent intervening days, he had been unable to claim. Included in this private parcel were nearly a quarter million acres.

None of the land cleared of Iroquois claim by the establishment of the new boundary was yet open for settlement but other western lands were now available and there was a way to reach them—difficult but at least not interdicted by the Proclamation of 1763: open territory in Kentucky, not yet named, could be reached through either the southern mountains or by the waters of the Ohio.

Although exploring parties had crossed the mountains into Kentucky from the south and southeast—including Dr. Thomas Walker, of Virginia, whose interest in possible settlement had led him to the junction of the Red and Kentucky rivers in the southern mountains in 1750—and traders, coming usually down the Ohio, had covered a good deal of the territory near the old Indian trails, the first practical move toward settlement by white families resulted from the association of two men: the trader John Finley, who conceived the idea and got but little credit for it, and Daniel

Boone, who carried out the plan and became the demigod of the frontier.

Not that Boone was one to seek fame—quite the contrary; he was only a fearless man, living in an age which offered frequent opportunity to demonstrate that rare quality, and had fame thrust upon him by an earlier biographer than most of his contemporaries—John Filson, who first described Kentucky in print. Boone was one who simply took the challenges offered; but in his position as leader in the movement of a people to a wild new land these challenges were frequently for battle to the death, not only of the challenged but of whole communities of his people.

No one could have been more amazed than Boone, had he been able to imagine a figure such as that of Lord Byron, at his own later appearance in the poet's *Don Juan*—

> Of all men, saving Sylla, the man-slayer,
>> Who passes for in life and death most lucky,
> Of the great names which in our faces stare,
>> The General Boon, back-woodsman of Kentucky
> Was happiest amongst mortals anywhere;
>> For killing nothing but a bear or buck, he
> Enjoy'd the lonely, vigorous, harmless days
> Of his old age in wilds of deepest maze.
>
> Crime came not near him—she is not the child
>> Of solitude; Health shrank not from him—for
> Her home is in the rarely trodden wild,
>> Where if men seek her not, and death be more
> Their choice than life, forgive them, as beguiled
>> By habit to what their own hearts abhor—
> In cities caged. The present case in point I
> Cite is, that Boon lived hunting up to ninety;
>
> And, what's still stranger, left behind a name
>> For which men vainly decimate the throng,
> Not only famous, but of that *good* fame,
>> Without which Glory's but a tavern song—
> Simple, serene, the *antipodes* of Shame,
>> Which Hate nor Envy e'er could tinge with wrong;
> An active hermit, even in age the child
> Of Nature—or the Man of Ross run wild.

'Tis true he shrank from men even of his nation,
    When they built up unto his darling trees,—
He moved some hundred miles off, for a station
    Where there were fewer houses and more ease;
The inconvenience of Civilisation
    Is, that you neither can be pleased nor please;
But where he met the individual man,
He show'd himself as kind as mortal can.

—and Byron, in view of his distance from the scene, painted a very
fair likeness of the character of the colonel at that (it was *colonel,*
not general, by the way).

But it was certainly not Boone who conceived the plan and
manner of the exploration of Kentucky; those came from Finley.

John Finley had been one of those western traders already
mentioned who had spent some time and had some spirited adventures at Indian Old Fields in Kentucky, and had been robbed
of their goods and taken captive during the troubles preceding the
French and Indian War. When Finley returned to the settlements
he had enlisted, probably under George Croghan, in Braddock's
force. In the same army, engaged as a wagoner with the North
Carolina troops, was the young farmer and hunter named Daniel
Boone. Finley soon became acquainted with Boone, who must
already have cast a tentative eye toward the possibilities for hunting in the mountains to the west of his home. Boone listened avidly
to Finley's stories of the wonders of the country that extended
south from the Ohio between the Kentucky and Licking rivers to
those same mountains whose eastern slope he already knew.

After the war Finley returned to the licensed Indian trade but
he left it in 1768, prompted by a desire to set up business for himself in Kentucky—the memory of which was improved no doubt,
in his own mind, by his frequent rehearsals of its wonders to
others. With a pack-horse load of needles, pins, thread, and notions
he started south from the upper Ohio, trading with the isolated
frontier families in Virginia as he went. He arrived at the home of
his former comrade-in-arms, Daniel Boone, on the upper Yadkin
River in the winter of 1768-69.

Boone was not hard to convince that a tour to Kentucky was

in order. Finley spent the rest of the winter as his guest, together they enlisted some neighbors, and on May 1, 1769, the start was made.

How they crossed the mountains, the rivers, the rocky gullies; how Finley found the old Indian town he had known—now burned but still easily identified by parts of the stockade and the iron-hinged gateposts still standing; how Boone pushed on and finally reached a knob from which he saw the rich middle valley of the Kentucky River spreading before him—that is another story which has been told and retold. It is enough that now, at last, someone had set eye on Kentucky who had not only the vision to anticipate its settlement but also the guts to start the movement.

In the next winter Indians found the party trespassing upon their hunting grounds and took its guns, powder, hides, horses—everything worth taking. In a heroic pursuit Daniel Boone and a companion recovered a few necessities, returned to camp, and presently had the satisfaction of greeting Squire Boone, Daniel's brother, coming west with additional supplies from home. The Boones and some of their companions stayed on for another winter to try to recoup their losses by more hunting but John Finley, plundered a second time in Kentucky, had had enough. He returned to the employ of the Pennsylvania Company and, except that he was reported in 1772 as having been robbed of £500 worth of goods by Senecas, no more is known of his life.

Efforts to settle the Ohio itself along several approaches were renewed at the end of the French and Indian War. The agent who had been sent to England by the Ohio Company of Virginia had been unsuccessful in renewing its grants. In 1770 the stockholders exchanged their claims for two shares in Walpole's Grand Ohio Company, directors of which they believed to wield enough political power to accomplish their purpose. Eventually it proved to be of a character as ephemeral as any of its contemporary grandiloquent schemes but in the meantime stockholder George Washington, remembering the rich bottom lands he had seen on the Ohio and its tributaries while on his surveying and diplomatic missions, took time off from his now extensive operations as a Virginia planter to make a tour of the upper river. The Zane brothers took up lands on the south bank of the Ohio, which had been salvaged from the original grants to the Ohio Company, and founded the town of Wheeling downstream from Fort Pitt. The Indians invited the German Moravians to establish a mission on the Big Beaver River north of its junction with the Ohio and two years later the Delawares asked representatives of the sect to set up another mission still farther west on lands along the Muskingum.

Thus permanent settlement began on the Ohio River below Fort Pitt through the rather divergent interests of land companies grandly conceived, of frontiersmen in search of hides and pelts and lands to farm, and of Christian missionaries.

James Harrod came down the Ohio to Kentucky in 1773— the first permanent settler in the lower Ohio Valley who is recorded as having actually emigrated by the river—and after the turn of 1774 he and his men laid out the first Kentucky town, Harrod's Burg (Harrod's Town, it was also called in the early days, but it eventually became Harrodsburg, spelled as one word). They had not yet finished building the first shelters when they were frightened out of the country by wildly exaggerated reports of Indian hostility which emanated from the Fort Pitt neighborhood where a certain Dr. John Connolly, instructed by Lord Dunmore, the governor of Virginia, was trying to make capital of the exposed situation of the river with a view to making protection appear to be a necessity. Harrod and his men later returned, however, and completed their work in time for their town to become the first important base of settlement in the then far west.

In 1773 the result of Boone's inspection of the country and the observations of the Long Hunters led to further developments downstream. Daniel Boone sold his farm on the Yadkin River, gathered five neighboring families with him, picked up a party of forty men on the eastern side of the mountains, and pressed forward with the intention of settling permanently in Kentucky. Indians attacked them before they reached the mountains and, after losing six men, they turned back and encamped on the Cumberland River.

That year Virginians, including men of such important names in future Ohio River and Kentucky history as Bullitt, McAfee,

Taylor and Drennon, came down the Ohio by boat. They explored the banks of the Ohio, surveyed the Kentucky River valley where Frankfort now stands and laid out a town (future Louisville) on land claimed by John Campbell and the same Dr. Connolly who was exhibiting so much solicitude for the welfare of Fort Pitt. At the same time a General Thompson was on the Licking River surveying lands for Pennsylvania interests and during the summer of that year four hundred families were reported to have passed down the Ohio by boat on their way to settle at Natchez, on the east bank of the Mississippi.

A missionary project was being considered in 1772-1773 by the Rev. Mr. David Jones of New Jersey, who made an exploratory tour of the upper Ohio in search of a likely congregation. One of the first Protestants to feel the urge to minister to the spiritual needs of the western Indians—always excepting those selfless, zealous, but withal practical Moravians—Jones was discouraged by the lack of either Indian or white interest in his efforts. What preaching he did upon his journey must have been entirely ineffectual but he had the hardihood to make the effort and his lack of success was no more marked than that of his later Protestant fellows—again excepting those Moravians—until, forty years later, the Negro Methodist, John Stewart, "emphatically God's missionary to the Wyandots" settled among the Ohio Valley Indians and the good and sympathetic Baptist, Isaac McCoy, came upon the scene.

From the very beginning of exploration, of course, some of the Roman Catholic fathers had made real, sincere, and lasting converts (although not, perhaps, to the extent they fancied) and had shepherded and guarded their flocks in temporal as well as spiritual matters. It is to be feared that the Protestants of British and American colonial background generally shared the belief of their government—that Indians had no souls worth saving.

The Rev. Mr. Jones, then, contributed little to the spiritual betterment of the Ohio Valley but he did keep a journal of his tour which is interesting for its view of the Ohio country at the time just before settlement really began, and because of its introduction of certain people who were shortly to be of considerable importance to the nation.

For instance, Jones arrived at Fort Pitt on June 4, 1772, and found "At this time the fortification was remaining, but somewhat impaired. Here were about eighty soldiers and one commanding officer. It is said the erecting of this fort cost the crown £100,000 sterling: by some orders in the fall, it was demolished and abandoned. East, at about 200 yards distance, by the Monongahela, there is a small town chiefly inhabited by Indian traders, and some mechanics. Part of the inhabitants are agreeable and worthy of regard, while others are lamentably dissolute in their morals. . . ."

Later he floated downriver in one of trader John Irvine's canoes commanded by James Kelly: "It was 60 feet in length and at least 3 feet in breadth . . ."

He saw the budding of great industries in the region of the Big Sandy's mouth: "In this country also are seen alumn mines, as the people call them; but some of them, from a chymical experiment, appear to be rather a mixture of vitriol with alumn . . . great abundance of stone-coal may be reckoned as one advantage, especially in process of time. [Something of an understatement, that.] The blacksmiths about Redstone use none other . . . spent some time in getting poles for the canoe—the wood used is called paupaw, it is very light, and bear a kind of fruit in shape resembling a cucumber, but too luscious for some stomachs."

David Jones was either naïve even beyond the limits expected of the clergy or his trader friends had developed a technique for guarding themselves against hostile Indians not reported elsewhere, for he reported that "we passed a place where some rude Indians were, who had behaved insolently. . . . Our canoe-men, understanding the disposition of Indians, for their safety, made themselves half drunk and as they passed the Indians, made such a horrid bustle, that the Indians were afraid to molest us, as they said afterwards.

"I thought at first this was only an excuse for excess, but was afterwards convinced that the Indians are extremely afraid of any person intoxicated; for they look on such as mad. . . ." Pastor Jones had probably better trusted in his first judgment.

Although he had no means of knowing it at the time, he had begun his tour from Fort Pitt in distinguished company—"with Mr. George Rogers Clark and several others who were disposed to make a tour through this new world. . . ."

When, on his journey, he encountered a large gathering of Indians he tried to secure an invitation to speak to them and received one of those polite brush-offs at which the Indians were so adept—perhaps in further evidence of their Asiatic origin. He had to content himself with preaching to a few small gatherings and, of course, always through an interpreter. Like many of his fellow ministers of later years, he made no concession to the possible inability of his audience to grasp the abstruse tenets of his faith, nor allowed for any possible limitation of his translator to reproduce nor of the Indian language to express his remarks. Upon one occasion he "addressed them on these subjects, viz. 1. The state in which God created man. 2. His fall. 3. The promise of a Savior; his coming and sufferings. 4. The work of God in renewing our souls to qualify us for heaven. . . ."

One wonders what might have been the reaction of these good men of the cloth—Protestant and Catholic alike—could they have understood the connotation of a literal translation of their words into Indian, or the liberties which their frequently impious and seldom literate interpreters must sometimes have taken with their sermons! Even those interpreters who wished to do service to Christianity—and there were certainly many who had no such pious aim—must have been put to it to translate the language of the faith into the poetic but inexact Indian tongues.

In late summer David Jones returned to Fort Pitt. He found no gathering of Delawares—some of whom had promised to meet him there with other potential converts—and he soon returned to New Jersey.

He was able to raise but little money to support another visit but he came out anyway and on his second trip was accompanied by an elderly minister who wished to join in the work. The Rev. Mr. John Davis was old and worn, and made the journey against the advice of Jones. Sure enough, the old man fell ill upon his arrival at the Ohio and died. Jones's remarks upon what fol-

lowed point the difficulties of frontier life upon even the most
thickly settled portion of the upper Ohio in that day:

> He departed this life, and left me his remains to commit to the
> earth. *My distress was not small on this occasion for materials to make
> a coffin, and a spade to dig the grave. Was relieved by hearing that in
> a cabin at some distance there were some sawed boards, and a spade
> could be had in going about eight miles. . . .*

Parson Jones soon returned to the East, where he ministered to
white congregations—who in truth were likely in as great need of
salvation as any Indians—served through the American Revolution, and lived a useful life.

George Croghan was one of those who spent the years between
the Treaty of Fort Stanwix and 1772 in attempting to open western
lands—his own, acquired in the Iroquois purchase. It is not too
surprising that, once the immediate excitement over the news of
this great purchase had died down, it became evident that Sir
William Johnson, the Indian agent, might have an interest in
them; such things had happened before, at Indian treaties, and
would happen again. But neither Croghan, Johnson, nor any associate they may have had, nor least of all the purchasers of the
land, made any permanent profit from the transactions, for within
a few years whatever title Croghan held was to be invalidated by
a change in government which not even Sir William Johnson
could prevent.

In June, 1774, Parliament passed the Quebec Act. It was another step in the procession of impositions that would eventually
lead to a change of sovereignty. While not actually one of the
"Intollerable Acts," it was considered by the colonists to be of a
pattern with them. The Quebec Act was designed to give the
country north of the Ohio a capital more accessible than Williamsburg—and at the same time to pacify the French Canadians by
permitting free exercise of the Roman Catholic religion and by
re-establishing French civil law in the predominantly French-in-
habited region from Lake Ontario to the Gulf of St. Lawrence.
None of these projects was unreasonable, as far as that immediate
area was concerned, but the British government, bungling as

usual, specified government of the whole by an appointed governor and council, with no elected legislative body of any sort! The French Canadians were partially pacified, but the Anglo-American colonial population seethed with anger. Charter grants of Virginia, Massachusetts, and Connecticut were abrogated and the lands north of the Ohio were effectively closed to immigration by any people interested in free institutions; the hopes of colonial citizens who wished to promote land development and speculation were again brought to nothing.

There was still plenty of hard feeling among the Pennsylvanians and Virginians over the ownership of the land around Fort Pitt and the head of the Ohio. The question was an old one, originating in the overlapping charters the British crown had granted to William Penn for Pennsylvania, to the Calverts for Maryland, and to Virginia. In the East friction over the exact definition of the border between these two colonies had begun early but had been settled in comparative peace when a surveyed line establishing the border between Pennsylvania and Maryland was begun in 1763 by Charles Mason and Jeremiah Dixon, who thus set up one of those pairs of names connected by an ampersand which have been so important to American culture—Mason & Dixon, liberty & justice, ham & eggs, doughnuts & coffee, even Sears & Roebuck.

Mason and Dixon required four years to run the line to the western border of Maryland, beyond which lay Virginia in a similarly questioned relationship to Pennsylvania. Seventeen-sixty-nine had come before the line as completed was ratified by the crown; then Virginia questioned the western terminus of Pennsylvania (that aforementioned matter of Pennsylvania's western line duplicating the meanderings of the Delaware River) which might or might not allow Virginia to claim the upper Ohio.

Geographically the claim of Pennsylvania to the head of the Ohio at Fort Pitt appears to have been logical, but Virginia had certainly taken the lead in first fortifying the point, in trying to protect it from the French, in regaining it, once it had been lost, and in furnishing troops to secure it when Colonel Henry Bouquet's Pennsylvania militia deserted. Perhaps Virginia's interest in the protection of the Ohio had been sometimes overt rather than eleemosynary but it was at least constant.

The quarrel over whether Braddock's Road through Virginia should be used or a new one built for Forbes's expedition in 1758 had been decided in the interest of Pennsylvania, the claim of the old Ohio Company of Virginia to the upper river had been abrogated and a continuation of the Mason and Dixon line west to the river would certainly put the disputed area into Pennsylvania.

Eventually a compromise was agreed upon, after Pennsylvania and Virginia had become states instead of colonies, which placed Fort Pitt and the head of the Ohio safely in Pennsylvania, after 1784, and gave Virginia a panhandle strip to the west and thus an additional stretch of shore line on the river. Virginia would lose this, after a time, in the course of a dispute not generally anticipated in colonial days, but all these things were yet in the distant future: Virginia, and particularly her current governor, had no thought of waiving the claim in 1773.

In the person of Lord Dunmore, Virginia had a governor who was not at all averse to adding honors to his name or to increasing the size and importance of the colony he governed. Probably with the full approval of most Virginia leaders, he moved to settle the question of control of the Forks late in 1773. For his special emissary he chose Dr. John Connolly, old George Croghan's learned nephew, a man who, second only to General James Wilkinson, succeeded in inspiring the most distrust in those who knew him.

Connolly's first move, spreading rumors of threatening Indian trouble, has already been reported as affecting James Harrod and his men in Kentucky. After he had established himself at Fort Pitt, where he arrived on January 6, 1774, Connolly posted notices informing the inhabitants and the garrison that Lord Dunmore, governor of Virginia, had instructed him to erect a new Virginia county to include the Forks and the settlement and that he had been appointed captain of militia and was empowered to muster men for the new force. Fort Pitt was renamed Fort Dunmore, for rather obvious reasons, and new justices, mostly associates of George Croghan, were appointed.

The justices in the nearest organized Pennsylvania county

who had previously held jurisdiction over the disputed lands caused Connolly to be arrested and jailed at Hannastown pending his appearance at court.

Connolly gave bail and was released. He returned upon the day set for his trial, April 6, 1774, but he brought with him a force of one hundred and fifty armed men and he, in his turn, arrested the Pennsylvania justices, marched them back to the Forks of the Ohio, and sent them off to Virginia to be tried.

Word of this disorganized state of affairs at the Forks naturally excited the hopes of the more turbulent of the Indians in the vicinity—an object at which Dunmore and Connolly had probably aimed. The Moravian missionaries succeeded in quieting the Delawares but the Shawnees, always looking for trouble, took advantage of the lawless situation to begin new raids on the outlying settlements. Eager for an excuse to recruit a larger military force around Fort Dunmore, and to show the necessity for his opportune arrival, Connolly made every raid count; he enlarged upon the damage done by the Indians and continued to originate rumors of an impending general uprising which, by his account, would make such previous incidents as Pontiac's War appear to be holiday picnics along the Ohio by comparison.

Several occurrences either coincidental to or resulting from Dr. Connolly's war scare have come down in history as of very little credit to the whites.

There was "Cresap's War," for instance, a subsidiary conflict in the early spring of 1774: Michael Cresap's action might even receive whatever credit or blame attaches to the actual beginning of Dunmore's War; credit in the dubious case that the campaign was necessary to protect the Virginia frontier or blame if, as appears more likely, the sole object was to justify Dunmore's seizure of the Ohio and to exalt that gentleman's name.

Cresap was, or had been, a Maryland man of substance. His family had held stock in the Ohio Company of Virginia and he owned land near Redstone Old Fort. He was naturally irritated by the series of misfortunes that had rendered the investment in the Ohio Company well-nigh worthless and he had long been a professed Indian-hater. Cresap's troubles had aggravated a characteristic which would later mark George Rogers Clark—then, according to a letter Clark himself wrote later, one of his active young followers: Cresap had long since shown himself able and willing to fight for his country when the necessity arose but if there arose no necessity for fighting his bellicose nature was all too prone to invent one!

He had already gathered and armed some irregulars on the Ohio—not all of them of the most stable or reputable element in the neighborhood—and on April 26 there came the trouble he hoped for, or at least expected. Some Indians passing up the river, later identified as Cherokees, either fired at or were fired upon by one of Cresap's riflemen, an employee of trader William Butler. Whoever began it, Cresap's follower was shot and an incident had occurred.

The men who rallied around Cresap at least included some colorful characters: besides venturesome George Rogers Clark, there was Simon Girty—that former Indian captive and future renegade white whose deeds would be exploited by frontier stories as the antithesis to those of Boone. Simon Kenton was among them, although, owing to the fact that he believed he had recently killed a rival for the love of a Virginia maiden, he then called himself Simon Butler. There were others, too; Daniel Greathouse, for instance, whose savagery soon came to amaze the savages themselves.

Greathouse presently found some other Indians upon whom the whites could avenge themselves for the killing of William Butler's man. The Indians Greathouse found happened to be Mingoes rather than Cherokees and were in no way concerned in the matter, but they were Indians and they thus would serve the purpose. Greathouse and his detachment saw the Indian family encamped at the mouth of Yellow Creek on the north shore of the Ohio, invited them across, gave them liquor, and murdered them all in cold blood.

This should have been a minor matter, only one of many such incidents, except that these Mingoes happened to be the family of the eloquent half-breed Logan—and audiences at elocutionary recitals have not yet heard the last of the repercussions of the event.

No one even then thought very highly of the fact that some of Cresap's men—including Clark—prepared for one of their forays by indulging in an Indian war dance complete with scalps, paint, liquor, yells and all, and there were other such episodes, not entirely disreputable but in no way commendable. The whole of Dunmore's War and the actions that brought it about were, from any viewpoint, a shady business.

As a result of Cresap's performance, of Connolly's cries of calamity echoing up and down the valley, and of the retaliatory Indian raids that naturally followed Cresap's assaults, a general exodus of peaceful settlers from the frontier as far west as Kentucky soon began. By June 10 Governor Dunmore had plausible enough excuse to call out the militia from southwestern Virginia. The companies rallied under General Andrew Lewis and an attack upon the Shawnee towns north of the Ohio was ordered.

Governor Dunmore came to the Forks to take personal charge of this, his own private war. One of his first acts was to open an office for the sale of neighboring lands; that important matter disposed, he began to organize his military force by appointing officers—including George Rogers Clark, who may have served on his staff.

In July and August, 1774, General Angus McDonald with four hundred Virginia militia crossed the Ohio and raided the town of Wappatomica on the Muskingum: this was a move of special interest because in it were first associated George Rogers Clark, William Harrod, Leonard Helm, and Joseph Bowman—four young men who were to see much of each other during the next decade.

Next Governor Dunmore called out the neighboring militia to join General Lewis and his eleven hundred Virginians at the junction of the Ohio and the Great Kanawha. General Lewis marched his force to the appointed spot and on October 6 camped at Point Pleasant, the triangle formed by the meeting of the two rivers. There the army awaited the coming of its rear guard of some two hundred and fifty men.

But the Shawnees did not wait to be invaded. A large force under the great chief Cornstalk crossed the Ohio from the north during the night of October 9 and attacked the following morning. This was a desperate and bloody fight, man to man, in frontier fashion, unhampered by European military tradition or protocol. By afternoon it became apparent to the Indians that they could not drive the whites into either of the rivers—their original aim—and toward evening they withdrew across the Ohio. The losses of Lewis were about fifty killed and a hundred wounded; the total of Indian casualties was unknown, since it was Indian custom to carry their dead and wounded from the field, when possible, in order to conceal losses. They may be assumed to have suffered at least as much, for they soon sent word to the effect that they were ready to treat for peace.

This engagement ended the current hostilities. The Shawnees were reduced in power and the Virginia and Pennsylvania militiamen had proved themselves competent to hold their own without the support of regulars or British officers.

Governor Dunmore's meeting with the Shawnees under Cornstalk was held in Camp Charlotte, which the governor set up on the Pickaway Plains near the French-built Upper Shawnee Town. The resulting treaty provided that the Shawnees give up all prisoners, that they refrain from hunting south of the Ohio, and that they trade only with and respect trade regulations of the British, certainly not terms too rigorous for a conquered people.

The great Logan failed to appear for the preliminaries of this treaty meeting and his old acquaintance Simon Girty was sent after him. When they met Logan is believed to have made the speech to Girty who, being unable to write, must have memorized it before he came back and dictated the words to John Gibson (husband of Logan's Indian sister-in-law), who wrote the speech down and read it to Dunmore and his staff.

Unquestionably one of the three who had a hand in its origin and transmittal—Logan, Girty, or Gibson—was a master of prose style: the reader may take his choice. Logan's speech certainly ranks only a bit below Washington's Farewell and Lincoln's Gettysburg address.

The British government at last began to realize the delicacy of its position in America, and Lord Dunmore compiled a list of "Persons Well Disposed to His Majesty's Government" before he and Connolly returned to Virginia's capital. Simon Girty's name was included and that eventually proved to be correct but there were many errors. All those who served with Dunmore were not his Majesty's friends—including a large delegation of young officers who immediately returned to Point Pleasant after the treaty and formed an organization of veterans: George Rogers Clark was one of them and all became officers in the colonial forces after the Battle of Lexington, Massachusetts—these were distinctly *not* "well disposed" toward his Majesty.

At the end of Dunmore's War that ambitious and now for the time successful administrator appointed George Croghan president of the justices of the new Virginia county of West Augusta—surrounding the Forks of the Ohio. Croghan's presidential career was obviously short, for his name did not appear as a justice after 1775. Perhaps that was no reflection upon his conduct on the bench, for early in that year came distractions which overshadowed Lord Dunmore's amateur warfare and gave pause to all his appointed officeholders.

News could travel from Concord to the Forks of the Ohio even in that day and on May 16, 1775, a "Meeting of the Inhabitants of that part of Augusta County which lies on the west side of Laurel Hill" was called. A Committee of Safety for the district was chosen, with George Croghan as chairman, and resolutions were passed approving the radical but spirited actions of fellow colonials of Massachusetts in "opposing the invaders of American rights and privileges." The people at the head of the Ohio River had endorsed the American Revolution.

As was to be anticipated, the outbreak of British-American hostilities soon caused the eyes of the ubiquitous John Connolly to turn again to the Ohio. He proposed to Lord Dunmore that he be allowed to go to Detroit, gather British troops and Indians, recapture the Forks from the rebels, and move to Richmond, Virginia, to help Dunmore put down the rebellion. General Gage agreed and Connolly started west. But General George Washington, forewarned, had him taken in Maryland and imprisoned. Connolly's plan was potentially effective and, had it been put in practice, the horrors of Indian war in settled Virginia might easily have early turned the tide of the approaching conflict.

Lord Dunmore by now had plenty of trouble nearer to the gubernatorial palace. He retreated to the British ship *Fowey* in ostentatious order, issued a proclamation emancipating the Virginia slaves—in the hope of arousing them to counterrebellion—and thereafter disappeared from the Ohio Valley scene.

To the west, and probably unaware of the developing revolt behind them, the nine Carolinians who had organized what was later called the Transylvania Company—including, among others, Judge Richard Henderson, John Williams, and the brothers Thomàs, Nathaniel, and Captain David Hart—had held a meeting with the Overland Cherokees in March, 1775. They had either purchased from the Indians or secured rights to settle a vast tract of land in central and western Kentucky and Henderson, David Hart and some others came up to inspect their purchase and to plan its settlement. At their meeting with the Indians, after the transaction was completed, a phrase often to be applied to Kentucky was first made known to whites when the Cherokee chief Dragging Canoe told the Carolinians that they had acquired "a dark and bloody ground." Most of Kentucky had certainly been a no man's land for centuries, maintained as a neutral ground between northern and southern Indians and crossed only with reluctance by either; it would maintain some of this character—although the Cherokee chief had no way of knowing it—off and on until 1864!

The governors of both Virginia and North Carolina refused to recognize this sale, and the Transylvania Company was to remain without clear title to western lands until 1778, when the two states granted the company 200,000 acres at the junction of Green River with the Ohio and the same amount in Powell's Valley, then North Carolina, now eastern Tennessee.

Back in the spring of 1775, however, Daniel Boone, who was still on the upper Cumberland where his party had camped after having been driven back from Kentucky's margin almost two years before, organized an expedition under the direction of Judge Henderson and in the interest of the Transylvania Company crossed the mountains and erected on the Kentucky River the stockade and cabins to be called Boonesboro (near present Winchester, Kentucky). Shortly Harrodsburg, which had been begun in 1774 and abandoned at the time of Connolly's rumors of Shawnee attack, also was re-established. There was a scattering of other units of settlement and when Henderson arrived on April 25, 1775, there were sufficient settlers on Kentucky (still, then, Virginia) soil to enable him to call a conference of Kentuckians to discuss mutual interests and policy.

That meeting convened at Boonesboro on May 23 and deliberated through four days; it was totally ineffectual, as far as legal action was concerned, since the two state governments shortly declared that the Transylvania Company had no rights, but it was, nevertheless, the first deliberative assembly of American settlers in the West.

Seventeen-seventy-five was a great year for land speculation and for settlement. William Murray, of Philadelphia, organized the Wabash Land Company and a bit later the Illinois Land Company: both were chartered and received grants from the British crown—but that, shortly, became a source incompetent to deliver a fair and merchantable title to land on the Ohio and development of the specified tracts was forced to await a more auspicious occasion.

Chapter Six

THE REVOLUTION IN THE WESTERN COUNTRY

NEXT year, 1776, the American colonies had many problems more worrisome than land-company charters on the remote middle Ohio. The American Revolution was under way; there was a government to be organized, a war to be financed, and munitions and supplies of all kinds to be wrung from an economy largely agrarian.

Most Easterners even forgot the speculative financial possibilities of the Ohio—certainly they had no time to fear for the safety of a few tiny western settlements which, hemmed in a country bounded upon north, south, and west by foreign powers, might fall to Indian or white foes. True, the Continental Congress early deliberated the possibility of an attack upon Detroit, and General Washington suggested that Indians might be employed as fighters on the project, but the fact soon became evident that the Indians' partisanship, where there was any, favored the British and affairs immediately on the Atlantic seaboard, and obviously of prior importance, intervened anyway. The plan of an attack on Detroit was dropped.

Dropped officially, that is; to redheaded 24-year-old George Rogers Clark the protection of his adopted Kentucky and the driving of the British from the West was still a live project—with the capture of Detroit by troops under his command to be the ultimate and glorious triumph. Clark defended his Kentucky nobly and he drove the British from the western country south of the Great Lakes but that longed-for crowning achievement, the capture of Detroit, was not to be his. Disappointment over that fact cast a shadow over his entire life and helped to embitter the ungracefully conceded honors that came to him in the end.

George Rogers Clark was a type of young man—although a singularly distinguished specimen—not uncommon upon the frontier. Dashing, fearless, alert, restless, they appear as men of the hour when battles are to be fought and rash heroism displayed, but the same qualities that make them great in time of stress may lead them into indiscretions when no emergency is to be faced. In the lower orders of their kind such indiscretions may easily result in wanton excesses; among superior individuals, though the indiscretions may amount to little more than disgusting foolishness, relatively harmless, they are still likely to rob the participants of the respect due for great works accomplished.

Unquestionably George Rogers Clark was a great man; unquestionably, also, had he included a mite more of continence within his makeup he could have been a greater one. He was born in Albemarle County, Virginia, on November 19, 1752, one of the six fighting sons of John and Ann Rogers Clark—both of whom were well connected in the Old Dominion. Clark seems to have received a reasonably thorough schooling for the place and day. The letters and journals he wrote in later life are obviously the

work of an educated man, marred only by the vagaries of spelling
and capitalization common to the eighteenth century.

Practically nothing is known of his boyhood. At the time when
his biography would ordinarily have been written; when the
memoirs of more fortunate men, even of lesser achievements,
would have been jotted down for posterity, George Rogers Clark
was suffering from a disrepute resulting from peacetime manifestations of some of the very qualities that had made him the great
soldier of the wild frontier. Later still, when before his death his
true stature as a patriot came to be appreciated by a few, he was
too much a wreck in mind and body to remember much of what
had occurred in the distant years before glory, misfortune, and injustice had come to him.

Like many of his contemporaries, he studied surveying in his
youth—perhaps following the example of his grandfather, John
Rogers, who as a surveyor had located and secured for himself
some fine Virginia lands. When he was about nineteen and had
mastered the subject, Clark set out across the mountains to explore
the Ohio River around the mouth of the Kanawha. Whether he
returned to his home in the meantime is not known, but he was
certainly at Fort Pitt on June 9, 1772, when he set out on that
canoe trip with the Rev. Mr. David Jones and the other young men
"inclined to make a tour in this new world."

The missionary and his companions had gone downstream as
far as the Indians' town located on the Mingo Bottom, site of
present Steubenville, Ohio, had bought horses and returned overland. Since even Jones forgot his evangel long enough to take particular note of the lay of the land and the quality of the soil it
may be assumed that Clark did likewise. He presently located some
land and built himself a cabin near the mouth of Fish Creek, on
the south bank of the Ohio about twenty miles below modern
Wheeling.

He gave his address, when he wrote to his brother Jonathan
on January 9, 1773, as "Ohio river, Grave Creek township" (it was
Virginia, of course), and his letter reported that his land had increased in value, that he had "a great plenty of provisions," had
raised some corn and made "a good deal of cash by surveying."

But evidently the secure life of the husbandman had no great attraction for him even after such an auspicious beginning, for he left his corn unplanted that spring and took an exploring trip farther down the Ohio. During the summer he returned to the settlements to visit his father.

George Rogers Clark was back on the Ohio when, early in 1774, the passing Cherokees killed trader Butler's man and, welcoming adventure, he hastened to join the Virginians rallying at Wheeling under Cresap.

Much that shortly happened around Wheeling was not praiseworthy—as has been noted—but certainly the operations in that neighborhood offered valuable training to a young man who would presently direct the defense of the frontier. In Cresap's guerrilla campaign and in Dunmore's only slightly more orthodox war Clark held his first commands.

When Dunmore's young officers had met and decided where their fealty would lie in case of a break between the colonies and the mother country, Clark, Harrod, Helm, and Bowman found themselves in agreement and decided, while waiting for the expected trouble, to try their fortunes in the Kentucky country.

Clark had gone down the Ohio and was at Leestown, seventy miles up the Kentucky River, by the 1st of April, 1775. Vastly impressed by what he saw—"A richer and more beautiful country than this I believe has never been seen in America"—he resolved to make it his home. During the ensuing winter he returned to Virginia to settle his affairs and to try to persuade his father and brothers to accompany him to the west.

He was back in Kentucky again in the spring of 1776 and in June helped to organize the council of Kentuckians that was held at Harrodsburg for the purpose of setting up a government and planning defense. Clark urged that efforts be made to secure a closer connection with Virginia and to lay up arms and stores for the settlements. His ability as a leader must have been readily apparent, for the council selected him and John Gabriel Jones to serve as delegates to the Virginia General Assembly.

The two young men set out overland by the Wilderness Road

to Virginia, and after a most difficult journey arrived at Williamsburg early in November,—to find the Assembly adjourned.

Jones returned to the western part of Virginia for a time but Clark stayed on at the old capital and determined, if possible, to make some progress in bringing the plight of the Kentucky settlers to the attention of officialdom. To this end he called on Governor Patrick Henry, then ill at his home in Hanover County —and the governor, immediately impressed by Clark's patriotic zeal and bearing, recommended that the executive council give him an order for five hundred pounds of powder from the stores that New Orleans and Philadelphia trader Oliver Pollock had secured from the Spanish and which Colonel John Gibson had brought up the Mississippi and Ohio to Pittsburgh. This recommendation was gracefully accepted by the council—with the understanding that Clark himself should be financially responsible, should their action prove unlawful, and that Clark, personally, should pay for the transportation of the powder to Kentucky in any case!

Clark agreed to these stipulations and thereby laid the foundation for the troubles with the unappreciative and capricious Virginia, colonial, and federal governments that would plague him to his death—and his heirs for a half century thereafter.

Finally, upon this same trip, he secured the recognition of Kentucky as a county of Virginia and he and Jones started west by way of the Ohio. They left Pittsburgh with a boat, seven men to row it—and the precious powder.

They passed through the rockbound upper river, where every mile of underbrush, fallen rock, and drift-lined channel provided cover enough for a thousand hostile Indians, and reached the mouth of Limestone Creek (present Maysville, Kentucky) safely. They knew the next hundred miles would be even more dangerous and, daring as George Rogers Clark would presently prove himself, he had walked too many miles and pleaded with too many comfortably situated Virginia legislators to take unnecessary chances with that powder. Wrapping it in buckskin parcels the men concealed it in the underbrush of a series of islands above the creek and stopped for rest at a deserted cabin: all but Clark, that is; his mission was not yet accomplished and he pushed on to recruit a force in his new "Kentucky County" strong enough to assure safe delivery of the powder to the settlements.

Shortly after Clark's departure a party of ten Kentuckians, apparently on an independent expedition up the Ohio after supplies, came upon John Gabriel Jones and his Fort Pitt boatmen resting in the cabin on Limestone Creek. Colonel John Todd was in command of the travelers and he and Jones decided to recover the powder from its hiding place and, with only Todd's men, to try to run it through to the settlements. It was Christmas Day, 1776.

Fortunately for the future safety of Kentucky, they did not even reach the cache. Indians attacked them on the river and the scalped bodies of Representative John Gabriel Jones, William Graden, and Josiah Dixon were left to the already well-fed Ohio River buzzards. The survivors were carried as captives north into the Indian country.

One of the captives was Joseph Rogers, cousin of George Rogers Clark, for whom was waiting a more dramatic fate than even the torture he probably anticipated. As was common when a prisoner was to be saved from the stake, he was adopted into an Indian family and painted and dressed as an Indian. Four years later, still in Indian garb, young Rogers saw his opportunity to escape from his captors as they retreated from a fight with a troop of Kentuckians. Probably forgetting in his excitement that, in Indian clothing and after four years of living in Indian camps, he could not be recognized as a white man, he dashed toward the Kentuckians and was met by a fusillade from the squirrel rifles. He died in the arms of the colonel commanding the campaign, his cousin, George Rogers Clark.

At the time Todd and Jones made their unwise decision about the powder, however, Clark was making his way alone between the steep tree- and brush-covered banks of the Ohio, unaware of what was going on behind him. He was pursued by Indians at times but he reached Harrodsburg safely at the end of a month—just before the arrival of a survivor of Colonel Todd's party with word of the defeat. There was no knowing whether the Indians had found the powder but there was also no question as to the next move: a troop of thirty volunteers was raised to try to recover it.

As future events would prove, the volunteers under the command of James Harrod included several men well qualified for this or any other desperate mission—Simon Kenton, Leonard Helm, Silas Harlan, and Isaac Hite among them. They found the hiding place, brought back the buckskin parcels, and the Kentucky stockades soon had powder and easily recognizable leadership for their future defense in the persons of the men who had delivered it.

Clark's terse journal describes both his own activities in the ensuing months and the hell of dread, danger, and death in which the British-incited Indians were keeping the people upon the lower Ohio:

Dec . . . 29th. A large party of Indians attacked McClelland's fort and wounded John McClelland, Charles White, Robert Todd and Edward Worthington—the first two mortally. 30th. Charles White died of his wound.

January 6, 1777.—John McClelland died of his wound. 30th. Moved to Harrodsburgh . . .

March 5. Militia of the county embodied. 6th. Thos. Shores and William Ray killed at the Shawanee Spring. 7th. The Indians attempted to cut off from the fort a small party of our men. A skirmish ensued . . . four men wounded and some cattle killed. We killed and scalped one Indian and wounded several. . . . 18th. A small party of Indians killed and scalped Hugh Wilson, about half a mile from the fort. . . . 19th. Archibald Neal died of his wounds. . . . A large party of Indians . . . killed and scalped Garret Pendergrest; killed or took prisoner, Peter Flin . . .

All this, mind you, within yards, or a mile or so at most, of the principal settlements in the lower Ohio Valley! All this when the only possible retaliation was the occasional stealing of a few Indian horses from the north side of the river or a rare opportunity for a pot shot at one of the besiegers grown careless!

Two good reasons existed for this continued harassment: British official Henry Hamilton, in headquarters at Detroit and with outposts at Vincennes, Kaskaskia, and Cahokia, was paying a fancy price for the scalps of rebellious frontiersmen, while some of the American militia on the upper Ohio added to the Indians' fury as they continued the barbarities that had begun under Cresap and had resulted in the murder of Logan's family and other similarly disgusting episodes.

It was in 1777, for instance, that Cornstalk, an exceptionally intelligent Shawnee chief who had worked to maintain peace with the Americans since his defeat at Point Pleasant in 1774, came to Fort Randolph at the mouth of the Great Kanawha to warn the commander that his control over his Indians was slipping and that trouble might be expected.

With what must be admitted to have been treachery typical of the time and place, the commander imprisoned Cornstalk and his companions as hostages; presently the prisoners were joined by Cornstalk's son, who came to see what had delayed the chief's return. While they were in the guardhouse a soldier was killed outside the fort by skulking Indians—the very eventuality against which Cornstalk had warned—and the undisciplined militiamen of Captain John Hall's company rushed to the guardhouse and shot three of the Indians who had sought to warn them. The fourth, according to a witness, "was shamefully mangled, and I grieved to see him so long in the agonies of death."

The murders infuriated hitherto friendly and hostile Shawnees alike; the tribe never forgot. In the next seventeen years it led in the attacks upon the western settlements; under its hero, Tecumseh, it incited the troubles of 1811; and it became an effective ally of the British in the War of 1812. Cornstalk's martyrdom ceased to be a threat to peace only, in fact, with the virtual annihilation of the Shawnees as a tribe.

The Kentuckians naturally suffered from the results of both British and Fort Randolph diplomacy in '77. Clark's journal continued:

April 7. Indians killed one man at Boonesborogh and wounded one . . .

Even in such constant danger, however, affairs civil and of the heart were not neglected:

19th. John Todd and Richard Callaway elected burgesses. James Barry married to widow Wilson. . . .

One wonders if the lady could have been the relict of Hugh, killed an even month before, but, wondering, should consider what must have been the lot of a lone woman, perhaps with children, in the time and place. In any case, life was being lived at such a pace upon the frontier that a month must have equalled at least a decade of normal mourning.

24th. Forty or fifty Indians attacked Boonesborough, killed and scalped Daniel Goodman, wounded Captain Boone, Captain Todd, Mr. Hite and Mr. Stoner. . . . 29th. Indians attacked the fort and killed Ensign McConnell.

May . . . 23rd. A large party of Indians attacked Boonesborough fort. . . . 26. A party went out to hunt Indians; one wounded Squire Boone. . . . 30th. Indians attacked Logan's Fort; killed and scalped William Hudson, wounded Burr Harrison and John Kennedy.

June 5 . . . Daniel Lyons . . . we suppose was killed going into Logan's Fort. John Peters and Elisha Bathey we expect were killed. . . . 13th. Burr Harrison died of his wounds. . . . 22nd . . . Barney Stagner, Sen., killed and beheaded half mile from the fort. A few guns fired at Boone's . . .

Small parties of reinforcements now began to arrive from the settlements farther south but the situation was still serious, though enlivened by "Lieutenant Linn married—great merriment . . ." and, on September 23, a report "that General Washington had defeated Howe. Joyful news, if true . . ."

The entire population of Kentucky was about right, in that month, to fill the guest list of a modest wedding reception or to gather for a neighborhood celebration of General Washington's victory. Captain John Cowan took a census and found that it' totaled 198, "Men in service, 81; Men not in service 4; Women, 24; Children over ten years, 12; Children under ten years, 58; Slaves above ten years, 12; Slaves under ten years, 7 . . ."

During 1777 Spanish Louisiana had done much to alleviate the shortage of munitions among colonial troops in the west. Governor Gálvez was said to have furnished supplies to a total value of $70,000 (in violation of Spain's neutrality) and the merchant Oliver Pollock had purchased more with his own funds. Gálvez had reasons for his generosity. Colonel George Morgan, commanding Fort Pitt, had planned to send a force down the Ohio and Mississippi to New Orleans for an attack upon Mobile and Pensacola, then in the hands of the British. Spain already had her eye on the towns, which would presently become important ports of Spanish Louisiana, and no doubt Gálvez considered the supplies he contributed as a premium sufficient to ensure a reasonable attitude on the part of the Americans toward his government's plans and against a possible future invasion of New Orleans by American troops.

Governor Gálvez was probably correct; his $70,000 was well spent. At any rate there was a new comparative plenty of powder and lead at Fort Pitt and the colonial government was cheered by Burgoyne's surrender. George Rogers Clark took advantage of this auspicious season of hope to present to the Virginia government his plan to capture Kaskaskia and Vincennes, thus to control the Ohio and its northern tributaries, and eventually to take Detroit, the western headquarters of the British. He first approached Governor Patrick Henry on December 1, 1777: shortly Thomas Jefferson, George Wythe, and George Mason were informed and gave their approval.

The scheme was presented only in part to the Assembly. Evidently in the belief that the body would flinch from an undertaking so extensive as that actually planned, its proponents asked and received authorization only for an attack against Kaskaskia, reported to be weakly garrisoned but the repository for a considerable quantity of potentially useful British arms. In this instance, as through most of his western campaigns, George Rogers Clark was to be under two sets of orders—one announced, the other secret—and while this arrangement proved efficient as a military measure it was to become a great contributing factor to the difficulties he later encountered in securing authorization for payments of the accounts of his forces; even for proper pension and pay rights for his men. But these troubles lay ahead and—fortunately for the West—Clark was always a man to win his battles first and to consider the problems of his paper work afterward. Now commissioned colonel in the Virginia forces, he was ready for action. He left Williamsburg on January 18, 1778, and made his headquarters at Redstone (present Brownsville, Pennsylvania) on the Monongahela.

Colonel Clark's first move was to dispatch recruiting parties to canvass the back settlements from Fort Pitt to the Carolinas. Requirements for volunteers were not exacting: a man should have two legs but of no specific length so long as they reached the ground; he needed at least one good aiming eye, but if he was not naturally so equipped the chances were against his having survived upon the frontier anyway; a finger to pull the trigger was required but it need not be a forefinger—any digit which served the purpose was acceptable; front teeth with which to bite and twist off the end of the wadding paper around a bullet were required for enlistees in the regular army (that requirement survived in some branches of the United States armed service until the 1940's, although the musket had been abandoned some time before) but recruiting agents on the frontier were not particular as to teeth— any coinciding pair with which a man could bite off a chew of tobacco would serve as well on wadding. As to age, that was the volunteer's own business. If a boy was persuasive enough to get his parents' consent or resourceful enough to run away from home, he would probably make a good soldier and if a man *felt* young enough to enlist, he didn't need to consult his grandchildren, did he? Even so, recruiting was a slow business.

Five hundred men were considered necessary for the campaign but securing them was no sinecure. Many leading men of Virginia opposed the western move, discouraged enlistments and even concealed those whom they could persuade to desert. At length Clark, weary of waiting for volunteers who did not come and of watching what men he had melt away into the woods, set off on the Monongahela from Redstone and continued down the Ohio with about a hundred and fifty volunteers. They made an easy voyage to the Falls.

The force was joined by several emigrant families who sought protection and whom Clark, well knowing the danger they were in, could not turn away. The whole party encamped on Corn Island, then well out in the rapid water at the Falls, and upon it they built a blockhouse and sheds for their supplies. Clark hoped that the swift water would deter desertions—especially at the critical moment when he should inform his men of the true extent of the expedition into the enemy country—but in this he was disappointed: "Kept strict Guards on the Boats, but Lieutenant Hutchings, of Dillard's Comp'y, contrived to make his escape with his party after being refused leave to return, luckily a few of his Men was taken the next day . . ."

Whether or not Clark had advance information of the event (Thomas Jefferson was much interested in the anticipated phenomenon and might well have warned him) he selected the very moment of the great eclipse of the sun of June 24, 1778, to shoot the Falls and begin the campaign. The effect of the eclipse, three-fourths to nine-tenths total, can be imagined as likely to impress even the more sophisticated of his men with the importance of the occasion. The eclipse impressed his own men and Clark was likewise armed with something to affect the French residents of the Illinois country—news of the Franco-American Alliance of 1778.

Four days later the boats entered the mouth of the Tennessee River and the force encamped. Scouts soon brought in a party of not unfriendly hunters from Kaskaskia. Clark, interviewing them, got what information he could about the state of arms at that place. He best showed his generaliship, however, when he instructed the Kaskaskians to emphasize only their town's weaknesses in answering the questions of his own men. Thus when his forces were allowed to talk to the prisoners there began to develop in the camp a far greater enthusiasm for the campaign than had hitherto been evident. That evening, before a night could cool their ardor, Clark ordered his men to run the boats across the Ohio and push them up Massiac Creek, a mile above the ruins of the old French fort on the north bank. In the morning they began the march toward Kaskaskia across the ridges and intervening meadows of what is now called the "Egypt" of southern Illinois.

The country was strange to these troops from the wooded highlands, where travel usually followed a creek or river course, and they had a reasonable doubt as to the good intentions of their guide. He was John Sanders, one of the Kaskaskia captives, who had agreed to conduct them toward that town. Soon even he was obviously lost and Clark's men, skittish enough at best, began to mutter. Clark did not hesitate: "I was in a moment determined to put the guide to Death if he did not find his way that Evening; I told him his doom . . . He accordingly took his course, and in two hours got within his knowledge."

(It should be noted here that John Sanders, far from holding Clark's threat against him, became his firm admirer; that he followed Clark through the campaign and back to the Falls, where he became a trader and later carried on what may be considered the first commercial banking business in the infant village of Louisville.)

How Clark took Kaskaskia, on the river of that name then near (but now beneath) the waters of the Mississippi, how he won the support of the largely French population of that place, and how he secured the capitulation of neighboring Cahokia and later of Vincennes through the good offices of the friendly Kaskaskians, are stories of other rivers—the Mississippi and the Wabash. The summer saw these things accomplished and a council held with the western Indians at which Clark explained in terms understandable to them the causes and progress of the Revolution.

During the fall Clark, at Kaskaskia, heard that British General Henry Hamilton, the "hair-buyer," lieutenant governor of Detroit, was moving south with a large force to retake Kaskaskia, Cahokia, and Vincennes. But he was also informed that American General Lachlan McIntosh was passing down the Ohio from Fort Pitt to establish Fort McIntosh at the mouth of the Beaver and to garrison it with five hundred or more men—that further McIntosh would attack Detroit in the winter. Thus the approach of the enemy was more welcome than otherwise. If Hamilton came to the Illinois country, even though he might recapture the posts there, he would have necessarily left Detroit with little defense. He would have Continental forces on both sides of him—McIntosh at Detroit to his northeast and Clark on the Ohio and in the Illinois country to his southwest. Detroit must certainly fall to the Americans, and that was still Clark's major object.

But another one of those frustrations was to confront Clark: McIntosh built his fort and garrisoned it with the largest American force thus far gathered in the west during the Revolution—but he and his troops had been ordered to Georgia before Hamilton reached the upper Wabash. Not only would Detroit remain in the hands of the British, but Clark lacked even the means to protect the Ohio.

Hamilton's expressed aim, at the time his campaign was planned, was to dislodge Clark from Vincennes and the Illinois, to seize the Forks of the Ohio and the mouth of the Mississippi and to build a fort at the junction of the Ohio and Mississippi. He left Detroit with about 33 officers and men of the King's Regiment of Foot, 66 Detroit militiamen (mostly French), 45 regular Volunteers of Detroit, and 70 Indians. He recruited more Indians along the way and had, according to his diary, 223 men under his immediate command and about 300 Indians commanded by their own chiefs as he reached the lower Wabash.

The general was unopposed until he reached Vincennes. There the British commander found a defensive force ready to resist: it consisted of the entire garrison Clark had been able to leave at the place—Captain Leonard Helm, Private Moses Henry, and two unidentified Virginia volunteers. Captain Helm's command stood by a charged cannon in the open gate of the stockade—all four of the personnel—Helm with a lighted match in hand while the enlisted men waited to reload.

"Halt!" ordered Captain Helm.

Hamilton's army halted and the general demanded surrender.

"No damn' man will enter till I know the terms!" replied the captain.

General Hamilton conceded the "honors of war" and, that seeming reasonable enough under the circumstances, Helm and his army of three marched out in good order. Certain letters in General Hamilton's own hand indicate that he added somewhat to the number of the defenders whose capitulation he had received— multiplying them, by one account, to a total of 217 men and furnishing a list of their names!

Vincennes was thus retaken and the Illinois towns to the west lay unprotected, for Clark's detail there had degenerated to almost nothing through fatigue, illness, necessary scattering on guard and messenger service, and that constant bane of commanders of the individualist frontiersman, desertion. Clark was totally unable to defend Kaskaskia, Cahokia—or for that matter the Ohio or the Kentucky country—had Hamilton chosen to attack. Possibly because, as Hamilton confided to his diary, "the South side of the river appears like a lake," he did not so choose.

It was at this perilous moment that the resourcefulness and daring of George Rogers Clark most fully demonstrated themselves: being unable to defend his own position he decided to divide his minute force, recruit some Frenchmen from Kaskaskia, and attack the enemy. Upon this decision and its accomplishment rests his undying fame. Even General Hamilton, later contemplating Clark's campaign in the quiet of his cell at Williamsburg prison, wrote that it was a military feat "unequalled perhaps in History."

Clarks own diary describes the progress in detail:

I had a large Boat prepared and Rigged, mounted two four pounders 4 large swivels Manned with a fine Comp. commanded by Lieut. Rogers. She set out . . . with orders to force her way if possible within ten Leagues of St. Vincents [Vincennes] and lay until further orders . . .

The boat would go, of course, down the Mississippi to the Ohio, up the Ohio to the Wabash, and up the Wabash to the point mentioned.

The troops to be used by land were now ready. They included two companies of volunteer Frenchmen, Captain Bowman's foot company, and William Worthington's light-horse troop. The population of Kaskaskia gave them Godspeed, Father Pierre Gibault "after a very suitable Discourse to the purpose, gave us all Absolution," and Clark led his army off across the prairie toward Vincennes, 240 miles east. The whole army, including the forty-six men who manned the boat, numbered a few more than two hundred! The date of starting was February 5, 1779.

The prairie, soaking in the early thaws as only flat land which has grown a tangle of matted grass for century after century can soak, was far from easy marching at its best. Clark's "greatest care was to divert the Men as much as possible in order to keep up their spirits." There was certainly a fife and drum, Clark's favorite instruments (in his old age he lay fully conscious while his leg was amputated to their music), and there was certainly the batting back and forth of jests and personalities blending the bawdy wit inherited by the Virginians from the day of the Virgin Queen of England with the happy vulgarity of the frontier French. No difficulty greater than the men had long known arose until, two-thirds on the way toward their destination, they encountered the twin streams that join in the forks of the Little Wabash.

> Although three miles asunder [wrote Clark], they now make but one, the flowed water between them being at Least three feet deep. Being near five miles to the opposite Hills. . . . This would have been enough to have stopped any set of men that was not in the same temper we was.
> But in three days we contrived to cross, by building a Canoe, ferried across the two Channels, the rest of the way we waded; Building scaffolds at each to lodge our Baggage on until the Horses crossed to take them . . .

They finally reached the River Embarrass—so named by the French in testimony of its being a hindrance to travel under optimum conditions—and found it in deep flood, its overflow joining that of the Wabash nine miles away. They marched down the Embarrass to its junction with the Wabash, about ten miles below Vincennes. Here Clark built a small canoe and dispatched a messenger down the larger river to meet and fetch up the gunboat. The force was now separated from the town by the Wabash and ten miles of flooded bottom land; their provisions had been entirely exhausted two days before.

Of the methods by which Vincennes was reached, Clark himself said, in a letter written later (November 19, 1779) to George Mason, one of his Virginia sponsors:

> If I was sensible that you would let no Person see this relation, I would give You a detail of our suffering for days in Crossing those waters, and the manner in which it was done, as I am sure that You wou'd Credit it, but it is too incredible for any Person to believe except

those that are as well acquainted with me as You are . . . I hope you
will excuse me until I have the pleasure of seeing you personally. . . .

Fortunately one participant, less fearful of aspersion than Clark,
left an account of his impression of the crossing. Young Captain
Joseph Bowman, under no necessity of reporting to Virginia elder
statesmen, kept a journal, after the pleasant fashion of the day, for
his own amusement. Captain Bowman's account, taken up at this
point, reports:

20th. Camp very quiet but hungry; some almost in despair; many
of the Creole volunteers talking of returning. Fell to making more
canoes, when, about 12 o'clock, our centry on the river brought to a
boat with five Frenchmen from the Post, who told us we were not yet
discovered, that the inhabitants were well disposed towards us &c . . .
news to our favor, such as repairs done the fort, the strength &c. They
informed us of two canoes they had adrift . . . One of our men killed
a deer, which was brought into camp. Very acceptable.

Next day they crossed the Wabash River's channel, although
miles of flooded bottom land still lay between them and the town:

21st. At break of day began to ferry our men over in our two canoes
to a small hill. . . . The whole army being over, we thought to get them
to town that night, so plunged into the water sometimes to the neck,
for more than one league, when we stopped on the next hill . . . being
no dry land on any side for many legues. Our pilots say we cannot get
along. . . . Rain all this day—no provisions.

22nd. Col. Clark encourages his men, which gave them great spirits.
Marched on in the waters . . . we came one league farther to some sugar
camps, where we stayed all night. . . . No provisions yet. Lord help us!

23rd. Set off to cross the plain called Horse-show Plain, about four
miles long, all covered with water breast high. Here we expected some
of our brave men must certainly perish, having froze in the night, and
so long fasting. Having no other resource but wading this plain, or
rather lake, of waters, we plunged into it with courage, Col. Clark
being first. . . . Never were men so animated with the thought of aveng-
ing the wrongs done to their back settlements . . .

About one o'clock we came in sight of the town. We halted on a
small hill of dry land . . . where we took a prisoner hunting ducks, who

informed us that no person suspected our coming at that season of the year. Col. Clark wrote a letter by him to the inhabitants . . .

Clark's letter informed the French citizens that he was within two miles of Vincennes; that, as in the past, those friendly would be protected but any who might find themselves attached to "the King, will instantly repair to the fort and join the *Hair-buyer General* and fight like men." Clark then released the prisoner and concealed his men as best he could. Capt. Bowman's account continues:

In order to give time to publish this letter, we lay still till about sundown, when we began our march all in order, with colors flying and drums braced. After wading to the edge of the water breast high, we mounted to the rising ground the town is built on about 8 o'clock. Lieut. Bayley, with fourteen regulars, was detached to fire on the Fort, while we took possession of the town . . .

Hamilton could not believe he was being attacked. He wrote later that he attributed the shots to a "drunken frolic of the inhabitants . . . but going upon the parade heard the balls singing." As his men began to fall, Hamilton saw his error.

Bowman continued:

Reconnoitered about to find a place to throw up an entrenchment. Found one, and set Capt. Bowman's company to work. Soon crossed the main street, about one hundred and twenty yards from the first gate . . . The cannon played smartly. Not one of our men wounded. Men in the Fort badly wounded. Fine sport for the sons of Liberty.

Of this first action there was early related one of those yarns for which Kentuckians were famous throughout the West. In view of the obviously hard-bitten nature of its chief characters and of the frontiersman's known facility at making himself what amusement he could, it may as well as not be true.

One of Clark's men—it must be recalled that there was little social distinction between officers and enlisted men in such an army —had made a special request as they entered the town. He told Clark that he wanted permission to locate the quarters in Fort Sackville in which their old friend Captain Helm was confined so that he could shoot some clay chinking from the fireplace into the apple toddy that he knew Helm would be heating in the embers at this hour in the evening! A sort of welcoming and reassuring ʒesture to the captain it was to be; designed to show him that friends were nigh.

Helm had been playing piquet with Hamilton himself when the attack began and some of the first shots, evidently aimed from a tree through one of the portholes in the room, knocked Helm's toddy cup full of mortar. Leaping to his feet he called to Hamilton that Clark's men were about them; that they would all soon be prisoners—"but the damned rascals have no business to spoil my toddy!"

Next morning, says Bowman:

24th. As soon as daylight, the Fort began to play her small arms very briskly. One of our men got slightly wounded. About 9 o'clock the Colonel sent a flag with a letter to Governor Hamilton. The firing then ceased, during which time our men were provided with a breakfast, it being the only meal of victuals since the 18th inst. . . .

Clark's letter demanded immediate surrender. It was rejected and firing was resumed but again with no serious casualties among the Continentals and, now that they had discovered the efficacy of firing in through the portholes and between the pickets, wounds aplenty for Hamilton's men. Hamilton now proposed a truce of three days: Clark refused but agreed to meet the governor at the village church.

About this time there arrived a party of Indians, led by "Francis Masonville" (said Hamilton), a Frenchman who was also in full war paint. General Hamilton later maintained these men had been ordered out on a scouting expedition but they had with them two Kentucky prisoners and several scalps; that was enough evidence for the Virginians at the moment. Sight of these bloody trophies marked the turning point in the siege: these might be anyone's scalps—possibly of children, brothers, fathers, wives, or mothers— of Clark's very men: the effect upon the Americans was terrible to see. The Indians were tomahawked (by Clark himself, according to Hamilton) before the gates of Fort Sackville, their bodies were thrown into the river, and Hamilton—probably within hearing and sight of the episode—fearfully signed Colonel Clark's terms of unconditional surrender.

It must be here recorded that the *Willing,* the boat Clark had armed with the cannons and swivels and launched at Kaskaskia, failed to arrive until well after the surrender. Thus ineffectual proved the first show of American naval power in the lower Ohio Valley.

Within three weeks the British privates had been released on parole, General Hamilton and the officers had been started to the Falls on their way to Virginia, Captain Helm had been made civil governor of Vincennes, Lieutenant Brashers commander of the fort, and Clark had sent peace talks to the neighboring Indians.

"Everything having the Appearance of Tranquillity," he wrote, "I resolv'd to spend a few weeks in Divertions which I had not done since my Arrival in the Illinois, but found it impossible when I had any matter of importance in view, the Reduction of Detroit was always uppermost in my mind . . ."

Bowman, back in Kentucky, sent word that as a result of the inspiring victory he could now raise three hundred men in Kentucky for an attack on Detroit; others promised men as well. Clark set June 20, 1779, as a time to rally at Vincennes for the expedition.

But July found only thirty men arrived at Vincennes under Colonel Montgomery, with word that Bowman, now a colonel, had led his force of one hundred sixty north across the Ohio and against some Shawnees at Chillicothe, on the Little Miami River, where they had suffered a rather humiliating defeat. As a result Bowman's men were in no mood for further action of the kind Clark proposed. In this unhappy situation Clark took the next best course. He realized that the Shawnees could not safely be left to enjoy their triumph over Bowman and the minor victories they had recently enjoyed on the upper Ohio. Late in August he returned to the Falls to raise troops to chastise those whom Bowman had only annoyed.

As matters developed, even this wise action had to be postponed. Upon his arrival at the Falls, Clark found orders which instructed him to establish a fort at or near the mouth of the Ohio.

Immediate action was impossible because of a "lowness of the Ohio which is so remarkable that it would be worth Recording, few being able to navigate it with the smallest Canoes for several months past." But the waters of the Ohio rose, as they always do with the beginning of the new year, and Colonel Clark moved down the river with the small but exceedingly loyal remnant of the original force which had survived early desertions and later hardships. Upon arrival the men began to build Fort Jefferson on the Kentucky shore of the Mississippi. Orders had been to establish it "where the line of latitude of thirty-six degrees thirty minutes strikes the east bank" but it was actually established about twelve miles downstream from the mouth of the Ohio, below the present town of Wickcliffe, Kentucky.

Wise enough in its broad concept of defending the mouth of the Ohio and protecting the line of supply from New Orleans, the plan for establishing a fort there and then was impractical in its details. The Kentucky lands between the Tennessee and Mississippi rivers (present western Tennessee and southwestern Kentucky) belonged to the Chickasaw Indians who had been good and useful friends of the Americans throughout the Revolution. The Chickasaws naturally resented this unauthorized incursion and so demonstrated, eventually, by an attack. Anyway, even from the beginning, Clark had no hope of constituting a proper garrison—not even a Clark-trained regiment could be divided to cover the entire West— and Fort Jefferson was farther from his sources of supply than had been even the far-enough Kaskaskia and lower Wabash. Fort Jefferson was abandoned within little more than a year after its completion.

In spite of the weakness of the Continental Congress, which resulted in countermanded orders and changes in tactics, progress had been made in the West during 1778 and 1779. The posts north of the Ohio had been taken, Vincennes had been lost but brilliantly recovered, and the British plan to penetrate Kentucky and make contact with their forces in Virginia was thus rendered hopeless of achievement for the time being. The campaign of McIntosh against Detroit had been diverted and, even though Bowman's raid on

Chillicothe had failed, small parties of independent volunteers had made effective reprisals for Indian raids.

The winter of 1779-80 gave the Westerners a new ally—the weather. That was a long-remembered "hard winter" and the intense cold that kept the Ohio water at low level also kept the Indians huddled in their smoke-filled huts and reduced their larders to the danger point. This situation, plus the harassment carried on by the settlers, had a tendency to cool the warlike ardor of the Indians, especially on the upper Ohio, and persuaded them to sue for peace with Colonel Brodhead at Fort Pitt. There followed, as far as Indian trouble was concerned, a comparatively tranquil period upon the upper Ohio for nearly a year; even upon the lower river there was respite for a few months. During this time, it is pleasant to note, civil progress continued in the west, cut off as it was almost entirely from contact with the world: the town of Louisville was chartered in May, 1780, with George Rogers Clark as one of the planners.

The calm was not for long: the very month following Louisville's elevation to chartered existence Kentucky was the scene of an invasion, by way of the Ohio and its tributaries, the Miami and Licking, which was the most dramatic of all the attacks upon the western country.

Governor Hamilton had originally hoped to use Vincennes as a base from which to assail the Kentucky settlements and even Spanish Louisiana. Clark's recapture of the town not only rendered this plan impossible but had put Hamilton in the Williamsburg military prison—a rather impractical headquarters from which to command such a comprehensive campaign. The next move of the British, made in June, 1780, threatened a far more efficient settling of the Kentuckians than raids from nearby Vincennes could ever have effected.

Detroit was well situated to control the Ohio, except only for its distance from the river. It was located at the head of well-established trails and waterways which led through the Maumee and Great Miami valleys to the Ohio near present Cincinnati, by the Maumee and Wabash rivers to Vincennes and the Ohio River below, and by Lakes Huron and Michigan to the St. Joseph River

and the portage to the Kankakee or the Chicago River and ready
connection with the Illinois and thence to the Mississippi and the
mouth of the Ohio. Governor Hamilton had come down the
Maumee and Wabash to Vincennes and had failed. Someone now
presented a plan for seizing the Ohio between the Great Miami and
the Falls after an approach by another route—down the Great
Miami valley.

Obviously the British control of that stretch of the Ohio River
(from the modern western suburbs of Cincinnati to Louisville)
could stop both shipments from New Orleans to the upper Ohio
and the movement of troops and supplies downstream from Fort
Pitt and Wheeling to the Kentucky country. Such control would
put an immediate end to the effectiveness of George Rogers Clark's
projects and would prevent any effective co-operation between his
command and the upper Ohio posts; his only remaining contact
with the Continental Congress or the government would be by
the Wilderness Trail over the mountains—a slender thread indeed
when arms, supplies, or even any significant number of men were
to be moved. Why the advantages of such an attack had not
occurred to the British before is hard to understand; now, in the
spring of 1780, its possibilities became apparent.

Contemporary accounts of the expedition that the British finally
launched from Detroit down through the Maumee and Great
Miami valleys toward Kentucky are conflicting and confusing.
Owing to the secrecy in which it was planned and the manner in
which it was carried out, neither the defenders of Kentucky nor the
early historians understood its true object until, almost a century
later, the papers of British General Haldimand became available
to American scholars. Only then did its significance and the reasons
for its failure become known.

By the British plan, attacks were to be made simultaneously
by British and their Indian partisans against all readily assailable
points in the west—Kaskaskia and Cahokia, Vincennes, Spanish
St. Louis (for Spain and England were now at war), the new Fort
Jefferson that Clark was building on the Mississippi and the Falls
of the Ohio, key to all river activity. The plan even extended to

New Orleans, which was to be attacked by British troops then stationed at Pensacola.

Early in the spring the settlements heard rumors of threatening activity at Detroit but Colonel Clark, with the only regular troops in the West, was busy at Fort Jefferson and even Daniel Boone was out of the country. The only defensive action taken was a casual alerting of the Kentucky militia and some minor strengthening of independent community stockades. Truth to tell, there was a large emigration to Kentucky in process at the time and land speculation was more in the shortsighted public eye than was defense.

The rumors were correct. Early in June, 1780, British Captain Henry Bird assembled an army of between five and seven hundred whites and Indians for the principal object, the attack on Kentucky and the seizing of the Falls. All in all it should have been an effective force, even without regard for the fact that it was one of the largest bodies of fighting men yet moved in the western country. The expedition was well equipped, the British having only recently supplied their Indian allies with new guns and tomahawks, and it even carried a few pieces of artillery. Three of the Girty brothers were included in the personnel—and those three themselves constituted a fear-invoking army on the frontier—there were several prominent Indian chiefs and even the upper Ohio River Tory (then deputy Indian agent) Alexander McKee. The only weakling in the expedition, apparently, was its commander, Captain Bird. He was a British regular and his weakness, already demonstrated in a previous command over Indians and whites of the Girty stripe, consisted in common decency and a pronounced distaste for the practices of scalping, maiming, and burning captives at the stake, for the slaughter of noncombatants and wounded, and for indiscriminate looting. It was probably this squeamishness of the captain's that saved Kentucky and the west.

Bird's force moved south along the Great Miami to the Ohio and stopped to hold council at the junction of the two rivers. It was there that Captain Bird's troubles began. Instead of the quick dash down the Ohio to the Falls (the original plan—which must certainly have been disastrous to the Americans), his Indians insisted that they be first permitted to cross the Ohio, proceed up the Licking River to the small stockades located upon it and to raid them and the settlements that lay in the Bluegrass before moving on the principal objective. Bird had no choice: his Indians outnumbered his whites at least three or four to one and the Indians wanted loot and prisoners. The advance up the deep-cut Licking River valley began.

As evidence of the state of Kentucky defenses, it should be noted that this large body was able to cross the Ohio unobserved, and to spend twelve days on the march up the Licking River to the first major settlement, Ruddle's Station, chopping out a road as it proceeded, while the inhabitants of that station carried on their daily business as usual!

Ruddle's was a rough stockade enclosing several cabins and planned to shelter the outlying families of the neighborhood in time of danger. The total population it was expected to serve was probably less than three hundred men, women, and children. Isaac Ruddle, founder of the settlement, was its commander, and it may be assumed that his force of able-bodied men was below the average; at least some of the youngsters must certainly have been away on service in Clark's and other armies.

During the night of June 21, the little stockade was surrounded by an advance party of Indians under McKee and firing began at dawn. By noon Captain Bird came up with the main body of troops, cannon fire opened a breach in the log wall, and official interpreter Simon Girty was sent in under a flag of truce to demand surrender.

Resistance was hopeless but doughty old Isaac Ruddle refused to be intimidated either by odds of ten to one or by the bloody Girty in person. Ruddle would surrender, he said, but only on his own terms: to the British officers and with the understanding that, as British captives, his people would be protected from the Indians. Captain Bird agreed to this provision and the gates were thrown open.

The captain issued the proper orders and Simon Girty translated them to the Indians—but to the horror of Bird (if not, perhaps, of McKee or the Girty brothers) the Indians, uncontrollable, swept in to seize those individuals who caught their fancy and to tomahawk any who resisted. In Bird's own words, "They rushed in, tore the poor children from their mothers' breasts, killed and wounded many." Cattle and hogs were wantonly and purposely shot down and all supplies were burned in the stockade and cabins.

The Indians divided the prisoners and booty it had pleased them to save—ornaments, weapons, kitchen furniture, and clothing —and, delighted as they always were under similar propitious circumstances, demanded to be led to the next station. Bird agreed to this after the chiefs of the various bands in his force promised to take only the plunder as their share, leaving further prisoners to the officer's discretion.

Martin's Station, smaller than Ruddle's but next most important in the district, was surprised as completely and surrendered as promptly. In this case, however, the promise of the chiefs held good and the prisoners were taken by Bird's whites while the Indians looted the settlement.

The Indians now wished to move against Lexington, largest settlement of the Bluegrass, but Bird refused, sick with disgust at what he had seen at Ruddle's and aware that the expedition could not now expect to reach the Falls of the Ohio in any case. The high water that had enabled his men to bring their cannon up the Licking had begun to fall. Provisions were running low, with all the prisoners to feed as well as the men of the expedition (the ample stores at Ruddle's and Martin's had been destroyed by the savages for amusement), and Lexington was known to be well fortified. His expedition a military failure, Captain Bird ordered a retreat by the route he had come.

Bird's redskin allies soon forgot their disappointment; many of them had prisoners from Ruddle's, available as slaves, as objects of torture and fireside amusement or for adoption into their families, and there was a windfall of loot. As was usual with them, this immediate plenty distracted their thoughts from further gains.

Knowledge of the raid was confined to its victims in Kentucky until John Hinkston, whose farm gave name to the stream that enters the Licking at Ruddle's Station and who was one of the prisoners of the Indians, escaped during the first night. Running to Lexington next morning he brought the first word of the disaster on the Licking and of what Lexington had been spared.

Once back north of the Ohio, most of the Indians dispersed to their own towns. Only the Lakes Indians continued with Bird to Detroit, which place they reached on August 4, when their prisoners were turned over to the British. More than three hundred Kentucky men, women, and children were delivered to the commanding officer at Detroit—all of the people from Martin's Station and those from Ruddle's who had been so fortunate as to have been captured by representatives of the Lakes tribes. What happened to those whom the Delawares, Shawnees, and Miamis had seized may only be conjectured.

This British campaign was certainly the most destructive of the Revolutionary War period in the Ohio Valley and, but for the Indians' love of destruction for its own sake, possibly encouraged by a perverted translation of Bird's orders by the American-hating Simon Girty, the results might have been much worse. If Girty did give the Indians an incorrect translation of Bird's commands at Ruddle's Station the renegade only succeeded in defeating his own purpose, for, well fed on the beef and corn stored there and restrained from their more wanton excesses, the Indians and Canadians could easily have decimated Kentucky and given the Ohio to British control within another two weeks.

The other parts of the British plan—the attack upon St. Louis, New Orleans, and the Illinois towns and the capture of Clark at Fort Jefferson—were thwarted, and Clark, as soon as he heard of Bird's expedition, hurried east to central Kentucky. He was, of course, too late—for Bird had flown, and precipitately.

Back at the Falls Clark found a letter from the governor of Virginia recommending a campaign against the Indians on the Great Miami (who had participated most recently in Bird's raid, though the governor did not know of it when he wrote) and particularly against the British trading post known as Loramie's Store, located on the portage between the upper Maumee and Miami rivers in western Ohio. Clark immediately called for volunteers to supplement his veteran troops, but even the horrors at Ruddle's Station had not entirely diverted the attention of the

Kentuckians from their current mania for land speculation: in order to raise the minimum of six hundred volunteers needed he was forced to hamstring business by closing the land offices by martial order.

Then he got his men. Just under a thousand, including the regulars, rallied at the mouth of the Licking (present Covington, Kentucky) with some artillery brought up the Ohio by boat from the Falls.

The expedition was a success, though not a brilliant one. There was only one skirmish, at Piqua, where each side lost seventeen men, but the Indians retreated and Clark's men were able to burn towns, fields, and stores. By December George Rogers Clark was back in Kentucky and again planning to assail Detroit. There seemed to be hope, this time, for reviving that old dream; General Washington believed the move was practical and agreed with Governor Thomas Jefferson of Virginia that the Continental command should furnish supplies and arms if Virginia could raise, through Clark, men enough to carry out the project.

But again there was to be only disappointment. Clark went up the Ohio to Fort Pitt, where Jefferson had arranged for Baron von Steuben to release Colonel John Gibson to act as Clark's second in command and was trying to persuade Colonel Brodhead to permit Gibson's regiment to accompany him. Brodhead refused. The Continental government was pitifully weak, the best of the troops were worn and discouraged, and something very like a state of anarchy existed upon the upper river.

Blood of innocent and noncombatant women and children shed upon the Ohio in 1781-1782 was by no means confined to Kentuckians—nor whites. Through the winter hostile Indians aimed desultory raids against isolated settlers on the Pennsylvania and Virginia shores and a retaliatory move was made in March, 1782. Typical of the purposeless and occasionally vicious conduct of the whites in that region, it again struck at peaceful Indians who had had no connection with the previous troubles and whose worst offense consisted of entertaining British partisans—as they did travelers of all persuasions who applied to them.

On the 4th of March one David Williamson, colonel of

militia from Washington County, Pennsylvania, gathered a force of upper river militia at the Mingo Bottom (a riverside meadow below present Steubenville, Ohio) and marched north toward Gnadenhutten, the Moravian mission town peopled by Christian Delawares. Williamson had no orders; his object and that of his men was said to have been chiefly to create a safe and amusing diversion for themselves which might save them from being ordered to participate in the more dangerous campaign George Rogers Clark was planning against Detroit. Most of Williamson's men were such as had already distinguished themselves by bloodthirsty and savage excesses between Wheeling and Fort Pitt on the Ohio. Of them their former commander, General William Irvine, had written to George Washington: "No man would believe, from their appearance, that they were soldiers; nay, it was difficult to determine whether they were white men!" Now, in the attack on the Christian Indians, they excelled their previous records.

Colonel John Gibson, in temporary command of Fort Pitt and of the western department, sent a messenger to Gnadenhutten to warn of Williamson's unauthorized sortie as soon as he heard of its object but the messenger arrived too late. Williamson's ruffians had already reached the town, had found about a hundred and fifty unarmed men, women, and children at work in the fields, and had slaughtered some ninety of them before the rest escaped into the woods.

The killing required only a little time and a small expenditure of strength, for there was no resistance; the Moravian mission Indians, having only a few years' experience in living under the teachings of Christ, took His admonitions seriously and naïvely trusted that others of His professed followers would do likewise.

The massacre of the Moravian Indians, as even a rather prejudiced local historian of the nineteenth century felt impelled to remark, "did not allay the excitement upon the frontier; it was now prevailing all along the border!"

More friendly Indians were killed—some of them even bearing arms for the Americans—and Colonel Gibson's life was openly threatened by whites when it became known that he had attempted to warn Gnadenhutten of Williamson's approach. Shortly General

Irvine returned to the command of the department, and fortified by rank and some rather positive ideas on the subject of discipline, began to regulate the wild and mutinous troops by the only means their animal minds seemed able to comprehend. After a brief season of court-martials, executions by firing squad and noose, and a liberal dealing out of the particularly impressive "hundred lashes, well laid on," some semblance of discipline was achieved.

On May 25, 1782, from that same Mingo Bottom at which Williamson had rallied his murderers, Colonel William Crawford led four hundred and eighty Pennsylvania and Virginia volunteers north on another campaign. This time there was a legitimate military objective, the Wyandot and Shawnee towns on the upper Sandusky River. The campaign was almost as bloody as Williamson's expedition—but the blood was not Indian.

Crawford's force was surprised on the march by a party of British and Indians under Captain William Caldwell. The ensuing battle had lasted through two days before Crawford was compelled to order a retreat—with the results that were to be anticipated from frontier volunteers. The upper river men, never exactly amenable to discipline, could no more be controlled in defeat than in victory and this time it was unquestionably defeat, and this foe was not composed of unarmed mission-converted men, women, and children handicapped by the influence of New Testament principles; this was a body of warriors efficiently commanded.

Colonel Crawford's men ran through the woods like hunted rabbits. Seventy of them were killed or captured; and the lucky were the killed.

Crawford himself, captured, was awarded as a prize to the Delaware members of the attacking party. These were unconverted Delawares but the Christians who had fallen at Gnadenhutten were their relatives and these, not bound by a gospel which taught that vengeance was the monopoly of the Almighty, were free to take full measure of revenge upon their captives for the wrongs done their people. On Colonel Crawford they expended all their native ingenuity plus what refinements they had learned from association with some of the border whites.

William Crawford was a competent officer, a lifelong friend of

George Washington, and by every evidence a good soldier and a responsible citizen. These virtues did not save him now, for he had commanded a body of volunteers, many of whom had seen action with knife, club, and ax at Gnadenhutten: that was enough for the Delawares.

The story of Crawford's torture has been told and retold—how he survived whipping, roasting by a slow fire, seventy rounds of powder shot into his skin at close range, maiming by knives, scalping; how he begged Simon Girty, who was with the Delawares, to shoot him and how Girty laughed at him. How finally, as the sun set, he died—paying, as had the Christian Indians of Gnadenhutten, for a crime he did not commit.

The Indians, inspired by their victory, moved on into Pennsylvania, occupied a part of Westmoreland County and talked with the western tribes of a large-scale attack upon Kentucky. The odds in favor of the complete success of such a move were long indeed but the Indians' native lack of concerted or long-continued purpose again saved the western whites. Diverted by temporary success, they let their opportunity slip.

Amid conditions such as these which prevailed upon the upper Ohio George Rogers Clark had tried to gather the additional troops necessary for an attack on Detroit. This last effort resulted in frustration—as usual—and a pathetic tragedy as well.

Back in August, 1781, Clark had laboriously collected four hundred men at Wheeling and—hopeless of further accessions from recruiting—had decided to drop down the Ohio. He waited beyond the date set to embark hoping, but probably by now no longer really expecting, to be joined by a promised force of Pennsylvania volunteers under Colonel Archibald Lochry. Five days passed with no report of the Pennsylvanians and Clark set out. He left boats and supplies for Lochry, should he arrive.

Meanwhile Lochry, with good men but far too few of them, had come overland to Pittsburgh. He embarked on the river for Wheeling but reached that place several days after Clark had gone ahead.

Clark waited again at the mouth of the Kanawha, but his own men, many of whom were, as usual, unwilling volunteers, were deserting in alarming numbers. To discourage this defection Clark wisely decided to move farther from the settlements and he had dropped on downstream before Lochry overcame further delays and reached the Kanawha. The Ohio was low, Lochry and his men were inexperienced in river travel, he had no competent pilot, and his supplies were almost gone. He wrote a letter to Clark which detailed his condition and ordered Captain Shannon to push ahead with seven men in a light boat and deliver it to Clark. The captain, his men, and, worst of all, the letter describing Lochry's situation were captured by the Indians.

Clark had by this time given Lochry up and was moving toward the Falls as rapidly as low water permitted. Lochry was pushing downstream as best he could, expecting aid from Clark hourly. So the days passed until Clark was safe at the Falls, without knowledge that Lochry and his weakened men were still following.

Finally, moving slower and slower as their strength failed and their fears increased, Lochry's men passed the mouth of the Great Miami. They recognized it as the last navigable stream reaching north into the Indian country above the Falls and—even though a difficult passage still lay ahead—they probably believed they had cause to rejoice.

A few miles farther they passed the entrance of a small stream which flowed sluggishly through a deep-cut slit in the high terrace that here formed the north bank. Stagnant, dark, and drift choked, it was as dismal and forbidding a trickle of water as existed upon the Ohio and they scarcely noticed it. It was worthy of their attention, for, though it then had no name, it would thereafter be called "Lochry's Creek."

The men were desperately hungry. They realized by now that Captain Shannon's party must have been taken captive but their commander was either not familiar with Indian methods—did not realize that they might well have been followed for days by unseen Indians on both banks ready to pounce at the first opportunity—or he believed that his best chance lay in giving his men one good meal before undertaking a final dash for the Falls. What happened is plainly stated in Lieutenant Isaac Anderson's journal:

August 24th. Colonel Lochry ordered the boats to land on the Indian shore, about ten miles below the mouth of the Great Mayamee river, to cook provisions and cut grass for the horses, when we were fired on by a party of Indians from the bank. We took to our boats, expecting to cross the river, and were fired on by another party in a number of canoes, and soon we became a prey to them. They killed the Colonel and a number more after they were prisoners. The number of our killed was about forty. . . .

August 27th. The party that took us was joined by one hundred white men under the command of Captain Thompson and three hundred Indians under . . . Captain McKee.

August 28th. The whole of the Indians and whites went down against the settlements of Kentucky, excepting a sergeant and eighteen men . . .

So passed all hope of an expedition against Detroit; so, indeed, before his twenty-ninth birthday ended the distinguished military career of George Rogers Clark himself. Twice afterward there were flashes of greatness, but they were only flashes. The high purpose and the broad military concept were gone.

Clark, the man John Randolph of Roanoke would call "the Hannibal of the West," had lost his last chance for further military glory—through no fault of his own—but his misfortunes were themselves by no means at an end: they would continue to dog him throughout his life. From time to time he is seen, as he pledges his own lands to meet the obligations for arms and supplies which Virginia refused to recognize; as he sits—more often lies—in his cabin above the northern side of the Falls, his solitary bouts with his jug becoming more and more frequent; as, having burned himself by falling in his own fireplace, he lies still while his leg is amputated as his old fifer and drummer march round and round the house playing martial music to keep up the courage not even yet lost. He is seen again, a broken old man but a proud one, as he shatters the sword the state of Virginia had presented to him in 1779 (after buying it second-hand "of a gentleman who had used it but little") saying, "Damn the sword! I had enough of that—a purse well filled would have done me some service!"

But then back in 1781, those later disappointments were still in the future; in 1781 only Clark's dream of capturing Detroit was certainly ended. That place would remain a stronghold from which, through peace and war, the British would continue to incite the Indians against the frontier for another thirty-two years.

* * * *

George Rogers Clark reached the Falls late in August: he found trouble on all hands. The civil and military governments were in conflict at Kaskaskia and Vincennes and the citizens were objecting—with justice—because they were being forced to accept worthless Virginia currency for supplies purchased from them for the troops. This indignation of the citizens is readily understandable in view of the fact that Virginia herself generously estimated her paper money to be worth one thousand to one of gold or silver at this time. The woods were full of Indians, especially in present Indiana between the Great Miami and the Wabash. Clark proceeded to build Fort Nelson, at Louisville, the fall after Lochry's defeat, and the ordinary harrowing attacks on outlying cabins plagued Kentucky through the winter, but March, 1782, saw Indian warfare reopened all along the Ohio with greater fury than before.

First in Kentucky came the attack on Estill's Station, which resulted in a drawn battle. Then followed forays against half a dozen of the small settlements from the Kentucky to the Licking River. There was now no effective force of regulars between the Ohio River and Canada from Fort Pitt to the Wabash.

In early summer Simon Girty and a detachment of Canadian troops under Captain William Caldwell began to gather the warriors of the Shawnees and other western tribes for a grand attack on Kentucky. Soon they had five hundred men and during the first week in August this force crossed the Ohio. On the 10th it attacked recently constructed Hoy's Station and moved on toward Bryant's, the largest post on the eastern Kentucky frontier.

Bryant's Station consisted of about forty cabins erected in two rows, the ends of the lines closed by palisades to make a parallelogram. The spring that supplied water was outside the enclosure and the whole structure was lacking in eligibility for defense. Caldwell's Indians surrounded the fort in the night and their presence was discovered by the inhabitants. Before the attack began at sunup men of the station had been dispatched for help and the women of the fort had put into execution a piece of daring which is one of the proudest of Kentucky's historical episodes. Forming in single file, they had marched to the spring and returned, each with her two buckets of water, in full view of the besieging force!

Either the command was weak, the tactics were unsound or, more likely, the Indians were attacking a defending garrison now so experienced in their methods that their every move was anticipated, for Bryant's Station held out through all day against the vastly superior Indian numbers—even received some reinforcements within the stockade—and at sundown the Indians made one of their quick and inexplicable retreats.

Further reinforcements began to arrive at Bryant's the next morning: Colonel John Todd from Lexington, Lieutenant Colonel Daniel Boone from Boonesboro, and Lieutenant Colonel Stephen Trigg from Harrodsburg. Colonel Todd brought one hundred eighty-two men in all. Of course the Indians and Canadians must be pursued, but there was a serious debate as to whether to wait for Colonel Benjamin Logan, who was expected within twenty-four hours at the head of a large force, or to go ahead without him, so that the Indians would have less chance to cross the Ohio and scatter.

Kentucky blood was up and the pursuit began immediately. Daniel Boone, premier tracker of the wilderness, took the lead. Boone was nobody's fool, adventurous as he was; his life had too often hung in a balance of which his own discretion was the deciding weight. He had been a captive of the Indians upon several occasions—as had most members of his family—and his judgment was highly regarded by Indians as well as the experienced frontiersmen among the whites. As the Kentuckians proceeded, Boone began to see things he did not like; the Indians' trail was preternaturally clear, it looked to Boone as if they were going out of their way to make sure of being followed. If the invaders planned an ambush, Boone reasoned, there would be no better place than

Lower Blue Lick, located in the deep bowl where the Licking River made an almost complete circle.

Boone called this fact to the attention of his fellow officers but it was brushed aside. There was then in Kentucky—and would continue to be—a class of men whose fetish was bravado. Mostly of families imbued with the old traditions of the Virginia Tidewater, they had brought their carefully sustained reputations for daring into the wilderness and, although less competent in dangerous circumstances than were many such practical individuals as Kenton and Boone, they were far more facile in speech. Thus, their rashness was likely to inspire followers who, in their hearts, had more respect for the cautious woodsmen. One such bravo was Major Hugh McGary.

The first night out the Kentuckians camped within five miles of the Lower Blue Lick, which Boone feared. Next morning they moved forward, stopping for a conference before descending into the Licking valley.

Here the commander, Colonel John Todd, called publicly for Boone's opinion: Boone said that the enemy probably awaited them at the Lick; that they had best wait for Logan's reinforcements but, failing that, they had better divide into two forces and send one in a circuit over the highlands to advance upon the Lick from the opposite side. In any case, said Boone, the valley should be carefully reconnoitered. The words were no more spoken than McGary, in one of those wild bids for immortality which would still be the glory and the nemesis of his type of Southerner in the days of the Confederacy, raised a war whoop (presumably a rudimentary form of the rebel yell of the sixties) and "called vehemently on all who were not cowards to follow *him* and *he* would show them the enemy." Most of the men leaped up and followed him, less than half remained in council with Boone and Todd, and the whole army finally crossed the river. There Todd's orders, given at Boone's insistence, prevailed and a halt was called while scouts went ahead. Too soon they returned, having seen nothing. The men were ordered to move forward.

As the Kentuckians descended into the wide pocket surrounding the Lick they were met by a withering fire from the over-

grown ravines on the opposite side, where, as Boone had anticipated, Caldwell's Canadians, Girty, and the Indians were concealed.

When the battle was over at least a third of the Kentuckians
were dead; some of the dead were among the best men on the
frontier and of this best were several of the very cream: Colonel
John Todd, Lieutenant Colonel Trigg, Major Silas Harlan, and
one of Boone's sons. British records show, upon their side, one
white and ten Indians killed and fourteen wounded. It was a sorry
day for Kentucky.

As soon as possible, George Rogers Clark called a meeting of
leaders at Louisville; if Detroit could not be assailed, then again
the next best thing must be done, the Indian towns nearest the
Ohio's north shore must be once more destroyed so that there
would be no future place at which to base such invasions as that
just past. There should be no foolishness this time, the Kentuckians
decided; if volunteers were not forthcoming there must be a draft.
There would be no waiting for approval from the Virginia government or the Continental Congress; Kentuckians must furnish
their own arms, their own powder, their own beef. In September
a thousand men gathered at the point where the Licking River
enters the Ohio.

With Clark in command they struck north into the wilderness: the Indians, forewarned, escaped. But the towns on the Mad
River were burned again, rebuilt Loramie's Store was leveled and,
though few Indians were caught, they were at least driven back
to the north. Though it was not a campaign glorious in the Clark
tradition, it had its effect. After his disappointments of the past two
years, who could wonder if George Rogers Clark was falling somewhat below his early promise?

Indian troubles on the Ohio that winter were confined to horse
stealing here and a captivity or two there; by comparison with the
year preceding the river must have seemed to its people a very
haven of peace. So matters continued, too, through the spring and
summer. Immigration into Kentucky increased. Families from the
east coast, impoverished by the passing and repassing of colonial and
British troops, each making their levies on livestock and foodstuffs,

decided that perhaps even the threat of Indian raids would be preferable. So they crossed the mountains or took the longer chance and floated down the Ohio to the new, cheap lands. Finally, in November, 1783, provisional articles of peace were signed with the British, hostilities ceased, and, subject to the final ratification of the treaty (which would come on April 9, 1784), the Ohio River and its lands belonged once and for all to the United States—if the claims of the Indians were ignored; which of course they were.

Chapter Seven

BLOOD AND INTRIGUE IN THE BACK COUNTRY (1785-1815)

AT THE END of the American Revolution the Definitive Treaty of
Peace of 1783 launched a new country upon the troubled international seas.

The United States of America—American even to the extent
that it followed the pattern of the Iroquois League as at first a very
loose confederation of independent states—did not appear to be a
vessel too seaworthy.

That analogy of the vessel launched is perhaps uncomfortably
appropriate, for if ever a ship of state was completely at sea it was

that of the young Union. Even the first act necessary to its commissioning—that of gathering a quorum of war-worn members of Congress to ratify the treaty—proved a difficult business.

The delays, the disagreements, the misunderstanding, and the contention incident to the first five years after the surrender of Cornwallis in 1781 must have been a period of hellish anxiety for those patriots who foresaw the glorious future possible for the nation and realized the necessity for strengthening the Union immediately, regardless of their own exhaustion.

To this disorganized United States, along with some hundreds of billions of dollars' worth of other undeveloped wealth, belonged the Ohio River and its valley. In the weakened state of the government under the Confederation this soon appeared to be a piece of property almost too hot to handle.

The West was little considered. The United States-Canadian border had been established by the treaty as the center of Lakes Erie, Ontario, Huron, and Superior and thence on west by the Grand Portage route to Lake of the Woods. Baron von Steuben was appointed to receive the western posts from the British: but those posts eventually turned over did not include Detroit, the most important, and British and Indians continued to control both sides of Lake Erie and to threaten a good bit to the south and west of it until 1796.

Probably this was considered generally to be of little moment; there was plenty of room and the British-Indian threat from the north was no more fearsome than were those from many other directions, within and without. People were presently swarming to the Ohio, mainly disregarded and left to their own devices by a government which had no great enthusiasm for its job and plenty of other problems on its hands, anyway. Those who arrived early found the lusty three decades between 1785 and 1815 on the river a period as lively as any in the history of America.

After the French and Indian War the Ohio had been open for settlement upon the Virginia side of the headwaters west of Kentucky. Clandestine settlement upon the north shore over the present Pennsylvania state line in eastern Ohio had begun before the Revolution. Now, with war at an end, legitimate settlement of that country also was possible, once Congress should establish a plan and authorize it.

In June, 1783, General Rufus Putnam, erstwhile tavernkeeper and efficient, if somewhat unorthodox, Revolutionary officer, applied to Congress for a grant of land on the north side of the Ohio. The general's idea eventually developed into something much more important to the West than securing land only for his own use. In the same year other veterans acquired a none too welcome interest in Ohio River lands. The state of Virginia, alarmed by the fact that General George Rogers Clark's expenses (since he had been an officer commissioned in the Virginia militia rather than by the Continental Congress) would have to be paid by the state, if paid at all, withdrew his commission but granted him and his soldiers some lands on the north side of the Ohio just above the Falls.

In the year 1783 a military man of future prominence came down the Ohio to settle in person. Maryland-born General James Wilkinson, Doctor of Medicine and handsome and brilliant young officer out of a job, brought his beautiful and well-connected Philadelphia wife to Kentucky, where he expected to act as the western representative of merchants in his wife's home town.

The interest that these veterans—Putnam, Clark's men, and Wilkinson—now had in the Ohio Valley would have profound effect upon the activity in the locality during the next twenty years.

The year 1784 saw both England and the States ratify the treaty at long last. The Revolution had now been over, for all practical purposes, since the surrender of Yorktown three years before, but the enervated Congress had found this added chore almost impossible to complete—even now Virginia complicated matters by refusing to comply with the terms of the treaty and England made this the excuse to refuse to surrender the western posts.

During 1785 some recuperation from the strains of the war at last became evident; even some little attention began to be paid to the security of the Ohio Valley, to the promotion of law and order, and to preparations for organized settlement on the north bank of the river. A treaty was made with the Delawares at Fort Stanwix on January 21 which established "friendship"; an ordinance was passed which provided for a survey of lands on the Ohio at the mouth of the Muskingum and during the winter Fort Harmar was constructed there to protect the surveyors. Other posts also were ordered built in the west and a party of regulars, under command of General Richard Butler, was dispatched to make treaties securing land on which to establish them. This armed force, traveling under government orders, suffered an indignity which demonstrates all too clearly the weakness of the Confederation, the disregard in which its emissaries were held and the appearance of a new threat to peace which would shortly exceed that of the menace of the Indians on the Ohio. General Butler and his men were held up and robbed by white outlaws! "I find we are infested by scoundrels more unruly and unprincipled than the savages, and who wish to frustrate the treaty," wrote the general in understandable indignation.

Individuals of the common sort—probably some of those who had already entered and been driven back from Ohio lands before —were moving in to establish themselves as squatters without formality of title or payment. Unauthorized settlers were said to have attempted to seize lands at the mouth of the Scioto River in 1785-1786 and an organized colony moved across the mountains with the intention of settling in southern Illinois—a country unforgettably familiar to those Virginians who had served through George Rogers Clark's campaigns against the Mississippi and Wabash posts.

Kentucky, still a Virginia county, advanced a strong bid for independence of the Old Dominion, with a view toward eventual statehood. The chief arguments of the Kentuckians were their distance from the seat of government and the consequent inability of Virginia to supply either protection from Indian dangers or the means for enforcing civil law. Also there was the conviction, although it was not expressed in so many words, that the parent state did not, and could never, understand the problems and potentialities of the new country in the west—the same understandable intolerance that any normal 16-year-old child is likely to feel toward his 40-year-old parents. The Kentuckians had strong arguments on their side and they were ably presented by the eloquence of one of the newer citizens of Kentucky—General James Wilkinson, formerly of the Continental line.

On January 10, 1786, General Rufus Putnam—who, with Samuel Tupper, had developed a grander plan for settling himself on the Ohio than his first request for a personal grant had contemplated—advertised for fellow veterans of the Revolution to join in a company to promote colonization in the West. Eleven men met at the Bunch of Grapes Tavern in Boston on March 1, and the Ohio Company of Associates was formed. The eleven intended to sell one thousand shares of stock in their new company to raise working capital and the price per share in the currency specified casts further light upon the still distressed state of the nation: each share in the Ohio Company of Associates was purchasable for $1,000 in Continental certificates plus (in order that there should also be some *money* on hand) $10 in either gold or silver!

Little wonder that the organizers of the company insisted upon the gold or silver; actually the $10 in specie *plus* the $1,000 in Continental certificates made the price of each share only $18 or $19— thus far had the certificates depreciated. About that time a court, convened in Louisville, established the local tavern prices, in Virginia paper, figured in pounds and shillings:

The court doth set the following rates to be observed by ordinary keepers in this country, to wit: whiskey fifteen dollars the half pint; corn at ten dollars the gallon; a diet at twelve dollars per day; lodging in a feather bed six dollars; stablage or pasturage one night, four dollars.

At about the same time the property of a deceased citizen was appraised—

To a coat and waistcoat £250, an old blue do.
and do. £50.. 300
    To pocketbook £6, part of an old shirt £3........... 9
    To an old blanket, 6 s., 2 bushels salt £480.......... 480 6
                                                              ————
                                                            £789 6

—which was equal to $30 or so in minted coin! Fair enough, in view of the fact that both men's tailored clothing and salt sold high on the frontier in that day.

A year was required to sell all the stock in the Ohio Company of Associates. Then, with the Rev. Manasseh Cutler in the role of lobbyist, the petition for the purchase of land was presented to Congress. In the meantime the Scioto Company, with similar intention, had been organized by the citizens of New York and the astute Rev. Manasseh secured its agency also. Through his efforts a joint purchase by the two companies of two million acres, to be located around the mouths of the Muskingum and Scioto rivers, was eventually arranged, with payments deferred to suit what was hoped would be the convenience of the purchasers.

Clearing the title to the western lands of the claims of individual states had been a difficult business. Many of the original charters and grants of the colonies had been vague and uncertain as to western boundaries, because the British crown and those departments of his Majesty's government which made them had been innocent of exact geographical knowledge of the interior. There existed many conflicts such as that which had caused the dispute between Virginia and Pennsylvania over the jurisdiction of the country around Fort Pitt.

Having no western lands of her own, Maryland proposed that all other states pool their claims in the West and open the country to settlement under the direction of Congress. In 1780 the state of New York led the way toward this solution by waiving her claims to western lands and Connecticut and Virginia followed the good example in 1781, except that Virginia reserved the right to grant certain lands north of the Ohio as a bonus to her Revolutionary veterans.

Thomas Jefferson offered a plan, embodied in the Ordinance of 1784, for dividing the western land into districts, each of which might be admitted to statehood upon achieving a population of 20,000. This ordinance was later repealed but it had served a purpose in introducing the idea of admitting more states to equal participation in the Union.

An agreement was reached in 1785 which proved to be a most

important aid to orderly settlement of the new lands. The Ordinance of 1785 provided a method for easy establishment of bounds by the laying out of land in "legal townships" each 6 miles square, containing 36 "sections" of one square mile or 640 acres each. All were to be aligned upon north and south and east and west base lines and all, therefore, were capable of division without regard to such landmarks as trees which died, streams which changed their courses, or rocks which might be moved. It was an entirely new idea; one which was to be followed for all new national lands in the Northwest except for the few cases in which previous surveys existed, such as the Virginia Military Reserve in Ohio and the old French holdings in Illinois and Indiana. That and other provisions of the Ordinance of 1785, "for ascertaining the mode of disposing of lands," which provided this plan, opened the West for sale and settlement.

In 1786 Congress took another step important to the Ohio; it authorized an invasion of the Northwest for the purpose of reducing recalcitrant Indians and impressing the British at the northern border. Congress also called for a great council of the northwestern Indians to be held in December, 1786, at which so much of the future aim of the United States as seemed expedient would be explained to the red men.

The Ohio Valley was awakening to more than matters of land titles, politics, and military affairs: it gave welcome, in a rather lackadaisical way, to its first newspaper. John Scull and Joseph Hall, erstwhile Philadelphia printers, had hauled a hand press and some type by wagon over the mountains to Pittsburgh. On July 29, 1786, they brought out the first number of the Pittsburgh *Gazette and Manufacturing and Mercantile Advertiser*.

Virginia agreed to Kentucky's independence that year, young Wilkinson's talk being persuasive—as it would continue to be, to one purpose or another for many years to come—and with statehood in prospect for one of its sections the West gained importance in its own eyes, as well as in those of the Easterners.

But with all these happy portents there was trouble brewing; trouble which would involve the whole West but primarily Kentucky—and of Kentuckians, none so much as that persuasive James Wilkinson.

France had ceded Louisiana to Spain at the end of the French and Indian War to save it for the house of Bourbon. Spain, none too enthusiastic at receiving the gift of a territory well known to have been a source of great expense to France, took no particular interest in her unsought acquisition at first but, now that American settlement impinged, she began to reappraise her holdings.

Spain complained of the misuse of her frontier by United States citizens and in this she was certainly justified. Farthest from the seat of United States government, and thus least of its worries, was the east bank of the Mississippi, western limit of United States control. The United States side was rapidly becoming the haven of all the criminal, dissolute, abandoned, and degenerate element that was likely to find its greatest safety in distance from the settlements: Spanish territory, in which to take refuge in case of pursuit by law-enforcing powers, lay but a canoe ride or a float on a log across the river. There was no question but what the frontier was being abused.

Spain, proud old land though even then in decline, saw no reason to temporize with the raw young States and demanded an accounting. She was in a mood to dictate and the American government was in no position to argue. So fearful was the government, and so ignorant were even the majority of New England and Middle Atlantic congressmen as to the importance of the Mississippi that John Jay and the commissioners appointed to deal with Spain were authorized to go to almost any length to secure harmony, including the yielding of all rights of navigation of the Mississippi for a term of years.

Pamphleteers later quarreled for years over the question of just what John Jay may have said and done in private; certain it is that, ignoring the advice of his fellow commissioners, he ended his negotiations by leaving the control of the Mississippi, so important to the back country and thus to the nation, in a state undefined. Undefined, that is, in words; it was certainly defined for all practical purposes by the fact that Spain held the river's mouth, its waters for some hundreds of miles upstream, and commanded forces which were still sufficient to make effective such decisions as to commerce and navigation as she might fancy.

Spain soon demonstrated that her fancies included the complete closing of the Mississippi to American shipping. That closed the Ohio also, since its export traffic then moved only downstream.

Of course this blockade stirred up violent resentment upon the Ohio, whose citizens felt themselves slighted by the eastern statesmen and crippled as to future development. George Rogers Clark, whose truculence had been such an asset while the country was at war, added anger at this indignity to resentment of his treament by Virginia, and these feelings, because of the progressive moral dissolution of which he was a victim, led him to seize the property of the Spaniards at Vincennes, his former friends and long the local capitalists. His action, understandable as it was, naturally complicated the relations of the two countries.

Every neighborhood upon the Ohio River system which was sufficiently developed to spare a dozen hogs or a barrel of tobacco for export buzzed with new talk of the vicious ignorance of the eastern government and the shameful neglect of western interests.

In May, 1787, a conference of Kentuckians was called to consider the possibilities of reopening navigation of the Mississippi, the more prosperous citizens on the Ohio River and its navigable tributaries being most vociferous. In June, 1787, James Wilkinson undertook a project which was probably intended originally as a perfectly logical and reasonable test case designed to learn, for the benefit of the Ohio River shippers, just how far the Spaniards meant to go. He loaded flatboats with the tobacco, flour, and bacon he had received in trade at his store and had purchased with borrowed funds and sent them off down the Ohio and Mississippi toward the forbidden Spanish territory. Wilkinson followed his cargo in a smaller boat.

There was no cause for questioning his act; on the contrary, it was regarded as a constructive move to the public benefit.

Meanwhile the activities of Manasseh Cutler bore fruit as Congress established the Territory North-west of the River Ohio and authorized the sale of some of its lands on the river to the Ohio

Company of Associates and the Scioto Land Company. The ordinance, passed on July 13, 1787, which established the new territory made some novel and farsighted provisions which eventually proved to be of vast importance to the nation.

The Ordinance of 1787 provided for the temporary government of this territory (which included, on the Ohio's banks, all of present Ohio, Indiana, and Illinois and also Michigan and Wisconsin to the north of them) by agents appointed by Congress until the colony should include 5,000 adult free males, when a representative legislature should be established. It also provided that the territory might become a state when the population came to number 60,000 and that in the future the whole might be divided into not less than three or more than five states. This fulfilled the aim of Thomas Jefferson's plan of 1784 and was a new departure. Formerly colonies had been only that—now this colonial possession (for that is what the west was in fact) could acquire full status in the Union. The future greatness of the United States was assured.

The ordinance provided a covenant to safeguard the rights of the inhabitants in the matter of religious freedom, the prohibition of slavery, and the fundamentals of representative government, as well as other guarantees of what has come to be called "English liberty." Further, and it was a most striking innovation at that time, it included a statement to the effect that education, being necessary to freedom, should be forever encouraged. Some force was put behind this sentiment by providing that, of the lands sold or granted by the Ohio Company, the proceeds of the sale of one section out of each "legal township" should be reserved for financing schools. Another section was designated for religious purposes, three for later sale by Congress—as a measure designed to prevent monopoly of large bodies of land by speculators—and that, out of the whole purchase, two full townships were to be reserved for the purpose of establishing a university. This, indeed, was something new: the beginning of the "land-grant college!"

Within the pattern set by this ordinance, as has been said, came the development of the Midwest and of the vast, then scarcely dreamed of, country beyond the Mississippi (except for California, to which territorial status was refused but which was admitted to statehood, and Texas, which was an independent republic at the time of its admission).

In October, 1788, Congress granted to John Cleves Symmes a tract of land on the north bank of the Ohio between the Great and Little Miami rivers (surrounding present Cincinnati) and Symmes issued proposals for the sale of land to settlers. The registration of lands in the Virginia Military Reserve in Ohio began (it extended along the north bank between the Scioto and Little Miami rivers) and generous parcels were awarded to Virginia veterans of the Revolution.

General Arthur St. Clair was appointed governor of the Territory North-west of the River Ohio and some additional troops were ordered to support his brand-new government. St. Clair's force is worthy of notice, it being actually the beginning of the United States regular army. During the summer of 1788, forty-five hundred settlers were estimated to have passed Fort Harmar, at the mouth of the Muskingum, on their way down the Ohio to settle.

The West was thus being established, but those forty-five hundred did not pass down the Ohio without incident. In spite of the great land purchases, the well-considered plans of government, the newspapers in the valley (for publication of a paper in Lexington had begun by now), there still remained unprotected miles upon miles even on the upper half of the river.

Indian depredations were no longer committed so much by organized war parties as by small bodies of warriors in search of loot, but the effect was no less terrifying. At times the north bank of the river seemed to the emigrants to be lined with Indians, who often followed the slow, clumsy boats for days awaiting a propitious opportunity to strike.

James Finley, later a pioneer circuit-riding preacher in the Midwest, accompanied his father with a party bound for Limestone (modern Maysville) in Kentucky. The little fleet, which left the Virginia shore of the upper river in 1788, consisted of one boat carrying armed sentinels followed by two large flatboats filled with passengers and their goods. Finley remembered witnessing an effort to rob or murder his party by one of the most common of the techniques employed by renegade whites associated with the Indians or, even more likely, by parties of white "wreckers" which infested the river:

Just below the mouth of the Great Scioto . . . a long and desperate effort was made to get some of the boats to land by a white man, who feigned to be in great distress; but the fate of Wm. Orr and his family was too fresh in the minds of the adventurers to be thus decoyed. A few months previous . . . this gentleman and his whole family were murdered, being lured to shore by a similar stratagem. But a few weeks before we passed, the Indians attacked three boats, two of which were taken, and all the passengers destroyed. The other barely escaped, having lost all the men on board except the Rev. Mr. Tucker, a Methodist missionary, who was sent out by the Bishop to Kentucky. Mr. Tucker was wounded in several places but he fought manfully. The Indians got into a canoe and paddled for the boat, determined to board it; but the women loaded the rifles of their deceased husbands, and handed them to Mr. Tucker, who took such deadly aim, every shot making the number in the canoe less, that they abandoned all hope . . . and returned to shore.

After the conflict this noble man fell from sheer exhaustion, and the women were obliged to take the oars and manage the boat as best they could. They were enabled to effect a landing at Limestone . . . and a few days after their protector died of his wounds, and they followed him weeping to his grave . . . no stone marks the spot where this young hero-missionary lies, away from his home and kindred . . .

Nor were the troubles of ministers over after they arrived at the place of settlement. The elder Finley, himself a Presbyterian divine, had to take the same precautions as any other immigrant before he could mount the pulpit and open his Bible. The Finleys settled at Stockton's Station (present Flemingsburg, Kentucky) eighteen miles below Limestone. The younger Finley describes the home:

This was the frontier house of the settlement, there being none between it and the Ohio river. The house was built of round logs from the forest trees: the first story made of the largest we could put up and the second story of smaller ones, which jutted over two or three feet, to prevent anyone from climbing to the top of the house. The chimneys were built on the inside. The door was made of puncheon-slabs, six

inches thick, and was barred on the inside by strong iron staples driven into the logs on both sides, into which were placed strong bars. In the upper part of the house there were port-holes, out of which we could shoot as occasion might require; and, as no windows were allowed, they also answered for the purposes of light and ventilation. The house for our colored people was built the same way, and immediately adjoining the one in which the family lived.

The senior Finley did not hesitate to shoot Indians at sight, regardless of their aims and the probably unsaved state of their souls; but living as he did under semisiege and recalling, as he must, that scene on the boats coming downriver, perhaps this seeming inhumanity can be forgiven.

In 1788—the same year as the Finleys' descent—the Ohio River from source to mouth came for the first time under on-the-spot government.

On July 9, 1788, Governor St. Clair reached Marietta, on the Ohio Company's Muskingum Grant, only three months after that town had been populated by the arrival of General Rufus Putnam and forty-eight earnest followers. During those three months government of the community had been administered by one of its members, Return Jonathan Meigs, under a set of laws drawn up by the settlers and published by the simple and direct process of writing them in a fair hand and nailing the written sheet to a convenient tree.

Late autumn saw colonists arriving at Symmes Purchase. In December the first town, fancifully named Losantiville, on the north bank of the Ohio opposite the mouth of Kentucky's Licking River, began to receive settlers. Its proprietors, Denman, Ludlow, and queer John Filson, had shown considerable ingenuity and drawn upon several languages in compounding the rather musical name of the place, *L* for Licking, *os* for mouth, *anti* for opposite and *ville* for town, but it would presently be changed—to Cincinnati.

The first laws of the Territory North-west of the River Ohio had been published on July 25, the first court had been held at Marietta and all appeared salubrious upon the middle river—except that the notorious Dr. John Connolly, one-time owner of the site of Louisville, inciter of raids against the Indians and former captain of militia under Lord Dunmore at Fort Pitt, was in Kentucky. Officially he was a tourist but actually an agent in the pay of the British. There was another unfavorable circumstance also: Spanish Louisiana now had a governor, Miro by name, who was considerably above the average of Spanish officialdom in both enterprise and sagacity.

Neither Connolly nor Miro yet cast any perceptible shadow upon the Ohio's waters and the following year opened no less auspiciously for the West—although Kentucky, the only section with anything much to export, continued to fume over the restraint of commercial shipping on the Mississippi.

Governor St. Clair concluded two treaties at Fort Harmar on January 9, 1789: one with the Iroquois League (except the Mohawks) confirmed the last treaty at Fort Stanwix; the other with the Lakes Indians—Ottawa, Chippewa, Potawatomi, Sauk—and the Wyandots and Delawares reaffirmed a previous agreement to the effect that these tribes should act as a buffer between the United States settlements and unfriendly Indians. For a consideration of $6,000 they also relinquished their claims to lands already occupied by white settlers.

The Territory North-west of the River Ohio began its cultural progress that spring when Daniel Story, preacher and apparently the first American teacher north of the river to follow his profession along conventional lines, arrived at Marietta. John Filson, minor proprietor of Losantiville (his contribution consisted of surveying and probably supplying chief inspiration for that name), was a teacher, but he had taught only south of the river, in Kentucky: anyway he now lay dead in the woods, somewhere back of the new town, supposedly a victim of lurking Indians.

Completed treaties and the arrival of a teacher notwithstanding, all was not to be well in 1789: trouble was obviously coming as the summer ended. Some Indian conflict was to be anticipated. But there were also other threats of a mysterious and unsettling nature, all the aspects of which have not yet yielded their secrets even after a century and a half of historical research.

James Wilkinson had returned from his first flatboat venture into unfriendly Spanish waters and had held whispered conversation regarding his New Orleans experiences with those whom he considered likely listeners or who had influence of which he could make use.

Wilkinson's reception in that city had been as encouraging as he could have hoped. A markedly suave and personable young man, he had presented himself to Don Diego de Gardoqui, commissioner in America for Spain, and explained his mission as an unofficial but influential representative of western commerce. He told Gardoqui that the West had "nothing to hope for from the Union"—later he confirmed this belief by letter—and that the hope of the new land lay with Spain first or, failing that, England. He met Governor Miro and secured a practical monopoly on American trade down the Mississippi—providing that it be limited to trade with Spain. He became, without doubt, a paid agent of the Spaniards, expected by them to bring about a secession of the western states and territories adjoining Louisiana. Probably no one will ever know whether he employed these traitorous promises only, as he later maintained, in an effort to open the Mississippi for the good of his country; whether, as his friends believed, he sought only to increase his fortune at the expense of Spain or if he seriously contemplated treason. Certainly there is every indication that treason was his aim. Whether or not he worked actively for secession, he would probably have done so had it become necessary to his own interests—though had that situation arisen his love for intrigue and double-dealing would probably have prompted him to take an opportunity to deceive the Spaniards!

In the course of his dealing with the Spanish authorities he eventually took a secret oath of allegiance to Spain and received an annual salary known to have been $2,000 at first and later $4,000. This pay he continued to draw long after he had been commissioned an officer in the United States Army in 1791, and since he will appear from time to time as a valiant Indian fighter and leader in civil affairs, this is a fact to be remembered and at which to marvel.

Many of the leading men of the West were more or less associated with Wilkinson in the trade and in the plot; how much more, how much less is a still deeper mystery. Judge Harry Innes, Judge Benjamin Sebastian, George Nichols, Kentucky's Senator John Brown, Davis Floyd, and dozens of others were involved. Their activities were no secret in the West and reports of them must certainly have come to the capital at Philadelphia. Little was done, however, either to counter the plan—which was always nebulous enough—to bring the conspirators to justice, or even, until public indignation demanded, to remove some of them from the public offices they held. Chances are that Wilkinson enlarged to the Spanish upon the western interest in a possible revolt and to his American associates upon the extent of Spanish friendship; he must have done at least that to justify his salary. Even so, there was threat enough and the lack of interest shown by the national government seems to indicate that it was either more apathetic toward the problems of the West than even Westerners claimed or that men in high places had also succumbed to the smooth-tongued James Wilkinson's blandishments.

When public exposure came through stories printed in Kentucky newspapers in 1806 and in a book by Daniel Clark, a Philadelphian who had been a resident trader at New Orleans and had himself been rather closely associated with Wilkinson there, surprisingly little discredit seemed to attach to the conspirators.

Perhaps there was some feeling, even among those citizens who abhorred the idea of treason, that since the government and the older states were still inclined to neglect the West this harmless dress-up game of conspiracy might be a way as good as any to arouse a bit of interest in western problems. Certainly, when at last the intriguing ended, America did have a grip on Mississippi trade which could not be broken, Wilkinson and others had received many thousands of Spanish dollars in retainer's fees—and only Spain had suffered loss.

The excitement over the possible development of commerce in the West, and the manifest fact of the settlement by farmers and mechanics which was taking place along the north bank of the Ohio as far down as the Great Miami, alarmed the Indians. Quiet except for sporadic raids and minor captivities since Clark and Harmar's campaigns at the end of the Revolutionary War, they began to threaten in concert during 1789. This time the lead was taken by the tribes of the Miami Confederacy—the Miamis, Weas, Piankishaws, and Kickapoos, mostly inhabiting what is now northern Indiana and extreme western Ohio.

Congress empowered President George Washington to call out the militia, and Fort Washington was built upon Symmes' lands and occupied by three hundred troops under veteran General Josiah Harmar. In the spring of 1790 Governor St. Clair toured the old French posts above the Ohio and called the western militia to rally at Fort Washington in the fall.

On the 30th of September, General Harmar led these troops north to attack the Miami villages, which his advance force under Colonel John Hardin of Kentucky reached in late October—and soon his men were straggling back to the Ohio, badly, ignominiously defeated.

The fact that the settlements along the Ohio River, growing but still pitifully small, could not thrive with hostile *and* victorious Indians at their back doors, must by now have been evident to all observers and arrangements were immediately made for retaliation. Next spring two dashes were to be made to the north against the western Miamis in present Indiana, and what was intended to be a final, annihilating campaign was planned against the main body of these Indians on both sides of the present Indiana-Ohio line.

General Charles Scott's campaign in May and June did destroy Kethtippecanuck, a large and important Miami Indian town at the junction of the Wabash and Tippecanoe rivers, and Wilkinson's force—the general had temporarily deserted commerce and diplomacy for the military though his Spanish annuity continued—was successful against the Eel River towns of the Miami above present Logansport, Indiana. But St. Clair's great fall campaign met with another defeat—horrible, bloody, and again not entirely honorable—on November 4.

St. Clair's campaign had been planned carefully. President Washington had obtained authority to raise an army of three thousand men for his old comrade-in-arms; a force which should have been strong enough for the purpose and more. These units were to gather at Fort Washington and to march northwest toward the Indian town on the present site of Fort Wayne, Indiana, on July 10.

But the old bane of the western campaigns, the same which had haunted Clark and Harmar, interfered again on the side of the Indians. October had come before even two thousand men had been assembled and they were mainly the least competent of the frontiersmen, their ranks filled to even this slender total with the scum of eastern cities, "purchased from prisons, wheelbarrows and brothels at two dollars a month"—plus their rations, of course, which were supplied by a corrupt commissary department skilled in the arts of petty graft.

Arthur St. Clair himself was sick and at the time entirely incapable of the command of an operation on this scale.

The very movement of such an army through the woods was a major undertaking. Already held at Fort Washington three months past the time for which the campaign had been scheduled, the frontiersmen were excusably restless and slipped off to return to their farms and families as opportunity offered along the line of march. Although no specific record so indicates, the jailbirds and bums from the eastern cities probably ceased to desert or even to stray, once the march through the wild country north of Fort Washington began: product of the crowded slums of the cities, they must have suffered the same terrible fear of the wilderness and the Indians which had afflicted Braddock's enlisted men— though no season of rigid discipline assured that *they* would stand their ground as had the British. Every cracking twig by night, every falling leaf by day must have soon reduced them to a state of helplessness. Finally the ragtag army straggled to a camping place which later became the site of Fort Recovery.

The command was weak; the enlisted men unfamiliar with even the rudiments of a knowledge of border warfare: early on the morning of November 4 they were victims of a surprise attack by the very Indians they had come to conquer. Two hours of hopeless, undirected firing followed—then the frenzied survivors fled twenty-two miles back south and secured themselves in the blockhouses at Fort Jefferson.

Through some miracle they managed to take three hundred wounded with them: how many others they left on the field cannot be known, for once the Indians moved in with tomahawk and scalping knife, the wounded soon joined the dead—to a total of six hundred—whose bones lay unburied in the swampy woods for two full years after.

The fault was not all St. Clair's; again it lay partly in the casual treatment of the West by a national government which permitted criminal malfeasance in the service of supply, recruiting with no regard for fitness, the placing in the field of an army two-thirds the promised strength in number and far less than that in efficiency.

The settlements on the lower Ohio were dazed by this last catastrophe—most of all by the knowledge that, under both Harmar and St. Clair, their soldiers had been defeated by inferior forces. True, it was known that the Indians had the backing, and even the leadership, of British officers from Canada, but the defeats were no less humiliating for that. Humiliation alone could have been borne, had that been all, but the Indians, two such victories won, immediately became both more bold and more bloodthirsty.

John Finley, the preacher's son, in his father's new home in Bourbon County, Kentucky, remembered that "Many who retired for the night were surprised and murdered, and the glare of their burning habitations, shooting up amid the darkness, told the surrounding settlements of the work of death." Flames like these were to be seen, more or less, in every neighborhood of the middle Ohio Valley.

The worst danger still lay upon the stretch of the Ohio between the mouths of the Big Sandy River and Limestone Creek where the attack upon the Rev. Mr. Tucker's company had taken place and in the region where the Finleys had first settled. The countermeasure employed by the citizens, now hopeless of practical aid from their government, was a bold and dramatic one. According to James Finley:

To prevent this state of things . . . the choicest men of the country were selected as spies. Men of the greatest integrity, courage and activity

. . . well skilled in all the modes of savage warfare . . . among the number I will mention the names of William Bennet, Mercer Beason, Duncan McArthur, [those McArthurs were fighting men and strategists long before the present General won his stars] Nathaniel Beasley and Samuel Davis. These men were dressed like Indians. They were to guard the passes of the Ohio from Maysville to Big Sandy. While some of these were passing up the river between these two points others were coming down, so that it was almost impossible for the Indians, in any considerable numbers, to cross over from the Ohio side without discovery. When they did cross the river, the settlements were . . . put on their guard.

Around the headwaters of the Ohio a new and different and entirely insupportable atrocity was being perpetrated: the federal government, at the insistence of Alexander Hamilton, was attempting to tax whisky, long the principal product of the Monongahela branch of the river! It was indeed a year of woe on the Ohio.

The people of the upper Ohio and the Monongahela were conscious of the local importance of the distilling business. As early as 1787 a "trade at home" campaign, which anticipated by a century and a quarter that losing contest which the less competent local merchants waged against chain stores in the twentieth century, was announced in Pittsburgh. There, according to the Pittsburgh *Gazette,* at "a meeting of the magistrates, sheriff, prothonotary, attornies, and a number of reputable inhabitants of the county of Washington, together with some gentlemen from Ohio County" it was resolved that the importation of spirituous liquors to the west of the Allegheny mountains "tends to the prejudice and impoverishing of the western inhabitants." Fifty-one of these gentlemen "pledged themselves and such others as will cooperate, after March next, to buy or use no liquors not manufactured on western waters."

Upon a community already so conscious of the welfare of one of its principal industries, the effect of a threatened government tax upon its product may be easily imagined.

There has been a considerable amount of jesting about the Whisky Insurrection. Many of its aspects would have been ridiculous enough had it been a matter only of the tax on that estimable product of the combination of earth's golden grain and man's art. The episode actually was the result of ample real—and many fancied—indignities offered the people of the Ohio Valley by the government throughout the preceding decade.

Neglect of military protection, dilatory tactics in opening the lands for settlement, the prevalence of nonresident speculation, the total ignorance of western requirements exhibited in Jay's Treaty, all gave reasonable grounds for complaint. The belief that Alexander Hamilton was attempting to introduce government by an aristocracy and that the original states were inimical rather than only negligent may have been fanciful but it was no less irritating to the spirited Westerners.

The excise law of 1791, taxing the valley's best cash crop and most easily transported product—as well as its favorite beverage—only brought dissatisfaction to a head on the upper Ohio and Monongahela, then chief center of distilling. Indignation rose high as soon as word of the passing of the bill reached the country, and it continued during the three ensuing years, as will appear in due order.

The year 1792 was little happier on the Ohio than its predecessor. Besides the antiexcise demonstrations at Pittsburgh and continued fear of the Indians now grown daring and confident after their defeat of St. Clair, there existed an almost complete stagnation of trade from Marietta to Louisville. On the lower river, the Kentucky country, which still had most to export, was crippled by Indian threats from the north shore and was almost cut off from the headwaters by the tax disturbances there. The road from Kentucky to Tennessee and Virginia through the Cumberland Gap—the "Wilderness Road"—did well enough for emigration and for travel but it was as good as useless for either export or import in paying quantities of any products except livestock—and even cattle and hogs lost half their weight in being herded over the mountains.

Typical of the Kentucky raids from the Indian shore of the Ohio in this period was that on the farm of Daniel Ketcham, an emigrant from Maryland who had bought and cleared land about six miles from Squire Boone's old station in present Shelby County, Kentucky.

During the six years the Ketchams had owned the place, three men had been killed by Indians in its fields, and twice the family had sought protection at Boone's. Finally, in 1792, Daniel Ketcham himself was captured by what he called the "Tawa" Indians. They were probably Ottawas who were located near Detroit, where Ketcham's captors eventually took him.

Ketcham's experiences as he related them to his family were those of the casual captive of not too unfriendly Indians. His son preserved the account:

He was pursued by the Indians, who shot his horse from under him. He ran forty or fifty yards and was overtaken by one of the party with tomahawk in hand. He immediately surrendered, giving his hand to his captor, who took his overcoat from him and tied it around his own body, and led him to the company of . . . eleven Indians, who took up their line of march for the North . . . It would be natural for a man situated as Ketcham was, to meditate seriously making as speedy an escape as possible. To prepare the way for his return, in case escape could be made, he broke off the twigs and boughs of bushes and trees, as marks to direct the course of his returning footsteps. His sufferings soon became extreme. Being unaccustomed to exposure, and being deprived of some of his clothing, he became so severely affected by the rheumatism that he could proceed only with the greatest difficulty. To hasten his steps the Indians would punch him with their guns until he would fall to the ground, then would whip him with their gun-sticks to make him spring to his feet. At last, in desperation and despair, he told them to tomahawk him, which they seemed willing to do, and one of them seized his hand and raised a tomahawk. They then stopped to consult and told him to get up, and for some time after they aided him to travel . . .

They crossed the Ohio River [near present Madison, Indiana] where they had spent the winter and had collected many skins and furs, which they carried with them. After having been convinced that their unfortunate captive was not dissembling or feigning sickness, they were disposed to be very merciful, and did not compel him to carry any of their burdens. But a change soon took place in the state of his health. . . . Having been compelled to wade many streams, and to have much to do with cold water, Ketcham soon recovered from his rheumatic difficulty. This fact it was greatly to his interest most faithfully to conceal, which he succeeded in doing for a time. At length, on one unfortunate day, when crossing a creek on a log, he forgot to limp. This being observed by the Indians, they burst out into a most hearty laugh, and from that time to the end of their journey, laded him down like a mule . . .

As they drew near to their village, supposed to be not far from Detroit, they raised the war whoop, when all the inhabitants came out to meet them. One of the villagers, an old Indian man, fastened his eyes upon the unfortunate captive and advanced towards him, offering him his left hand with a very gracious smile, while with the right hand he gave Ketcham a blow by the side of the head severe enough to bring him to the ground . . . This was the first initiatory step to his new order, the first part of his introduction to his new home. A French trader stepped up and informed him that he might be thankful if he fared no worse, and the second step in his initiation soon showed that such was the truth. On a favorable day the whole community assembled, and poor Ketcham soon found that he was to be the hero of the occasion. They took him, blacked him, and gave him a looking glass that he might take the last view of himself, fastened him to a stake and prepared to burn him. Just at this juncture a daughter of the Chief, dressed in the most costly attire, and decorated with at least five hundred silver brooches, made her appearance and delivered a speech, of about thirty minutes in length, with an exceedingly rapid rate of utterance. At the close of her speech she advanced to Ketcham, (Pocahontas like) and released him, not one daring to gainsay or resist. Two women . . . then took him to the river to wash the black off him and the white blood out of him, that they might adopt him in their family, as a respectable Indian. After having completed that ceremony to their satisfaction they took him to their tent, and introduced him to his mama, who in the kindness of her heart offered him her hand, but who was so drunk as immediately to tumble off the seat where she was sitting. The duties assigned to him were, to carry wood for as many of the villagers as he could and to pound their corn, of the soup of which he was permitted to eat, but was always compelled to take the first and thinnest of it.

Soon becoming convinced that he could not long survive such treatment, he resolved that if he did perish, to do so in an attempt to escape. Having concealed a handful of corn and a small piece of squash, he departed from the village in the night, but was hotly pursued the next day, and for many days thereafter, and had it not been for the assistance of the French settlers, he would certainly have perished at the hand of his pursuers. On one occasion, after having traveled all night in a famishing condition, he, at the dawn of day, approached a log cabin with the hope of obtaining food and protection. As he opened the door of the cabin he saw the floor covered with sleeping Indians— his own relentless pursuers. Fortunately none were awakened by this occurrence but the Frenchman owning the house, who immediately signed Ketcham not to come in, came out and offered his assistance. They departed a short distance, when the guide said he would return and get some food for their journey . . . the man soon came again, with some food and stimulants, conducted him to a river near by, put him into a canoe and paddled for life. It is supposed that this was the Detroit river, and that Ketcham was landed in Canada, where he was cared for by the French. Ultimately he came to Detroit, where he hired himself to a French priest, who paid him with an old beaver hat, a second hand scarlet vest, and two dollars in money. With this liberal supply he succeeded in getting to his native place in Washington county, Maryland. Tarrying there until rested, he came to his family in Kentucky.

Reasonable as this narrative appears to be, one notes that even the sober Daniel Ketcham could not forbear the insertion of the Pocahontas motif, which was characteristic of such a large percentage of the captivity reports after Captain John Smith's happy experience. Perhaps the captive returned from such dangers should be permitted a bit of latitude but one wonders why at least a few of the unfortunates could not have been preserved through the interference of, for instance, a chieftain's spinster aunt, or, this circumstance impossible, why there might not have existed at least one Indian princess who was a girl of average or even homely mien, dressed in plain but serviceable clothing, with only a dozen or so silver brooches?

The force of arms having failed so signally north of the Ohio, commissioners were appointed to treat with the Indians during the year of Ketcham's captivity. General Rufus Putnam made one treaty at Vincennes: it was not an illustrious gathering of Indians who signed—they were mostly petty chiefs in search of the customary treaty gifts, leaders of bands whose total strength could hardly have attacked a good-sized farmhouse successfully. General Putnam must have known that most of the important chiefs were absent and that even those present affixed their marks with their fingers crossed.

The only hope for significant improvement lay in the knowledge that the government had at last begun to give serious and intelligent attention to the Indian threat against people of the Ohio Valley: General Anthony Wayne—the unbeatable "Mad Anthony" of Revolutionary days—had been ordered to the West.

Upon this unhappy scene now entered a party of colonists who were, in short order, to be the victims of disillusionment and tragedy as sorry as had happened upon the river since the murder of the Moravians' Christian Indians. The French who had bought lands of the Scioto Company arrived to settle Gallipolis in February, 1792.

Totally unfitted for frontier life, promised a mild climate and a ready market for the products of the fine artisanship that had been their chief source of livelihood in France, these people had sold everything they owned in order to purchase small parcels of land and to pay for passage to America. Arriving at Alexandria, Virginia, they learned that they had no title to the land for which they had paid. The Scioto Company had defaulted upon its contract with the government and had never owned the land it sold. Most of the French who still had money for the passage went back to France; the others, with no choice, continued to the Ohio to see what they could salvage. Arriving in the dead of an upper Ohio winter, their situation was pitiful indeed. Their children, their old and weak, died by the dozen and, but for the charity of their few neighbors, the emigrant colony might well have been wiped out entirely. Who was the gold-brick salesman who had traveled through France making the promises and taking the cash? He was none other than a gentle preacher-poet by the name of Joel Barlow —later to become a professional champion of the "common man."

Another year showed promise, but the people of the Ohio had seen promising portents before. General Wayne had reached Fort Washington with fresh troops, had rallied others there, drilled and seasoned his men for weeks, and had then moved north, building or rebuilding stockaded forts in a line toward the heart of the

Indian country. His progress seemed slow indeed to the people of the valley.

Still another year, 1794, and there was hell to pay on the headwaters of the Ohio. Around Pittsburgh and along the Monongahela the still-owning citizens had worked themselves into a terrible furor over Alexander Hamilton's whisky tax. More than twenty-five hundred stills were operated in the United States and the majority of them were in the Pittsburgh area. The first still in Kentucky had been set up in 1789, the Kentuckians made use of corn mash to distill a delightful beverage since become known as bourbon whisky, and the business was booming. Kentucky erected two thousand stills during its first twenty-one years in the business but the old center was yet the larger; "Monongahela" was still the hallmark of prime rye drinking liquor in 1794 and Pittsburgh, closer to the seat of government, was much more susceptible to governmental supervision and interference.

The people on the headwaters felt that not only had their economic welfare been endangered by a tax on their most readily marketable product but that the very foundations of democracy had been assailed. The home of John Neville, collector of the excise, was burned in July and self-appointed "regulators" robbed the outgoing mail and read letters regarding the enforcement of the measure. The contents of these communications irritated them still further.

Though President Washington wisely ordered a meeting of delegates of the government with representatives of the insurrectionists, no understanding was reached and the militia from surrounding states was ordered out to occupy the western Pennsylvania counties in November. The real object of the government in this move had little to do with whisky; Wayne's campaign was about to begin and the presence of a large armed force at the head of the Ohio was desirable in case the British should abandon secrecy and actually take the field to support the Indians. Nevertheless, the presence of the militia had a quelling effect on angry distillers. The trouble was over—especially after it was noted that those who had been arrested and sent to Philadelphia for trial were usually acquitted, pardoned, or dismissed. This fact somewhat reduced the cause of complaint and robbed the more recalcitrant of material with which to arouse their neighbors. Everyone growing probably a little weary of the cause anyway, order was gradually restored. This was the first test of the power of the federal government to act. Hamilton and the Federalists had won an important point and even the Pittsburgh region profited, for, with the return of stable government, outside capital began to move in and commerce was immediately accelerated.

Downstream General Wayne continued to move methodically north extending his line of forts from present Cincinnati along what would become the Indiana-Ohio border. The stagnation under which the lower end of the river had existed during the past two years was broken by a new wave of intrigue. Both French and Spanish emissaries were now in Kentucky, again whispering into the receptive ears of General Wilkinson about a possible break from the United States for Kentucky and the western part of the country north of the Ohio, and about a confederation with those European countries which so dearly loved and appreciated the Westerners. The general's enthusiasm, shown in return for his annual stipend now raised to $4,000 a year, was still infectious.

In the fall of 1793 Wayne's army had reached a point eighty miles north of the Ohio on the present site of Greenville. There Wayne established a base and from there he sent a detachment north to build Fort Recovery, christened in hope and located upon the site of St. Clair's horrible defeat. Before the men could begin work on the fort they found it necessary to carry out a preparatory work equally in the interest of morale and sanitation: the burial of the rotting carcasses of St. Clair's hundreds of dead, which had been scattered by scavenging Indians and animals during the two years since their slaughter.

Through the following winter and spring Fort Recovery was occupied by Wayne's army, which was augmented by sixteen hundred men who arrived on July 26. This was now a force well prepared by drill and seasoning to meet Indians—even Indians, as Wayne suspected, under British leadership.

The waiting campaign was shortly ended; on June 30, 1794, about two thousand Indians began an attack on Fort Recovery, which they continued, to their sorrow, for two days. Wayne's insistence upon proper training bore fruit. The discipline of the Americans was nearly perfect and by the second day the attackers' losses were so heavy that the northern Indians, who made up a substantial part of the attacking force, returned to their villages. Of necessity the Indians of the Miami Confederacy and those Shawnees and Delawares who participated also withdrew north.

The battle over, Wayne moved his troops to the Maumee River—of great importance in the period of early settlement— which heads at present Fort Wayne, Indiana, and flows northwest to Lake Erie. It was called, interchangeably, "Miami of the Lakes" or "Maumee." There he built Fort Defiance, at the site of modern Defiance, Ohio, and the point then farthest north to be reached by American arms in the Ohio country. The fortification well begun, Wayne continued on down the Maumee toward Lake Erie and finally, at a patch of tornado-leveled woodland known as "Fallen Timbers," he caught up with the remnant of the Indian forces and defeated them decisively; as, in fact, Indians had not been defeated before in the West.

Other expeditions had usually stopped after either the initial victory or defeat but Wayne still pursued the fleeing Indians down the Maumee to Fort Miami, which the British had only just built upon the north bank. The Indians sought protection in the fort— adding to the evidence given by the presence of Detroit militiamen at the original attack on Fort Recovery to indicate that the whole campaign had been British sanctioned; perhaps British led.

General Wayne did not attack the fort; action such as that must come by governmental authority, but he could and did burn the British trading post and the Indian houses about it before he marched back up the Maumee toward the spot where the St. Joseph and St. Mary's join to form it.

There, at the head of the Maumee, he built Fort Wayne—and the United States at last had a post to back its claim to the lands immediately below the Great Lakes.

Provisions agreed upon at the ensuing Treaty of Greenville, signed on August 3, 1795, extinguished the Indians' claims on southern and eastern Ohio but recognized their right to the rest of the Territory North-west of the River Ohio—with the stipulation that lands within it might be sold only to the United States. The Indians also granted sites for forts and posts and the right to navigate streams and to use portages throughout the territory.

The Treaty of Greenville was not the only one made in the years 1794-1795 nor were the Indians the only people who made concessions affecting the western country. Chief Justice John Jay's treaty was signed on November 19, 1794, and weak as it was in some respects, it at least secured the northern border of the Territory North-west of the River Ohio as the center of the Great Lakes and resulted in the abandonment, a year later, of that insolently conceived British Fort Miami. And there was also Thomas Pinckney's Treaty of San Lorenzo, on October 27, 1795, by which Spain (being in a tight spot in European affairs, what with the repercussions of the French Revolution and one thing and another) forgot her prejudices and gave American citizens freedom to travel and to ship goods on the Mississippi and to deposit them at New Orleans for resale or transshipment.

The intrigues of France and Spain with various Kentuckians did not end, however—these activities continued with motives since interpreted variously and in direct contradiction. It is entirely possible that, owing to the chaotic state of affairs in Europe and the delay that distance caused between the issuing and delivery of orders from home, the emissaries of the two countries in America did not themselves know what their current purposes were supposed to be: it is also possible that M. Genêt, the French minister, and the Baron de Carondelet, who had succeeded Miro as Spanish governor of Louisiana, may have taken a leaf from Wilkinson's book and have operated independently toward the consummation of entirely private aims.

Certainly de Carondelet had Judge Sebastian as well as Wilkinson in his pay at this time (Wilkinson is supposed to have been dropped from the roll the next year) and it was Sebastian who went west to meet an emissary of de Carondelet at New Madrid and discussed a separate Kentucky treaty with Spain, providing that the state withdraw from the Union, which would allow exclusive privileges of trade.

Word of the signing of Thomas Pinckney's treaty put an end

to talk of this particular scheme but the treaty otherwise seemed to have the effect of increasing tension between the people of the western United States and the Spanish of Louisiana. The conspiracy, whatever its actual purpose, continued to drag on.

It continued even after General Wilkinson had been commissioned by his forbearing, foolish, or conniving superiors to command the western forces of the United States after the death of Anthony Wayne: but this time it was surely Spain that received the double-cross. In June, 1797, Thomas Powers was dispatched north by de Carondelet. His intention was to induce Wilkinson to postpone occupation of the Mississippi posts, supposedly with the purpose of testing Wilkinson's current loyalty to Spain and learning what the disposition of his associates in Kentucky might be. Powers was instructed by Carondelet that "If a hundred thousand dollars, distributed in Kentucky, would cause it to rise in insurrection I am very certain that the minister would sacrifice them with pleasure . . . and you may promise them—with another equal sum to arm them . . ."

But the crisis of Kentucky's dissatisfaction was past. Powers went to Detroit to see Wilkinson but the latter refused to be moved in such a course as Powers suggested and, in a further complication of intrigue, told Powers that some of the Kentuckians had gone so far as to "propose to raise three thousand men to move against Louisiana, in case a war should be declared between the United States and Spain . . ." To further clear himself with his superiors of the suspicion of having resumed his old role, he had Powers arrested in Tennessee and taken, under guard, out of United States territory.

That phase of the "Spanish Conspiracy" was over.

There had begun to be positive evidences of the progress of civilization in the valley: a paper mill was set up and began to operate in western Pennsylvania, and Kentucky, with a population of some seventy-five or eighty thousand, had become entirely too crowded for the comfort of Daniel Boone, who packed up his belongings, spent a while in the Kanawha valley, took a trip to the south to see the possibilities of Spanish Florida as a future home, but finally moved across the Mississippi to Missouri in 1795. Roads were being opened upon the upper Ohio—other than those early ones, Forbes's, Braddock's, and the Wheeling Trace—to feed immigration to the stream. In 1796 Congress authorized the marking of a road through the Territory North-west of the River Ohio which would cut across a great southern dip of the river and connect Wheeling and Maysville. That fall one of the redoubtable Zane brothers, Ebenezer, earned the promised fee by blazing the trees along the line. It is now the highway that passes from Wheeling, West Virginia, by way of Zanesville, Lancaster, and Chillicothe, Ohio, to Maysville, Kentucky—the section from Wheeling to Zanesville being a part of the National Road.

According to James Finley's account, "Soon great companies passed over Zane's trace, and settlements were made at the Muskingum River, where the town of Zanesville now stands, and also on Wills Creek . . . William Craig was the first man who drove a wagon and team to Chillicothe over Zane's trace." The country around Chillicothe was not a particularly happy place to be that year for everyone was down with what became the curse of the pioneer in the rich lands of the Territory North-west of the River Ohio—chills and ague following the first plowing. "Bilious intermittents," Finley called them, "supposed to have been caused by the effluvia arising from the decomposition of the luxuriant vegetation which grew so abundantly . . ."

Danger of the "ager," as the name of the scourge was universally pronounced, did not deter one new settler who intended to introduce the life of the landed gentleman to the West, for he had no intention of plowing; his aim was ease and the philosophical pursuit of culture. He was a well-educated, visionary, aristocratic young Irishman named Harman Blennerhassett, possessed of a beautiful wife and more money than was good for him, and in 1798 he bought the upper end of Backus Island in the Ohio below the mouth of the Little Kanawha River. He began building what was probably the most elegant residence then west of the Alleghenies; it augured no good for the West but at the time it probably seemed an evidence of great progress to the neighbors.

Upon the people in the valley peace—or such comparative peace as any booming frontier country was likely to achieve—had descended for a brief spell. They had a few years' time off from militia duty and Indian raids in which to build houses, plant crops, and improve themselves and their communities in facilities for more gracious living. But they were still a restless lot and the profits one of them had made in intervals between Indian raids and rebellions against the government got him in trouble in 1798.

Zachariah Cox, who had speculated successfully in Ohio River lands, was bitten by the same bug which had once infected the Kentucky conspirators—and would presently attack Vice-President Aaron Burr. Cox enlisted a private army of eighty men, armed them, and embarked at Smithland, Kentucky, bound for Spanish conquest.

General James Wilkinson—none more competent than he to sniff out a plot—warned Governor Sargent of Mississippi that Cox was about to attack a friendly power and Cox and the thirteen men who had not already deserted him were arrested at Natchez. He claimed he only intended to explore the Spanish country but the fact that he had felt free to do so without the formality of permission well demonstrates the contempt with which the frontiersmen were likely to regard all governmental restraints and procedures—including those of their own.

Nevertheless, by 1799 representative government was functioning on the Ohio not only in old established Virginia, Pennsylvania, and Kentucky, but also in the new Territory North-west of the River Ohio. On September 4 representatives of the settlements in that vast and infinitely varied land met in Cincinnati as a territorial assembly and young William Henry Harrison was appointed delegate to Congress.

Those territorial representatives were a colorful lot—graduates of the New England and southern colleges and frontiersmen who were long on common sense but scarcely able to read or write; mannerly Virginians, reared in the Tidewater, and men who had never owned a linen shirt. They were, as a body, an entirely competent set of men, experienced in all the ways of life likely to develop in this new country and fully aware of its problems and its possibilities.

In 1800 the approximate boundaries that would eventually mark the state of Ohio were decided upon, and Indiana Territory (which consisted of present Indiana and all the rest of the now defunct Territory North-west of the River Ohio) was set up. The following year William Henry Harrison was appointed governor of this Indiana Territory and presently established the territorial capital and built his fine brick home, "Grouseland," at Vincennes. There was some complaint, shortly, from the only other thickly settled center in Indiana Territory. Residents of the Kaskaskia neighborhood— who must have numbered more than a thousand by this time—petitioned Congress for relief from the necessity of traveling clear across what is now Illinois to pay their taxes and go to court. No complaints were recorded from the country to the north, since the occasional Indian trader or squaw man in that locality had no intention of paying taxes anyway.

In spite of the derogatory statements that aging and querulous Governor St. Clair had made about the abilities and character of her people, Congress agreed in 1802 that Ohio was ready for statehood: and in that year a constitution was drawn up under the provisions of the Ordinance of 1787.

Downstream and on the Mississippi there was another recurrence of Louisiana trouble. By some complicated European horse trading, Napoleon and Talleyrand had reacquired Louisiana for France on March 21, 1801, but the territory had not yet been formally transferred and the acting intendant at New Orleans in 1802 was still a Spaniard, Juan Ventura Morales. For some reason— possibly because of the depredations of the lawless American boatmen on their visits to New Orleans—he revoked the American privilege of deposit at the port on October 16 of that year. When news of that came north in midwinter there began to be new mutterings on the Ohio.

President Thomas Jefferson, to whom the importance of the West had always been obvious, did not hesitate to take prompt action—foolhardy though it appeared to most of the Atlantic seaboard. He dispatched James Monroe to Paris to assist Ambassador Robert R. Livingston in making a final settlement of the troublesome problems of the Mississippi.

Monroe carried instructions to pay up to $10,000,000 for the port of New Orleans and for East and West Florida (present Florida plus a wide stretch of the Gulf Coast extending to New Orleans), which was assumed to have been returned to France along with Louisiana. Actually Spain had retained the Floridas and Livingston, evidently with an inkling of this, had taken great pains to enlarge to the French upon the value of them and the worthlessness of trans-Mississippi Louisiana.

When Monroe arrived and the two Americans approached Minister of the Treasury François Barbe-Marbois, they found that the entire Louisiana Territory, extending west from the Mississippi below the mouth of the Arkansas River to the Rocky Mountains and north to the Canadian border, could be purchased. Not all the horse traders were Europeans and within six months New Orleans and the Louisiana Territory had been bought by the United States for $11,250,000.

Spain agreed to the sale and, since the lands were still in Spanish hands, turned the enormous tract over to two American commissioners who had been appointed to receive it on December 20, 1803. One of these commissioners was William C. C. Claiborne; the other was none other than General James Wilkinson—playing his dual role to the last curtain!

Much angry talk filled the New England taverns that winter. Eleven million dollars for an unexplored waste which could never by any stretch of the imagination be populated! And this in addition to the Ohio Valley, already on the nation's hands and with scarcely a white inhabitant per square mile!

On the Ohio news of the purchase got a different reception, as may be imagined, and emigration and investment were promptly accelerated. Dreams of a peaceful and wealthy valley began to have a foundation of reality even to the casual and disinterested eye and prosperity and a degree of comfort soon became rather usual in the longer-settled communities on the river. All continued quiet in 1805, too, except that a native-born Westerner found himself becoming quite famous and a famous Easterner visited the West for the first time.

The eastern visitor was Aaron Burr, aristocratic, ambitious, ruthless former vice-president of the United States: the native was Tecumseh, one of the triplet sons of a Shawnee couple. For generations busybodies have indulged in speculation as to who, other than the husband of their mother, might have fathered the three Indian sons, but whoever he might or might not have been he had reason to be proud of at least two of his getting; they became distinguished men by any standard.

Aaron Burr, of New York, vice-president in Thomas Jefferson's first administration, who had himself missed election to the Presidency by one vote, was a rather foppish, handsome little man, with a fierce pride, a burning ambition, a glib tongue, and with no solid earthy virtues such as truth or loyalty or sense of duty to balance those superficial attributes. Throughout his term of office he had struggled with Thomas Jefferson—capable adversary indeed—for added power in his office. He had left it with his appetite entirely unsatisfied.

He was an unsuccessful candidate for the governorship of New York in 1804 and during the campaign was the object of pointed statements concerning his past and probable future probity which issued from an old rival, Alexander Hamilton. His defeat a certainty, Burr challenged Hamilton to a duel and Hamilton quixotically accepted and was killed. Under indictment in New York and New Jersey, Burr sought less squeamish fields in pursuit of a dream he had apparently conceived years before—though *exactly* what that dream was will probably never be known!

It must have had to do with the seizure of and "liberating"— perhaps in that term's World War II meaning—some part of Spain's holdings south of Louisiana Territory: perhaps there was planned a Burr-governed empire which might include present Texas, Arizona, New Mexico, California—even Mexico itself. Both Hamilton and Burr would once have welcomed a war with Spain in which each hoped for an opportunity to take the leadership in liberating Mexico; Burr had now made himself heir to both their aspirations.

While serving as vice-president Aaron Burr had gone so far as to approach the British in an effort to get their aid for such a move. Rebuffed, he had dropped the plan but, after his term was served and he had been defeated in the race for governor of his own state, had killed Hamilton and was himself a fugitive, such a scheme seemed the most direct road open to the aggrandizement he craved.

He came down the Ohio to the Mississippi and went back east overland, surveying the prospects for possible backing among the Westerners. Certainly he was entertained on Blennerhassett's Island as he passed down the Ohio, probably he talked to his old friend James Wilkinson and to such disaffected spirits as ex-Judge Sebastian. He seems to have contracted with one Andrew Jackson, sometime to be president but then only a prominent citizen of the upper Cumberland River, for the construction of flatboats to be floated down to the Ohio for the use of a prospective military expedition to the south, and he made sundry other arrangements with lesser figures.

Back east again he made efforts to gain the support of both France and Spain. Such support could scarcely be conceived as having any but treasonable aims toward the United States—doubtless involving, by this time, some plan which would revive Wilkinson's old one for the secession of the western states. But France and Spain were exceedingly busy with affairs at home; they had no support to offer at the moment.

Burr set about raising funds from private sources; some of them disgruntled politicians, some opportunists, and some—like Harman Blennerhassett—merely the bumbling visionaries of which the United States still seems to have an oversupply. How much money was promised, how much was actually paid in will never be known: after the affair was over no creditors were anxious to call attention to their participation through public suits for accounting or recovery.

Perhaps Burr had a definite plan in mind by the summer of 1806, perhaps he only intended to lead a private army down the Ohio and Mississippi, accumulating promised recruits and supplies along the way, and to move toward the Spanish country and see what opportunity for conquest might offer itself. At any rate, he wrote to General James Wilkinson—still in command of the American forces on the lower Mississippi—and informed him that the private expedition would soon start for the lower Mississippi from its rendezvous on Blennerhassett's Island in the Ohio.

Obviously Wilkinson had previous knowledge of the plan; undoubtedly he had given Burr assurance that he would either permit its action or connive in it—but Aaron Burr, devious as he was, had failed to reckon with the greater talents of Wilkinson.

The general now made the last of his important double-deals, this time, as it chanced, to the benefit of his country: Wilkinson had heard that news of Burr's activity had come to Jefferson and, fearing to continue his association, he ordered the men of Burr's flotilla arrested when they reached the lower Mississippi. When Burr arrived he, too, was seized. He escaped but he was retaken and was held at Richmond, Virginia, for trial on a charge of treason.

At the time of his arrest Burr's true aims were no less puzzling than they are today, after fourteen decades of scholarly investigation. At Cincinnati in October, 1807, Christian Schultz wrote a passage for his book *Travels on an Inland Voyage,* which gives the contemporary view and suggests, by a pointed lacuna, that either names too great to be casually mentioned were thought to be involved or that Schultz's publisher did not care to risk the challenge or horsewhipping that was then preferred over libel suits as a recourse for injured feelings.

The chief topic of conversation at present along the Ohio [Schultz notes] is Burr's late expedition, and his pending trial. Marietta was what may be styled the headquarters of this business; not that many of its citizens had embarked in his schemes, but rather as forming a kind of central point for the preparation and equipment of his flotilla. From every information which I have been able to collect, this affair still remains enveloped in a *cloud of mystery.* That Burr ever seriously meditated a separation of the western states is highly improbable . . . The seizure of the Spanish dominions, *without* the immediate aid of Wilkinson and the army, is equally absurd, as his whole force would not have amounted to more than three hundred men. And, lastly, his intended settlement on the Washita appears equally distant from the real object in view; for here it is a well known fact, that what little preparations had been made, were more for a military than agricultural expedition. In short I have conversed with some who were on board the *fleet,* who laughed at the idea of "leaving their friends and families, and a healthy country, to go and settle a *swamp,* in the most unhealthy part of Louisiana!"

The only rational conjecture, and which is the prevailing one in this country, is, "That as the prospect of an immediate war between the United States and Spain at that period appeared to be unavoidable, Burr's primary object was the seizure of Baton Rouge, and afterwards of Mexico, with the aid of Wilkinson's army; but matters having been mysteriously arranged with the Spaniards, Burr, notwithstanding, persevered in his plans; and had the counsel and . . ."

At this point either Christian Schultz or Isaac Riley, his New York publisher, became cautious or charitable: the printed page in Schultz's book is here interrupted by sixteen lines of asterisks, carefully enclosed in quotation marks, as is the preceding and following text, and occupying the space in which the original story obviously carried the names of those who were believed to have counseled Burr! These lines were evidently removed and the asterisks substituted after the type had been set and the pages made up ready for the press.

After this titillating break Schultz's quotation continues:

Until he was denounced to the government; and thereby frustrated a scheme which, had it succeeded, would probably have secured an empire (with out endangering our own) to one whose daring genius and towering ambition at least as well deserved it as Bonaparte, into whose hands it will probably soon fall.

When Burr was finally brought to trial either the old feud between President Thomas Jefferson and Chief Justice John Marshall had at least as much to do with the proceedings as did Burr's treason, or the chief justice was overzealous in guarding the constitutional rights of the accused. After considerable legal finagling Aaron Burr was acquitted.

Acquitted only in so far as the court was concerned, however: to the American people the name of Burr has been synonymous with treason ever since—to the almost complete popular eclipse of more persistent traitors such as Wilkinson and his associates.

The other man who came into prominence in the West at the same time as Burr was of very different character; far from a traitor to his people, Tecumseh was probably the most fanatically patriotic—and possibly the most capable—leader the Indians ever had.

The principal northern tributary of the Ohio River is the

Wabash, which rises near what is now the Ohio-Indiana line, slants across the upper half of Indiana and turns south to form the division between Indiana and Illinois until it empties into the Ohio River. The most beautiful tributary of the Wabash is the musically named Tippecanoe.

The neighborhood of the junction of these two rivers had been a favorite townsite of the Indians and an important trading post of the French throughout written history. After La Salle had aided the Illini and the Miami tribes to form an alliance for defense against the Iroquois the Great Wea Town had been established there and near-by the French had built Fort Ouiatenon, which would be held eventually by French, British, and Americans. Immediately at the junction of the rivers had stood Kethtippecanuck, the great Indian trading town, which General Scott had burned in 1791. To this eligible site came, in June, 1808, Tecumseh and his brother, Tens-Ka-Ta-Wah, or Elskatawa ("The Open Door"), called by the Americans "the Prophet." There they rebuilt a town which was peopled by the followers they had gathered from many tribes. Tecumseh was the governor, the Prophet was the spiritual leader.

The teachings of the Prophet—until he later began to add claims of personal miracle-working and sorcery to his gospel—were of the highest order of moral philosophy. Certainly no one but the most unreasoning Indian-haters could have taken exception to them, in so far as they affected the welfare of the Indians.

In an early "talk" which the Prophet sent to tribes throughout the Mississippi Valley and Great Lakes areas he set out the tenets of the faith he preached and, incidentally, made a curious and rather remarkable prophecy. With an introductory statement to the effect that he had been bidden to speak by a higher power (conventional device of prophets of all faiths) he began:

The Great Spirit bids me address you in his own words, which are these:
"My children you are to have very little intercourse with the whites. They are not *your Father,* as you call them, but your brethren. *I am your Father* . . . I am the Father of the English, of the French, of the Spanish, and of the Indians. I created the first man, who was the

common father of all these people, as well as yourselves; and it is through him, whom I have now awakened from his long sleep, that I now address you. But the Americans I did not make. They are not my children, but the children of the Evil Spirit. They grew from the scum of the great water, when it was troubled by the Evil Spirit, and the froth was driven into the woods, by a strong east wind. They are numerous, but I hate them. They are unjust. They have taken away your lands, which were not made for them.

"My children—The whites I placed on the other side of the *Great Lake,* that they might be a separate people. To them I gave different manners, customs, animals, vegetables . . . for their use. To them I have given cattle, sheep, swine and poultry, for themselves only. You are not to keep any of their animals, nor to eat of their meat. To you I have given . . . all wild animals, and the fish that swim in the rivers, and the corn that grows in the fields, for your own use; and you are not to give your meat or your corn to the whites to eat.

"My children—You must plant corn for yourselves, for your wives, and for your children . . . but plant no more than is necessary for your own use. You must not sell it to the whites. It is not made for them. I made all the trees of the forest for your use, but the maple I love best, because it yields sugar for your little ones. You must make it only for them; but sell none to the *whites.* They have another sugar, which was made expressly for them; besides, by making too much you spoil the trees and give them pain, by cutting and hacking them; for they have a feeling like yourselves . . .

"My children,—You may salute the whites when you meet them, *but must not shake hands.* You must not get drunk. It is a great sin. Your old men and chiefs may drink a little pure spirits, such as come from Montreal, *but you must not drink whiskey.* It is the drink of the Evil Spirit. It was not made by me, but by the *Americans.* It is poison. It makes you sick. It burns your insides. Neither are you on any account to eat *bread.* It is the food of the whites.

"If a white man is starving, you may sell him a little corn, or a very little sugar, but it must be by measure and by weight. .

"My children,—You are indebted to the white brothers, but you must pay them no more than *half their credits, because they have cheated you.* You must pay them in skins . . . But not in meat, corn and sugar. You must not dress like the whites, nor wear hats like them, but pluck out your hair, as in ancient times, and wear the feathers of the eagle on your heads. And when the weather is not severe, you must go naked, excepting the *Breechcloth,* and when you are clothed, it must be in skins or leather, of your own dressing.

"My children,—You claim that the animals of the forest are scattered. How could it be otherwise? You destroy them yourselves, for their skins only, and leave their bodies to rot, or give the best pieces to the *whites.* I am displeased when I see this, and take them back to the earth that they may not come to see you again. You must kill no more animals than are necessary to feed and clothe you, and you are to keep but one dog: because by keeping too many, you starve them.

"My children,—Your women must not live with Traders or other white men, unless they are lawfully married. But I do not like even this; because my White and Red children were thus marked with different colours, that they might be a separate people."

The Great Spirit, through the mouth of the Prophet, followed with advice on courtship, marriage, daily bathing, and further injunctions against cheating and getting drunk. He continued:

"Your wise men have bad medicine in their bags. They must throw away their medicine-bags, and . . . collect fresh and pure. You must make no feasts to the Evil Spirits of the Earth, but only to the Good Spirit of the Air. You are no more to dance the Wabano, nor the Poigan or Pipe-dance. I did not put you on the earth to dance these dances . . . You are to make yourselves *Paḳa tonacas* (or crosses) which you must always carry with you, and amuse yourselves often with that game [probably lacrosse]. Your women must also have handsome *Passa quanacles,* that they may play also: for I made you to amuse yourselves, and I am delighted when I see you happy. You are, however, *never to go to war against each other:* but to cultivate peace between your different tribes, that they may become one great people.

"My children,—No Indian must sell rum to an Indian. It makes him rich, but when he dies, he becomes very wretched. You bury him with all his wealth and ornaments about him, and as he goes along the path of the dead, they fall from him. He stops to take them up and they become dust. He at last arrives almost at the place of rest, and then crumbles into dust himself. But those who, by their labors, furnish themselves with necessaries only, when they die, are happy. And when they arrive at the land of the dead, will *find their wigwam* furnished with everything they had on earth."

The Prophet then told how the Great Spirit had opened a door and showed him a deer, and a bear, small and lean, and told him that these were the animals as they then were; through another door he was shown a great fat deer and bear—the animals as they had been created. The Prophet, no longer quoting the Great Spirit, continued:

Now my children, listen to what I say and let it sink into your ears—it is the orders of the Great Spirit.

My children,—You must not speak of this Talk to the Whites. *It must be hidden from them.* I am now on Earth, sent by the Great Spirit, to instruct you. Each village must send me two or more principal chiefs to represent you, that you may be taught . . .

The Prophet's rules for living—reminiscent as some of them are of early Christian doctrine—certainly covered the points most important to Indian welfare. And, since there is not evidence of anything but the best of faith on his part, we may assume that his interpretation was, if not inspired, at least a well-planned program of Indian reform. The wisest of the Indians must surely have seen that his was the only possible way to the salvation of their ancient culture and manner of life. .

Like many a religious leader before and since, Elskatawa made frequent prophecies of events to come. Some of them were so logical that their fulfillment was bound to follow but one, at least, from the preamble of the talk just quoted, foretold a catastrophe so utterly unpredictable even today that its occurrence on schedule should win even a modern audience to follow the Prophet's leadership.

There is no question as to the authenticity of either the talk or the prophecy. It was read by Le Maiquois—the Trout—a follower of the Prophet, to a gathering of Chippewas and other Lakes Indians at Michilimackinac on May 4, 1807, and was reported in full in a letter from Captain Dunham, commandant of that post, to General William Hull at Detroit in that same month.

In it the Prophet, referring to himself as "the first man created," says:

Children,—I was asleep, when the Great Spirit, addressing himself to another Spirit, said "I have closed my book of accounts with man, and am going to destroy the earth: but first I will awaken from the sleep of the Dead the first man I created; he is wise, and let us hear if he has aught to say." He then awoke me and told me what he was about to do.

I looked around the world and saw my *Red children* had greatly degenerated, that they had become scattered and miserable. When I saw this, I was grieved on their account, and asked leave of the Great Spirit to come to see if I could reclaim them. I requested the Great Spirit to grant, in case they should listen to my voice, that the world might yet subsist for the period of *Three full lives,* and my request was granted.

Now, therefore, my children, listen to my voice, it is that of the *Great Spirit!* "If you hearken to my counsel and FOLLOW MY INSTRUCTIONS FOR FOUR YEARS, THEN THERE WILL BE TWO DAYS OF DARKNESS, during which, I shall tread unseen through the land and cause the animals, such as they were formerly, when I created them, to come forth out of the earth . . ."

It was evidently the Prophet's plan to have this talk circulated among all the tribes during the summer and fall of 1807—four years and a small varying fraction before December, 1811!

And on the night of December 15, 1811, began the New Madrid Earthquake, greatest in the written or traditional history of mid-America. It shook the earth and darkened the sky throughout the whole Mississippi and Ohio valleys. The earth trembled, gigantic crevices, later to become lakes, appeared south of the junction of the two rivers, the Ohio ran upstream as far as the Falls, and dust clouds hung overhead not for days but for months.

Such a fulfillment must surely have convinced the Prophet's followers of his divine guidance—except that in the meantime he had made the error of teaching his warriors that he could make them impervious to white men's bullets and he had failed to heed the temporal advice of his brother, Tecumseh. By the time of the darkness the Prophet's own light had already been extinguished five weeks.

But back at the beginning, at the settlement of Tecumseh and the Prophet at Kethtippecanuck in 1808:

It required only two years for the brothers to become powers on the lush black Wabash lands. The Prophet, who first came to be well known, devoted his attentions largely to the Indians, following his own precept, while Tecumseh maintained contact with the whites. By August, 1810, Tecumseh's activities had become so extensive that he achieved the distinction of an invitation from William Henry Harrison, governor of the Indiana Territory, to come to Vincennes, the territorial capital, for a conference: it took only Tecumseh's closing remarks on this occasion to show Governor Harrison that he had to do with an exceptional Indian leader. Tecumseh's immediate complaint was that at the treaty signed in Fort Wayne the year before other tribes had disposed of lands to which the Shawnees had a claim. After he had stated his contention Governor Harrison said he would refer the question of ownership of the land to the President. To this Tecumseh made his famous reply, translated as:

"Well, as the Great Chief is to determine the matter, I hope the Great Spirit will put sense enough into his head to induce you to give up this land . . . he is so far off he will not be injured by this war; he may still sit in his town and drink his wine, whilst you and I will have to fight it out!"

They were two strong characters—Tecumseh and Harrison— both with a conviction that in positive action lay the best hope of victory.

Tecumseh evidently felt that no matter how much aid was to be received from the Great Spirit it would be well, also, to command as many men as possible when the clash came. In July, 1811, he went south to try to enlist in his forces the Creeks and Cherokees, who had befriended the Shawnees before.

He made a tremendous impression on the southern Indians —as he did on everyone, Indian or white. He held their respectful attention but he could get no definite promise of aid. Leaving the last council, however, he made a final plea of the eloquent sort of which he was so capable and closed it with a warning based on the Great Spirit's revelation to his brother: if the southern warriors did not start north to his aid upon a certain date in the late autumn, he said, he would "stamp his foot, the sky would be darkened and the earth would shake." Then Tecumseh started the long trip back to his own people.

Meanwhile Governor Harrison had not been idle. He had secured two companies of regulars, had received Kentucky volunteers and a few from the sparse settlements of Indiana Territory. He had moved up the Wabash and set up a small fort above the place called Terre Haute, and had marched on up the river to within a mile or two of Tecumseh's people at the Prophet's Town. There he had camped.

The arrangement of that encampment, on the night of November 6, 1811, was no masterpiece of military science: in the center were great fires which threw the surrounding men in clear silhouette, once they rose to their feet. Sentries were poorly placed and sleepy, Kentucky companies thought the Hoosiers green, Hoosiers considered the Kentuckians uppity, the regulars were dubious of the commanders' competence, and authority had been delegated more on a basis of expedience than ability.

The Prophet, seeing this weakness, either forgot Tecumseh's order that he was to avoid action at all cost until the latter returned from the south or actually had the full faith in the Great Spirit's guidance that he claimed. He ordered an attack shortly before the hazy Indiana dawn. It was a brief but bloody fight and the Prophet and his braves were beaten off: white men's bullets had not proved ineffective.

Days later messengers met Tecumseh on his way north but still below the Ohio. They told him of the disaster so that he might skirt the settlements and reach the Great Lakes where the British could be expected to assist him in rallying his people.

But before word of the defeat of the Prophet could reach the Creeks and Cherokees there came that phenomenon which, if reports are true, well-nigh wrecked the morale of certain chiefs of those tribes permanently. It began shortly after the date by which Tecumseh had ordered the southern Indians to dispatch the war party or suffer the consequences: the earth in the Mississippi and lower Ohio valley *did* shake—such shaking as was unknown to Indian tradition—the sky *was* darkened, and the sun glowed faintly red through those thick clouds of dust.

Immediately, by report, the southern Indians seized their weapons and started north—if this was a sample of Tecumseh's stamping, they probably reasoned, 'twere as well to make a propitiating gesture. Before they had gone far they learned that it was too late.

Actually the Battle of Tippecanoe set off the War of 1812 in the western country—the East's mind had been influenced long since by shipping interests and the fear of impressment of seafaring sons in foreign ports. Harrison's action certainly established the Indians permanently on the side of the British.

To the Ohio River itself the War of 1812 was no immediate threat, except for the harassment of lone farms and small settlements by Indians again directed from Canada, and for the fact that every man along the river who could walk and carry a gun was more or less engaged, during the next two years, in the fighting at Detroit, in southern Canada, and at New Orleans. Also, simply as a means of keeping them scattered and upon the defensive, frequent attacks had to be directed by the westerners against the northern Indiana, Illinois, and Ohio Indians.

Certainly not all of these campaigns by old men, boys, and amateur soldiers were effective. One of them, in fact, comes near to being a classic of ineptitude—it was that commanded by General Samuel Hopkins, who, with some vague idea of attacking Indian towns on the Indiana-Illinois prairies, moved north across the Wabash and the headwaters of the Vermilion in 1812, experienced mutiny when supplies ran low and the whisky ration had to be cut, and managed to lose himself and the entire detachment on the prairies for several days.

But the extent of the participation of the Ohio River people in the important campaigns was amazing. Not all men of the valley were seasoned frontiersmen at the time of this war—many were untrained, inexperienced farmers, mechanics, clerks and businessmen recently come from eastern cities—but they adapted themselves readily to wilderness warfare. Perhaps no such percentage of any regional population has been under arms in America before or since. How were their companies and regiments organized, how supplied, how were their campaigns carried out? A participant's account of the offhand manner in which these matters were handled may be interesting:

Nathan Newsom was an intelligent volunteer in a company raised at Gallipolis for the purpose of reinforcing General Hull at Detroit in 1812—a futile gesture, in view of his surrender before their arrival, but nonetheless heroic. Newsom kept a journal, as yet unpublished, which described his day-to-day adventures through the harrowing time of General Winchester's defeat at the River Raisin. After that event, during the time in which General William Henry Harrison was attempting to rally the American forces on the Maumee for the next summer's invasion of Canada, the diary ended abruptly. Whether Newsom sickened in the rigorous early spring and died, as did so many of his unhardened fellows, or whether the press of duty simply left him no time for journal entries is not known. Whichever the case his well-written diary gives a clear picture of the minutiae of volunteer life, of the organization—or frequent absence of it—of the troops, and of the natural complaint of the American soldier against both military authority over him and the attitude of the noncombatants he encounters along the way.

Preliminary organization and training was not a complicated matter for the Gallipolis volunteers: "On the 9th of August 1812 we met at Gallipolis under marching orders. Two meetings had taken place previous to this to elect the officers and to make such regulations as we thought prudent for our organization . . ."

It was as simple as that! No physical examination, no officer training, no screening, no instruction. An officer was supposed to lead, so any man popular enough to be elected could be expected to make a good officer; such was the rather logical reasoning of the day.

There was no unbecoming delay in *this* campaign:

"We left Gallipolis on Monday the 10th of August, and encamped at Major Blagg's the first night, supposed to be about 14 miles . . ."

Two days later they arrived at Chillicothe and a company already stationed there marched a half mile out to meet them. Here became apparent some of the glories of war!

The three companies . . . marched into Chillicothe in military order, attended by music and a numerous concourse of spectators. We all encamped in the Court-house. . . . Here we were first embarrassed in the pecuniary way, and our officers borrowed money to a considerable amount at Chillicothe at an exhorbitant rate of interest. Various things were promised us, when we marched, by our officers for our convenience, which were never complied with . . .

The United States government had still learned nothing from the unhappy fiscal experiences of George Rogers Clark. Still the officer must pledge his good name to feed his men and was frequently forced later to use whatever personal resources he had in repaying the loans; the patriotic citizen who advanced money without interest was still as likely to suffer. Even that "exhorbitant rate of interest" was understandable if not commendable—especially if those who made the loans had happened to hear that Francis Vigo's advances of money and stores made to Clark's force at Vincennes thirty-four years before were still unpaid!

Newsom's detachment marched on, up to twenty-four miles a day—but of course they were now seasoned troops of an entire week in the army—and they soon reached a rallying point at Urbana, Ohio.

Here they had first news of General Hull's surrender of Detroit, "The ever disgraceful and memorable act of Hull the traitor . . . The scene was horrid and pitiful. Some men who happened not to be convenient when Hull sold them, took their flight through the wilderness as they could, and passed Urbana in small parties in the utmost despair . . . in the most abject appearance and lowness of spirit. Anger and indignation were apparent in the countenance of every man at the infamous conduct of Hull . . ."

The troops encamped until October 1 and at last Newsom could note that "we were daily trained to military discipline." When they left Urbana, where "The merchants and mechanics . . . exhibited a great thirst to take the advantage of the necessity of the soldiers in demanding at least double the value for such articles as they thought the soldiers must have" they were about eight hundred strong. "Desertions were frequent, but with very little or no injury to the army, as none but the most trifling characters deserted."

Since the force was now on its way to the front, and was even now expecting attack by British Indians at any moment, Newsom's next comment is of particular interest:

> Most of our officers as well as the men are as yet destitute of the qualifications requisite in military life. Tho we had a capable adjutant who left us on the banks of the Mad River. The causes he left us is by some said to be, that he found his necessary expenditures nearly to amount to his income; and that he was dissatisfied with some of the officers. His name is Jacob Dunbaugh—as a disciplinarian he is exemplary, as a gentleman and officer highly respectable having merited the approbation and confidence of the soldiers, they viewed his departure with the utmost regret.

Side lights upon discipline are frequent:

> The army was drawn up . . . two prisoners were brought into the middle to hear their sentences read . . . Harris had deserted while on guard and Scott from the camp. The sentence was put into execution . . . Harris underwent every part of the ceremony prescribed by military useage a man should that had forfeited his life and sentenced to be shot; except he was not shot but fined $12. Scott was fined $12, and marched in front of the line with his hat off . . .

It was a democratic army and made much of the fact:

> Col. Sutton appeared as commander . . . he delivered a discourse stating his orders from his superiors and exhorting and admonishing his brothers and fellow citizens to obey him. . . . He dwelled a considerable time in explaining what would be the inevitable consequence of a disobedient, insubordinate army. . . . This discourse somewhat pacified the discontent. . . . On the 15th two men were punished. They had to ride a rail in front of the line. . . . On the 16th two men were marched in front of the line . . .

They were camped near Solomon's Town, a place where friendly and neutral Indians were gathered:

> The soldiers commenced killing chickens at Solomon's Town, and in some days after, hogs. They skinned the hogs and brought them into camp for immediate consumption. . . . The drawing of beef from the

commissary by the soldiers was not attended to, as they preferred pork to beef. The soldiers in small parties penetrated the neighborhood of the army every direction in quest of Indians, game, hogs, bees, &c, but the hog-hunters were most plentifully remunerated for their services . . .

The force moved on north toward Detroit, building breast-works and storehouses at strategic points along the route as it went. There were more desertions and those deserters who were caught had their heads shaved, were tarred, and received other unusual but not too cruel punishment. The lower order of individual stole his neighbor's blankets, fired at passing rabbits while on sentry duty and otherwise misconducted himself, but the orders in general were as well and promptly carried out as could be expected.

The men pushed on through the raw wet chill of November. No nominal dry land could have been wetter than that low north-western section of Ohio in the days before ditches had been dug and the timber cut. The watershed between the Great Lakes and the Ohio River, it had none of the characteristics usually associated with such a geological feature. Interspersed with glaciated muck-covered plains, little above the level of Lake Erie, this section was a vast swamp in which sluggish tributaries of the Wabash, Great Miami, and Maumee rivers rose within a few yards of each other to flow toward the Ohio and the Gulf of Mexico or Lake Erie and the North Atlantic.

Newsom did not complain particularly, nor apparently did most of his fellows. They were now reduced in number, something over six hundred and fifty of the eight hundred having volunteered for combat, while the others had been left behind to man the posts they had built. Newsom remarked that they could no longer move their one cannon through the mire; that three sick men had to be left by the trail because there was no way to carry them; that no provisions except flour were available after they passed Finley's Blockhouse; that their final day's march to old Fort Necessity was forty miles and that some of the men were now nearly naked.

But by the end of November, 1812, things looked bad even to Sergeant Newsom.

The men, most of them with only one blanket apiece, were dying by the dozens:

The army at this time far more presents an object of charity . . . every species of human misery . . . than the grandeur, dread and reverence which must necessarily mark an invading army which can reasonably expect to meet with success and honor. Nearly one half . . . sick, nearly the other half . . . almost naked . . . It is ardently wished by every naked man in camp that something would arrive where with he might cover his nakedness, keep his body warm and his feet dry so that he would be able to do his duty as a soldier . . .

Capt. Calvin Shepard died. He was an inoffensive, innocent young man . . . one little old blanket was the amount of all his bedding . . . James Jourdan died . . . he was a sergeant . . . a friendly, good-hearted man . . . left a widow and eight children . . .

And so they continued, a death within Newsom's immediate acquaintance every day. The whisky froze; shoes arrived—enough for all who were entirely barefooted; two hundred blankets were delivered for the garrison of almost seven hundred men and the feeling against the commissary service became daily more truculent. Word of General Winchester's defeat at the Maumee rapids came in and on the 31st of January General Tupper's Brigade—which included Nathan Newsom's outfit—was ordered forward to support Winchester's left wing, then cut off at the Carron River.

There Newsom's diary ends—and with it a participant's account of a body of troops typical of those who went forward under William Henry Harrison, very shortly, to invade Canada and to drive the British for the last time from the Great Lakes!

Those men of the Ohio were only human—very human in their small complaints and dissatisfactions—but they were ready to act when action counted and they seem to have possessed a fortitude well-nigh past believing.

Within eight months of the close of Nathan Newsom's diary the War of 1812 was won in the West by a glorious success in the naval victories on Lake Erie and by the invasion of Canada and the Battle of the Thames. Tecumseh, hope of the Indian, was killed in that engagement by one of two Ohio River citizens of the Kentucky side—whose partisans quarreled over the distinction for half a century following.

Wars with foreign powers, after the Treaty of Ghent was ratified by the United States on February 17, 1815, were over until a minor engagement to come in thirty-three years in which the United States itself was the shameful aggressor. There was also to be a major war, within half a century, which would be of great importance to the river but no foreign power would be involved— it would be strictly a family affair.

Chapter Eight

THE ADVENTUROUS,
THE MURDEROUS AND
THE RING-TAILED SNORTERS

IN ALL of history the only names that have survived much beyond their final publication in death notices have been those of the exceedingly good and the exceedingly bad.

The Ohio supported many in that first category, and of a great variety of goodness; there were more who could be classified as moderately good and a great body of people who, in the valley, would still be described as "so-so." The Ohio also had citizens of rather slack moral fiber, some who were downright ornery, and quite a number who were definitely bad—including a few really distinguished in their badness.

Of course it was the so-so element that really settled the country, fought the battles, did the work. The leadership of the excessively good is a necessity if a new country is to prosper and to grow in rectitude, but they are not as a generality too useful as hewers of wood or fetchers of water.

The badmen made two worth-while although entirely unintentional contributions: when their conduct became insupportable it caused their neighbors to bestir themselves and establish law and order, while their more notable villainies furnished themes for a frontier literature, first preserved in tales around log-cabin fires, later put in print.

A glance at the badmen is in order and no apology need be made for disturbing their dust: each of the more distinguished for infamy among them had enough of the showman in him to appreciate the inclusion of his name and exploits here—even should the history of the Ohio River be available reading matter in the nether regions.

The rascals came from all walks of life and almost every sort of family background. Some of them undoubtedly considered themselves gentlemen-adventurers, some had a social or religious crusading mania, some probably never thought of themselves at all.

It is difficult to classify these villains by the degree of their sinfulness: who, for instance, can say whether the Rev. Mr. Joel Barlow, when he defrauded the French artisans of their money in return for nonexistent titles to "subtropical" lands at Gallipolis, Ohio, was a lesser enemy than dissolute "Colonel Plug," who only pulled the calking from emigrant boats downriver in order to salvage their cargoes and dispose of an occasional passenger for the sake of acquiring his watch?

Many a man of gentle reputation at home on the eastern seaboard or in Europe succumbed to the absence of restraint and the example of the dog-eat-dog life around him and became something very different in the wilderness. Many another, with infamy in a secret past, responded to the challenge of a new life and became a worthy citizen, often a person of consequence.

Peculiarly enough, in many cases neither the Old World status nor the rise in the new land seemed to affect the progeny of these immigrants: the blue blood of one was likely to grow progressively more diluted with the passing generations, while the descendants of the man who first saw America as a deported criminal came to sit in high places. Even today the shanty-boaters on the Ohio and the most doless of the hillbillies back up in the mountains often bear proud names, and many a prominent family springs from an ancestor who, instead of emigrating to America, was "transported" to these shores.

Some characteristics were shared by all those who moved to the frontier of their own volition before the end of the first decade of the nineteenth century, were they good, bad, or indifferent: they all had curiosity, ambition of one kind or another, and a willingness to attack unknown problems or strange enemies for gold, glory, or only for excitement. It was the use to which they put these qualities that made the difference.

Of course some of the earliest of those who reached the Ohio Valley cannot be considered immigrants in the strict sense of that term; they were just emigrants—movers—motivated only by a desire to leave their previous haunts. They left little trace of their presence—which was often as they meant it to be.

These might be called "strays," since no other classification seems to fit their many characters and the varied stimuli that launched them on their wanderings. There must have been a great many before 1800, and the arrival of the first of them on the Ohio and its tributaries probably dates back to the years in which de Soto's Spaniards wandered along the upper Tennessee and middle Mississippi rivers. As their cruelty and arrogance aroused the hostility of the Indians the latter retaliated by armed attacks and, far worse for its psychological effect, stealthy harassment: some of de Soto's men became separated and others, half crazed by the strangeness and apparent hopelessness of their situation, deserted or simply wandered off into the wilderness. These men had seen the Tennessee, southern tributary of the Ohio; perhaps some of them returned to it or reached the Ohio itself and survived a while as Europeans before being assassinated or assimilated by the Indians.

Certainly there was an infusion of white blood, after their passing, in the Indian tribes which the men of de Soto's expedition visited.

There is no need to raise the delicate issue of what happened to Virginia, her parents Ananias and Eleanor White Dare, and the other possible survivors of Sir Walter Raleigh's Lost Colony at Roanoke. Whether they or their servants lived to flee northwest up Chesapeake Bay to contribute to the language of the Delaware Indians what travelers later got themselves laughed at by insisting were Welsh or Celtic words is a question far too hot to introduce— anyway the Delawares had other opportunities to welcome former citizens of the British Isles within the following century.

After 1600 English criminals were "transported"—some for the crime of as little as a dozen shillings' insolvency; some the very dregs of Newgate, marked with the scars of manacles and shackles on wrists and ankles or with ears cropped close. Many of them must have seen in the untamed back country a future in which they might either regain self-respect or avenge themselves upon a cruel society. Some such must have found a home in the woods for themselves and their part-Indian descendants after them, but certainly *they* were most anxious that the story of their origin *not* be preserved.

Many a man, between 1600 and the recorded exploration of the Ohio, must have spent time on its banks, leaving the settlements for no cause whatever except that the wilderness appealed to some instinct which awoke in him after he had learned its ways through a few seasons of hunting near the settlements. And what happened to all those runaway bondsmen and apprentices for whom the colonists advertised constantly?

What of those stories about Spanish miners, prospectors, and seekers after mythical Indian gold still wandering upon the Ohio's southern tributaries, the Kentucky, the Tennessee, and the Cumberland, in the early eighteenth century? *Someone* must have known of those Spaniards—fabricating tales of wandering conquistadors was a little beyond the imaginative powers of the first settlers in the Kentucky mountains.

And such matters as that brief glimpse of that party which had strayed from the Atlantic coast to the future site of Nashville? What of them? Timothy de Monbreun, who had come down from Canada to trade at that place before the last quarter of the eighteenth century, saw them as they wandered westward past the Cumberland: five men and a woman, they were heading west—destination unknown, even probably to themselves. There had certainly been other such wanderers before.

And the renegades, the whites who joined the Indians? Sometimes they were whites of substance who became useful and prominent members of Indian tribes—Jim Allen, former Tennessee lawyer who joined the Chickasaws after financial reverses broke his pride, was one—but more often they were crackpots or fugitives from punishment for capital crimes. The latter sort were usually accepted by Indians of social standing similar to their own—outcasts of the tribes—and thus they found their natural level. Of this stamp were the terrible Harpe brothers, whose only friends were a party of Cherokee desperadoes who had been outlawed by their own people. Similar were the Girty brothers, bogeymen of the frontier before the turn of the nineteenth century.

Infiltration of this sort had come from all the length of the coasts by which the whites invaded. The French from north of the St. Lawrence were more likely to "go Indian" than other whites. The majority of them who made this step did so entirely by preference and received no stigma from it. Dutch came west from New Amsterdam; more, doubtless, after it became New York. Englishmen made the change probably with a less easy adaptability but there must have been a great many English who were driven to the Indians by real or fancied persecution. The Spanish from the Floridas mingled with the Indians on the borders as freely and naturally as they had after the conquests in Central and South America but, being always few in number, they were a lesser factor.

Taking all this into consideration the fact becomes evident that there must have been a discernible amount of European blood mingled with that of most of the tribes east of the Mississippi by 1800—especially was this likely to be true of the lithe, yellow-brown skinned peoples such as the Cherokees, the Shawnees, and the Miamis who were most physically attractive to the whites. Some of it was blue, aristocratic blood; some the red blood of the healthy peasant; some of it yellow, of the scum of European ports, but a good deal of it must have been bad—if such a thing as bad blood really exists.

The story of the strays will remain a mystery, as the strays themselves intended it to be, but the histories of many of the undesirables are of full record; those of the worst of the Indian traders, for instance.

The directors of Indian trade were bold and enterprising men, although not too careful, usually, in their observance of the law of the land: such were Viele, Croghan, and Findley. Most of their employees shared only a few of their characteristics but in one of these they participated to a man and excelled. *They* had an almost complete disregard for law, civil or ethical, except in the matter of contracts and credits with Indians in fur transactions—which was, after all, a virtue forced on them by expediency.

Of the early travelers and diarists some disapprove of tavern-keepers as a class, others condemn the ferrymen, an occasional one criticizes the military—but they are unanimous in their opinion of the employees of the Indian traders. These, say all the people who met them in the wilderness, were a godless lot.

It was to be expected that the Rev. Mr. David Jones would find the traders' men upon the Ohio in 1772 to be "lamentably dissolute in their morals," but Christopher Gist, a man who had led no cloistered life, found "a Parcel of reprobate Traders" at Logstown and reports continued uniform until the day when Indian traders had become a profession extinct east of the Mississippi.

The aristocratic or prosperous conspirators against the welfare of the nation were mostly victims of the same lack of restraint on the frontier which encouraged moral dissolution among ignorant Irish traders: Sebastian, Innes, Brown, that very inept conspirator Zachariah Cox, and the accomplished James Wilkinson. Aaron Burr's doings were fully as nefarious, albeit upon a higher intellectual scale, as those of any of the double-dealing trading-post hands.

There is no reason to consider here the many men who simply

got into trouble of one sort or another in the lawless new country, and got out of it again: such men as George Rogers Clark, whose worst misdemeanor was that of roistering at the head of the same armed troops with whom he had chased the British from the Ohio Valley. Driven to bitter resignation in his later years by his Laocoönlike experience with the governmental red tape in which he became entangled, plus a taste for potent Kentucky whisky, he spent the years after he was tamed as a harmless man, old before his time, glowering from the door of his cabin at the Falls of the Ohio, his jug—his only intimate—by his side.

Thus the questionable characters all—strays, traders' men, conspirators, roisterers—were chiefly of only dubious repute; dissolute at least, traitorous at worst. After the American Revolution a new class began to develop upon the Ohio and the Mississippi. There was no question about this category; it was made up of as cruel, inhuman, and depraved a lot as ever preyed upon a countryside— the early outlaws were heartless, often purposeless, assassins.

Perhaps Simon Girty may be considered as one of the earliest of these, though Simon was at least occasionally loyal to his friends the Indians—whereas the outlaws upon the lower Ohio were steadfast to no connection. However, Girty was heartless and bloodthirsty enough upon occasion to qualify for the company.

As children he and his brothers, Thomas, James, and George, were being not very gently reared by their Irish-born parents at the time of that memorably unhappy housewarming at Chambers' Mill, Pennsylvania, in 1751. The occasion has already been described: "the Fish," an Indian guest, it will be recalled, killed their father, Simon Sr.; John Turner, a member of the household, killed the Fish and married widow Girty. Even then all might have prospered with the little family except that they returned to the forbidden Indian country and were captured during the French and Indian War by a party of Delawares, Shawnees, and Senecas.

As captives the Girty boys witnessed scenes which made the rather lax morality of their former home life seem staid by comparison. One of their fellow prisoners, a woman who tried to escape, was scalped but not killed, "next they laid burning splinters of wood here and there upon her body, and then cut off her ears and fingers, forcing them into her mouth so that she had to swallow them . . . the woman lived from nine o'clock in the morning until toward sunset, when a French officer took compassion on her and put an end to her misery. An English soldier, on the contrary . . . who escaped from prison at Lancaster and joined the French, had a piece of flesh cut from her body and ate it." The Girty boys also saw their stepfather, Turner, burned at the stake: "They tied him to a black post . . . made a great fire; and having heated gun-barrels red-hot, ran them through his body! . . . and at last held up a boy with a hatchet in his hand to give him the finishing stroke."

Indians, urged on by contesting French and British and well supplied with rum by both, were not in that day foster parents likely to develop the finer side of a growing boy's character; to this the Girty brothers' careers later testified.

James Girty was presently adopted by a Shawnee, George by a Delaware, and Simon by a Seneca. They learned the languages and within a few years were given complete freedom by the Indians to come and go as they wished. They wished to come and go considerably, and upon business usually bloody—to the lasting terror of the frontier.

The Girtys were often on the Ohio about Fort Pitt as young men, working as interpreters and as guides to the military and to traders. Thomas Girty married and settled down, more or less (although he was described in middle life as "a savage in everything but color"), and George became an independent trader. In spite of their patronymic which was likely to attract suspicion, George and even Thomas Girty bore a reasonably decent character for a time. Thomas reared a respectable family and George was of good repute until he deserted the colonial army to join the British during the Revolution. Both Simon and James continued more or less with the Indians, but since it was Simon who became most infamously known, only his career need be sketched.

Until the end of the War of 1812 latitude along two lines was extended to any returned captive of the Indians in the reporting of his adventures: he was permitted to report that he had been rescued by a beautiful Indian princess (the before-mentioned Pocahontas motif) and he was expected to claim to have seen Simon Girty, the White Savage. If Simon Girty sneered at the sorrows of all the captives who claimed this distinction he must have had time for few other activities. But there is reason for believing the claims of a good many of the captives, at that; by the unimpeachable authority of official British and American military records Simon Girty took part on one side or the other in the majority of disturbances involving Indians during more than forty years following his captivity!

Simon, a Pennsylvanian by birth, and a professed friend of the Indians, was a Virginia partisan in Dunmore's War. He was employed as a scout for the force of which George Rogers Clark, Michael Cresap, and Simon Kenton were members. Girty was apparently commissioned, for in Governor Dunmore's list of "Persons Well Disposed Toward His Majesty's Government" drawn up at the beginning of the American Revolution he is listed among the elect as "Lieut. Simon Girty."

At first he proved a disappointment to his lordship, for he joined the Rebellion as soon as news of it came to Fort Pitt. His service, at a salary of five shillings per day, was with William Wood upon a tour northwest to the Wyandot Town on the Sandusky River for the purpose of learning the temper of the Indians toward the rebelling Americans. The report of Indian temper was made to the Virginia House of Burgesses, accompanied by a petition for extra pay because of the hardships of the journey. Instituting its future policy toward petitions for funds, legitimate or not, the House accepted the report but refused the pay.

In April, 1776, the Virginia government appointed George Morgan agent for Indian affairs for Fincastle County (which included Virginia beyond the Big Sandy River and thus all of present Kentucky) and Girty became an interpreter for him. On May 1 he was appointed interpreter for the Iroquois at Fort Pitt as a salary of ⅝ of a dollar per day "during good behavior, or the pleasure of the Hon. Continental Congress" and after three months and one trip to the Indian country he was discharged for "ill behavior."

Simon Girty had helped to enlist men around Pittsburgh for

the Continental army, expecting to receive a captain's commission: he received only that of second lieutenant and was left behind, presumably as a representative among the Indians, when the company went on active duty. He resigned his commission in August, 1777, but stayed in Pittsburgh and was presently suspected of joining Alexander McKee and other Tories in a plot to seize the West. He was acquitted of the charge and was sent as emissary to the Senecas, who held him captive until he escaped.

Perhaps it was disappointment at his failure to receive a captaincy; perhaps it was indignation at being suspected of conniving with the Tory, McKee—perhaps it was only his innate cussedness: at any rate Simon Girty's gestures toward respectability ended shortly after this point in his career.

He accompanied Continental forces upon one more expedition —that of General Edward Hand, with five hundred men, against an Indian town on the Cuyahoga River. The Indian force in camp upon the arrival of this party was composed of four squaws and a boy. All but one squaw were killed and General Hand had one captain wounded and one man drowned.

Shortly after Girty returned to Fort Pitt, following this military coup, Alexander McKee did exactly what his neighbors had anticipated; he left his residence at McKee's Rocks on the Ohio to join the British at Detroit; as had also been anticipated, Simon Girty accompanied him. There was a difference in their two cases: McKee was a consistent Tory, while Girty was a traitor to a cause to which he was sworn. Girty was not alone in such turncoat action. In the early days of the American Revolution, loyalty was a matter in high question: was it more honorable to be loyal to one's king and one's established government or to one's neighbors and one's country? Many a better man than Simon Girty had trouble deciding, and many a man of lesser order simply sought what looked at the moment to be the "better 'ole." On April 20 following Girty's departure more than twenty other soldiers and civilians fled down the Ohio from Fort Pitt by boat. They were pursued and caught at the Muskingum, some were killed in the fight that resulted, two were shot as deserters, one was hanged, and two received one hundred lashes each.

Nothing like this happened to McKee and Girty's party. McKee knew the Indian country well and Girty knew it perfectly. They arrived at Detroit in June, and Girty, "giving satisfactory assurances of fidelity" to his Majesty's cause, was retained as a British interpreter. He was assigned to the Mingo tribe at a salary of sixteen shillings, York currency, and one and one-half rations per day—which was a substantial raise over his Virginia pay.

The Girtys (brother James joined Simon in this profitable employment) accompanied Indians on a raid in Kentucky in the fall of their first year with the British. The party brought back seven scalps and Mrs. Mary Kennedy and her seven children as prisoners.

On their return north from this expedition Simon Girty went to an Indian gathering at Wapatomica where Chief Blackfish and his Shawnees were holding a famous captive. Simon Kenton had been captured by the Indians while he was on a horse-stealing expedition north of the Ohio—that was about the only form of retaliation the scattered Kentuckians could then venture—and he had already been sent on his famous ride, stripped naked and tied to a horse which was whipped through the brush-filled woods. He had survived that and had successfully run the gantlet. When Girty arrived his next test was under consideration.

Many stories have been told of Girty's meeting with Kenton, some of them gaudy indeed. In a little-known volume of sketches by F. W. Thomas (then a Cincinnati newspaperman, but chiefly remembered as one of the few men who kept Edgar Allan Poe's friendship to the last) there exists an account which is unquestionably the true one, because it alone is Simon Kenton's own, given in an interview with Thomas at Kenton's home in Ohio in 1834. Thomas quotes 79-year-old Kenton, who was annoyed by a recently-published book:

"The book says that when Blackfish, the Indian warrior asked me after they had taken me prisoner, if Col. Boone sent me to steal their horses, that I said "No, *sir*" (here he looked indignant and rose from his chair) "I tell you, I never said 'sir' to an Ingin in my life; I scarcely ever say it to a white man."

Mrs. Kenton . . . turned around and remarked: "When we were

last in Kentucky, some one gave me the book to read, and when I came to that part, he would not let me read any more."

"And I will tell you," interrupted Kenton, "I never was tied to a stake in my life, to be burned; they had me painted black when I saw Girty, but not tied to a stake."

Kenton continued regarding Girty's arrival:

"He was good to me. When he came up to me, when the Ingins had painted me black, I knew him at first. He asked me a good many questions, but I thought it was best not to be too for'ard, and I held back from telling him my name; but, when I did tell him—oh! he was mighty glad to see me. He flung his arms round me, and cried like a child. I never did see one man so glad to see another yet. He made a speech to the Ingins—he could speak the Ingin tongue . . . and told them if they meant to do him a favor they must do it now, and save my life. Girty, afterwards, when we were at Detroit together, cried to me like a child, often, and told me he was sorry for the part he took against the whites; that he was too hasty. Yes, I tell you, Girty was good to me."

Thomas suggested that Girty's being good to anyone seemed rather remarkable but Kenton explained:

"No," he replied, quickly but solemnly, "It's no wonder. When we see our fellow-creatures every day, we don't care for them; but it is different when you meet a man all alone in the woods—the wild lonely woods. I tell you, stranger, Girty and I met [here Kenton referred to a former meeting, before the Revolution and apparently when he and Girty had first become well acquainted] lonely men, on the banks of the Ohio, and where Cincinnati now stands, and we pledged ourselves one to the other, hand in hand, for life and death, when there was *nobody in the wilderness but God and us.*"

Girty unquestionably saved Kenton's life, whether on account of the pledge and the past friendship or because of a whim of the moment no one can say. It was one of several such gestures in his life, and there were others who had reason to say "Girty was good to me." On the other hand, he sometimes seemed to enjoy the spectacle of torture as much as the Indians themselves, and he is said to have added some refinements to the ritual, from his superior intellectual endowment, which had not occurred to the red men.

Early in 1779 Girty took part in a raid upon a supply train from Fort Pitt to Fort Laurens (upon the Tuscarawas River) in which he captured some correspondence belonging to Colonel John Gibson, under whom he had served in Dunmore's War. One of these letters referred to him as a prime enemy figure, to be shown no mercy if captured. This correspondence probably heightened Girty's hatred for his own people. Back at Detroit, Girty urged that a British captain be sent to take charge of the several hundred Wyandots and Mingoes he reported as being ready to attack Fort Laurens. Captain Henry Bird (who commanded that celebrated raid on the Licking River settlements in Kentucky) volunteered, and well supplied with presents and ammunition he and Girty repaired to Upper Sandusky. After a successful campaign Captain Bird was able to report to his superior: "Girty, I assure you, Sir, is one of the most useful, disinterested friends in his department the government has."

To add weight to this testimony Simon Girty took about a hundred Indians to the Ohio in October and attacked a convoy of keelboats under the command of David Rogers which was en route from New Orleans with supplies. Girty's savages killed Rogers and more than forty of his men at a loss of two Indians killed and two wounded.

Sporadic raids against the western settlers continued. One of these was to be of special interest to Simon Girty, though he was far from the spot where it occurred. A party of Munsee Indians attacked a convoy of three boats loaded with immigrants to Kentucky. A few miles below Captina Creek on the Ohio one boat was captured, three of its occupants were killed and the remaining twenty-one, men, women, and children, were taken prisoner. Among these prisoners were the wife and children of Peter Malott, of Maryland, who, being in one of the other boats, himself escaped. The eldest child of the Malotts, and one of the prisoners, was their daughter Catherine, who would sometime become Simon Girty's wife.

Three of the Girtys were now at Detroit—George having deserted his post on the Mississippi as a lieutenant in Clark's force and joined the British—and all accompanied Captain Bird's previously described expedition in the raid on Ruddle's and Martin's Stations on the Licking River.

In 1781, Simon Girty was transferred from the Mingo tribe to the Wyandots at Upper Sandusky. From that point he led small parties upon raids to and across the Ohio, returning his prisoners and scalps to Detroit. He accompanied the British rangers and Lakes Indians who went to support the Wyandots successfully against Colonel William Crawford's attack. It was Girty who had the pleasure of informing Crawford that he was to be tortured "in part payment" for his earlier Indian raids. As has been told, Crawford was tortured for several hours and finally asked Girty, an interested spectator, to shoot him. Girty, laughing at his own coy jest, said he had no gun.

Simon and George Girty accompanied Colonel William Caldwell's Indians across the Ohio and against Bryant's Station in Kentucky—that Battle of Blue Licks, most disastrous in Kentucky history, ensued and "Never were so many valuable men lost so shamefully."

The British soon abandoned the offense on the Ohio—continuing to pay bounty for any American "hair" the Indians brought to Detroit of their own initiative—but not before Girty had led a raid for scalps and plunder to within five miles of Fort Pitt.

After the Revolution Girty retired on half pay and married the Indians' captive Catherine Malott in 1784. The bride was in her early twenties with suffering enough behind her, and the groom a leathery, hard-drinking forty-two. It is to be feared that little marital tenderness was evidenced at their home near Amherstburg, Ontario, across the river from Detroit. They had two sons and two daughters but after the birth of the fourth child Catherine Girty left her husband—whose favorite caress was said to have been a rap on the side of her head with the flat of his sword. She took their children with her, probably thus accounting for the fact that they grew to be respectable citizens.

When spineless General William Hull surrendered Detroit to the British in August, 1812, Girty celebrated by returning to the

Detroit side, drunk as usual, and proclaiming, "Here's old Simon Girty again on American soil!" That was his only contribution, however; he was too old, too worn, too dissipated to serve in any capacity away from the fireside.

After Perry's victory on Lake Erie and Proctor's defeats at Fort Meigs and Fort Stephenson, Canada was threatened with imminent invasion by General William Henry Harrison. Probably Harrison's army, made up partly of Kentucky, Ohio, Virginia, and Indiana volunteers, included not a single man who had not heard of Simon Girty, and a great many of them knew his infamous character very well indeed. Obviously the Canadian side of the Detroit River was no place for old Girty and he was advised to leave; he rejoined his friends, the Mohawks.

Girty returned to Malden in 1816. He was totally blind and crippled by age and old wounds, some acquired in battle, some while drunk. Now he spent much time at the home of his daughter whose husband, Peter Govereau, kept an inn in Amherstburg. The next year, feeling himself near death, he gave up liquor and listened to his wife, who now cared for him, while she explained to him how he might still go about the difficult business of saving his soul.

On the 18th of February, 1818, he died on his farm south of Amherstburg and was buried there when the snowdrifts had melted sufficiently to receive him in a grave which was, appropriately enough, unmarked.

Girty was not the only scourge of the frontier whose career was partially shaped by the internal conflict in the American colonies over the question of loyalty at the beginning of the Revolutionary War. That also has been sometimes advanced as one explanation of the maniacal doings of those precious brothers, Big and Little Harpe.

The historian, Consul W. Butterfield, was able to piece together the story of the Girtys after a thorough search of British and American military records. No such reliable source is available to document the myth-enshrouded lives of the Harpes. That is really no matter, for the tithe of provable stories of their doings is enough to admit them to any catalogue of ruffians. Scotch Tories in background, the Harpes were unquestionably the most ruthless and purposeless killers in American frontier history; the ladies of the traveling harem that accompanied them were, if not actively villainous, at least complacent of their misdemeanors and completely devoted to them.

The Harpes themselves, Micajah ("Big"), the elder, and Wiley ("Little"), the younger, were scarcely old enough to have formed any definite loyalty or prejudice during the Revolution when the ill-feeling between the Tories and the American rebels was at its worst. They were thought to be not beyond their late twenties when they first appeared in east Tennessee after 1795. If they were motivated by a desire to avenge the American rebellion it must have been at the instigation of their father who, an avowed Tory in 1776 and a volunteer with the British forces in the South, later tried to enlist in the Continental forces but was shunned by his North Carolina neighbors and became more or less an outcast.

The step from outcast to outlaw was an easy and logical one on the frontier; in fact a man who was not at least tolerated by his associates had little choice in the matter. In the ill-supplied back-country a family could scarcely exist honestly without the co-operation of neighbors. Perhaps it was to escape some such unhappy situation that Micajah and Wiley Harpe first left home for the West, where they joined a party of outcast Cherokees and spent considerable time with them. Probably the Cherokees and the Harpes were each able to add to the other's store of villainy; certainly the Harpes, in the association, developed a liking for the unrestrained life of the woods.

When, two years after leaving North Carolina, the brothers appeared near the west Tennessee settlements they were unquestionably a hard pair.

They brought two women with them from North Carolina, whose maiden names (if such a term as "maiden" could ever have been applicable to females likely to associate with the Harpes) were variously reported. By some accounts they were gently bred girlhood friends, kidnaped by the brothers; others make them sisters, Susanna and Betsey Roberts, who came along of their own free will. One of them claimed the dubious distinction of being Big

Harpe's wedded wife, while the other seems to have bestowed her affections upon either Big or Little, without discrimination.

The first recorded crime of the Harpes in Tennessee was the robbing of William Lambuth, a young Methodist minister whom they met near Knoxville, the women standing by while the men took his horse and money. Here was an error on the part of the Harpes; their technique had not yet been perfected, for they not only informed their victim that "We are the Harpes," as a sort of threat, but they neglected to kill him. Later, and increased in sagacity, they killed almost every lone traveler they robbed; eventually almost everyone they met. In this case they left the Rev. Mr. Lambuth a living witness, competent to describe them in the settlements when their description should become a matter of public interest.

The Harpes evidently intended to settle near Knoxville, then a community plenty wild enough to serve as an address for even such as they. The brothers built a cabin in the neighborhood and made gestures toward clearing and putting in a crop. Wiley, as was the custom of young men establishing a frontier home, soon went courting.

The object was Sally, daughter of a local preacher of the gospel named John Rice. Both father and daughter seem to have been either singularly innocent or broad-minded. They had both seen the Harpe entourage shortly after its arrival and it must have required an excessive naïveté to fail to note something rather unusual about the marital arrangements that prevailed. Besides, the Rev. Mr. Lambuth had stressed the fact that the whole party was notably dirty, unkempt, and ragged—as did, in fact, everyone who met the Harpes (and survived) then and later. The courtship of Miss Sally succeeded, however, and apparently the Rev. Mr. Rice offered no serious objection, since he himself performed the marriage ceremony.

The cares of agriculture and husbandry were not for the Harpes, however: there was a way to get hogs which was less burdensome than raising them and racing a horse was better fun than plowing with it. Presently suspicions as to their activities were aroused, but by the time the surrounding farmers were certain enough as to how their pork was being marketed to raise a posse and call on the suspects the Harpe cabin was deserted. Following their trail the posse came upon the Harpes, without their women, driving a herd of horses they had collected throughout the area. They were arrested without resistance, but before the party had returned far on the way to Knoxville the brothers leaped free and dodged off through the underbrush to join the ladies at some previously appointed spot. Their career was now definitely launched.

Presently they killed a guest at a trace-side tavern north of Knoxville; next a peddler on the Wilderness Road south of Barboursville in Kentucky, then two travelers on their way to Nashville, then a young Virginian who had invited their company for protection through the dangerous stretch of road south of Crab Orchard, Kentucky. The young Virginian was partially right in his judgment; three men and three women should have constituted a party of safe size, except that five of the six were Harpes. That made five murders in little more than twice as many weeks and the residents along the Wilderness Road became thoroughly alarmed.

Another posse was raised and, as before, its members had not far to look; the Harpes never seemed to hurry. The posse came upon the entire Harpe household seated on a log by the roadside—the men wearing some of the young Virginian's clothing. All were taken to Danville, Kentucky, to await trial.

Here begins a passage in the story which would be almost unbelievable were it not well authenticated by the court and county records. At the time of their incarceration all three of the ladies were in a delicate condition and the purchases of sugar and Hyson tea, then regarded essential fare for expectant mothers, formed a considerable item of public expense which was recorded in detail by the jailer. Eventually a midwife's services for Susanna also set the county back 18 shillings. The Harpes probably welcomed this period as a rest and change from the hard life on the road: there was no other reason for them to linger, for although the jail had been reinforced and manacles and new locks provided, the male Harpes departed through a hole in the jail wall shortly after Susanna's baby was delivered.

They were hunted, while the women remained comfortable in the Danville jail, presumably solacing themselves with an occasional glass of Hyson tea, but by report the hunt was only half-hearted. The brothers certainly left a well-marked trail, for they were even seen by one of the parties who went after them.

Their stay in the jail had not softened them, however. They killed Colonel Trabue's young son, whom they met returning from a mill, and they shot a man named Dooley and another named Stump as they moved north toward the lower Ohio.

Finally the women came to trial, but their apparent repentance moved the frontier court to sympathy. They were chivalrously acquitted and the goodhearted people of Danville gave them food, clothing, even a horse, so that the three might take the baby back to their old homes and be done with their villainous consorts. The hopes of the Danvillians were not well grounded, for the ladies Harpe went directly to Green River, traded the horse for a boat and floated off downstream, bound for the lower Ohio and a prearranged meeting with their menfolks.

By some accounts the meeting took place in the Kentucky cave country just south of Green River, and the whole party spent a waiting season in Mammoth or one of the other caves in the vicinity. Mammoth Cave, then scarcely known, would have made an ideal camping ground for such a party, the women newly outfitted by Danville's generosity and Big and Little Harpe still attired in the young Virginian's "fine linen shirts" (possibly showing signs of soil and wear by now but still superior to their accustomed ragged garments). They might well be imagined banqueting around a fire in one of the stalactite-hung underground halls—in a preview of the infernal scenes to which they were eventually consigned.

Perhaps it was here that Big Harpe took Susanna's baby by the heels and dashed out its brains against some convenient projection: he certainly did that to one of her offspring, for he said, before he died, that this was the only act he regretted!

The state of Kentucky now offered a reward of $300 each for the capture of the men, and in connection with the offer the only reliable printed description of the two is to be had: most people who met the Harpes, and realized the fact, were not likely to waste any time in taking mental notes. By the official notice

MICAJAH HARP alias ROBERTS is about six feet high—of a robust make & is about 30 or 32 years of age. He has an ill-looking, downcast countenance, & his hair is black and short, but comes very much down his forehead. He is built very straight and is full fleshed in the face . . .

WILEY HARP alias ROBERTS is very meagre in his face, has short black hair but not quite so curly as his brother's; he looks older though really younger, and has likewise a downcast countenance . . .

Micajah was, then, a low-browed individual, with curly black hair, wearing (had he ever owned a full suit) about a size forty-two, long. He was straight and broad of face but he preferred not to look the casual acquaintance in the eye. Wiley, apparently, could be described as rat-faced, with straighter hair and, since there is no mention to the contrary, apparently a more intellectual cast of brow. Neither could have been very attractive, although Big Harpe—Micajah—did appear to fascinate those he met by the strange flat black of his eyes in which the look of madness increased as time went on.

Whether or not the Harpe party hid out in the caves, they ultimately reached the Ohio River in the neighborhood of which Cave-in-Rock was then the center of activity. The stretch of the Ohio between Yellow Bank (present Owensboro, Kentucky) and the site of old Fort Massiac, below the mouths of the Cumberland and Tennessee—including Wabash Island, Shawneetown, Cave-in-Rock, Diamond Island, and Red Bank (now Henderson, Kentucky)—was already populated by a people who fancied themselves to be as desperate as any: robbers, wreckers, cutthroats in the night. But the Harpes soon proved themselves too depraved even for the taste of that company.

During their activities upon the Ohio the Harpes sought occasional shelter at Cave-in-Rock, where outlaws gathered in such numbers that no posse of honest citizens large enough to attack could be gathered, yet very shortly the Harpes are said to have perpetrated a jest so ghastly that it wore out their welcome among the murderers at the cave.

This prank followed a lucky day on which the outlaws had received a windfall. A flatboat containing two emigrant families had been waylaid, most of the emigrants easily killed and their gratifying ample property divided. The success was being celebrated by the drunken thieves around a fire on the river bank before the opening of the cave when suddenly there was a blending of screams—that of a horse and of a man—and man and horse crashed into the midst of the gathering from the brow of the cliff above.

It was only the Harpes' little joke: they had taken a survivor of the flatboat party and one of the stolen horses, had led both downstream and up the easy rise to the brow of the cliff, where they had tied man to horse à la Mazeppa or Simon Kenton, and had whipped the horse over the cliff. Their purpose had been only to enliven the evening, but such sport was too robust for the resident pirates, who drove the Harpes out of camp.

So they meandered back across Kentucky toward Knoxville. Chesley Coffey's son was brained for his rifle, William Ballard was killed in Tennessee, the Brassel brothers were waylaid; one escaped, one was killed. John Tully, the Triswold brothers, John Graves and his son, all these were known victims. There was no record of the casual travelers, not recognized locally, who may have fallen.

The Harpes began now to masquerade as itinerant preachers, and some sort of religious mania does seem to have been developing in Big Harpe's disordered mind. The women were in hiding and the Harpes started north, back across Kentucky, following the old trace from Nashville to Red Bank and the Ohio. One night they stopped at Moses Steigal's cabin near the Trade Water River Crossing. Now, Steigal was already a man suspected by his neighbors of ill-doing and, admittedly, both he and his wife had known the Harpes somewhere in the past. Steigal was away but his wife admitted them and sent them to sleep in the loft with another guest who was awaiting Steigal's return.

The temptation was too great even for men playing the part of preachers; as soon as the stranger was asleep the Harpes killed him, seized his valuables, and came down the ladder. Either the killing madness was upon them or Mrs. Steigal protested, for they dragged her outside, tomahawked her and her baby, set the cabin afire, and ambled away. They met two neighbors, possibly attracted by the fire, a short distance off; the Harpes killed both.

Steigal returned before his wife died and heard her story. Four neighbors joined him and he took the trail. As was usual with the Harpes, it was plainly evident and their progress was typically unhurried, though with a start of some hours.

There was something queerly wild in Steigal's manner, his companions thought, and since he eloped with a young girl of the neighborhood a short time later—and was shot by her brother—it was not necessarily the result of grief over the loss of his wife. Afterward the neighbors reached the uncharitable conclusion that Moses Steigal wished to kill the Harpes with his own hand so that he could be sure that his name would not appear in their possible dying confessions.

The second day of riding brought the little party in sight of the Harpes. They had rejoined the women and children (there had been those two other infant accessions anticipated at Danville; possibly more) and the whole party were standing in the road, the men talking to an intended victim. As Steigal's party approached and were seen, both men mounted their horses. Wiley dashed into the woods but Micajah rode straight ahead on the trail, pursued by the posse.

Samuel Leiper took the lead, his horse at a dead run. Riding with both hands free, aiming his rifle between his horse's ears, he shot Micajah Harpe square in the spine. Harpe did not fall but his horse veered from the road and presently slowed: Harpe, paralyzed from the shoulders down, rolled from the saddle.

Then Steigal rode up, apparently with a sudden thirst for revenge. His knife was drawn to finish Harpe but the others stopped him; the honor would be greater if Harpe could be brought in alive. An inspection showed that the outlaw was obviously beyond moving, so they let him lie. He talked of the murders he had committed; said they had been divinely inspired. He had no regrets about them but he was sorry for his hastiness in that matter of Susanna's baby. Finally his captors tired of waiting; they had

had little rest, little to eat, during the two days and two nights past. When Steigal took up Harpe's knife and proposed to cut off his head no one interfered. Harpe obviously couldn't live all the way back to the settlements anyway; why be squeamish? Harpe was still alive when the work began and criticized Steigal's awkwardness with the knife after the first cut, but it was soon over, and Harpe's head was stowed in a grain sack and tied to a saddle.

When they got back Steigal fastened the rotting head to a tree at Robertson's Lick on the Nashville-Red Bank Trace, where it stayed for years as a warning to evildoers—that place is still called Harpe's Head in Kentucky—and the terror of the Harpes between the Tennessee River and the Ohio was stilled—for no one expected Little Harpe to continue alone.

They were right; Wiley Harpe fled the country. The Harpe women and children remained in the neighborhood and at least two of them, Betsey and Sally, married and reared respectable families, one of which has been particularly noted, in its line of descent, for its great production of ministers of the gospel. Steigal soon met his end in the romantic encounter mentioned before. The whole business served to awaken Kentucky and Tennessee to the necessity for permanent arrangements for enforcing law.

Before leaving these interesting people, perhaps it is well to describe the remarkable coincidence through which Wiley Harpe eventually met his end, also by way of a severed head:

After his escape Wiley moved southwest, changed his name, and started professional life anew. But, lacking his older brother's guidance, he never achieved more than junior standing in it.

In 1803, after the rise of Samuel Mason's gang of robbers on the Natchez Trace had caused public indignation to grow great enough to place a substantial price on Mason's head, two men arrived in Natchez carrying a large ball of hardened clay which they broke to exhibit just that—Mason's head—and to claim the reward.

Some thought the two men, called Setton and Mays, might have been connected with Mason's gang at one time, but all were relieved by this obvious proof of his passing. Setton and Mays might not have been questioned, except that a Kentuckian, in town on a search for his stolen horses, recognized the horses they rode as his property and followed the men to court, where the two were preparing to collect their reward.

The Kentuckian entered the courtroom and saw the men. He told the magistrate that not only Mason's head was in his hands but that in the person of "Setton" he had Wiley Harpe. Captain Stump, the Kentuckian, had good reason to remember the Harpes; they had been in his neighborhood, and a relative of his had been one of their victims.

Many of the outlaws in the West were men who had served in the Revolutionary forces—sometimes with considerable distinction. Of the latter class was this Samuel Mason, a Virginian by birth, who had fought with George Rogers Clark.

Perhaps he had imbibed some of the unfortunate Clark's instability in peacetime, for when he settled at Red Bank after the war and began to prosper, was appointed a justice of the peace, and established as a leading citizen, he suddenly faltered in the paths of respectability and shortly deserted them entirely.

Red Bank was a charming spot (as its successor, Henderson, Kentucky, still is). Mason had first seen it, probably, while moving down the Ohio from Clark's Corn Island headquarters to old Fort Massiac, where the force had disembarked to begin the Kaskaskia campaign. It was a beautiful, fertile, and promising district, but at the time of Mason's arrival it was also exceedingly lawless, the home port of wreckers, robbers, and fugitives representing every section of the country. It was one of these gentry who was the immediate cause of Mason's fall.

This particular scape-gallows was a man who had left the southeast a jump ahead of the law. His occupation at Red Bank is not recorded, but he had friends and possibly business interests downstream on Diamond Island, then the headquarters of one of the more active of the bands that preyed on emigrant boats. Mason's daughter conceived an attachment for this eastern gentleman, possibly because he possessed certain social graces not of the common run. Without benefit of magistrate or clergy the two eloped to Diamond Island. Deeply affected by this blow to his dignity, Mason disowned his daughter, but after a time apparently he relented,

planned an infare (frontier equivalent of the reception at the home
of the bride's parents), welcomed her back to assist with the arrangements—and even invited her lover to drop in for the festivity.

The party began auspiciously. There was a fiddler, plenty of
solid refreshments, and ample drink. Each enjoyed himself according to his own lights, and it was probably quite late before anyone
noticed that the groom had gone outside and had not returned.

There was no need to wait up for him; he was not coming.
Mason's three sons and a friend of theirs had lured him out, under
Mason's direction, and killed him.

Whether the plans had extended beyond the killing and had
gone astray, or whether Mason's next act was the inspiration of the
moment, is not known, but next day, sacrificing the reputation he
had evidently sought at first to protect, Mason, his sons, and their
friend Henry Havard fled from Red Bank. Mason left his wife,
daughter, and property behind and the office of justice of the
peace vacant.

The original crime was complicated when a constable pursued
them and was shot down. There was now no turning back, the
next stop for the Masons was Cave-in-Rock. Havard rode home to
Tennessee, where, suspected of having shot the constable, he was
killed by a posse.

Cave-in-Rock was no sanctuary at the moment; no sooner had
the Masons arrived than word was brought in that a volunteer
army was coming down from Pittsburgh to wipe out the gang; yet
as a matter of fact the consternation spread among the bandits by
this threat materially assisted Mason in embarking quickly on a
large-scale career as chief of a desperado band of his own. He
already had his three sons and he was able to enlist several experienced hands from among those who were leaving the cave. With
this nucleus he set out for Wolf Island, in the Mississippi about
nineteen miles below the mouth of the Ohio. Here he camped and
perfected his organization.

Henceforth Mason ceased to be an Ohio River bandit. He
specialized rather in robbery upon the Natchez Trace (by which
Ohio Valley residents walked or rode home after selling their
produce in New Orleans), piracy upon the Mississippi, and Negro

stealing throughout the South. His followers were at one time numerous, his take quite large—and his end has already been recounted.

Colonel Plug had been some sort of military man also, back in his New Hampshire home before he came west to participate in what might be termed merchant marine activities. But his exact previous rank and the scene of his military experience are not of record. He is said to have introduced himself on the lower Ohio as Colonel Flueger (or Fluger or Pfluger)—it makes little difference. He was famed as "Colonel Plug" and, since every tenth man in Kentucky bore that title anyway, his claim was regarded as sufficient.

He set up on the Kentucky side in that murderous reach of the Ohio between the mouths of the Green and Tennessee, which was made fully as dangerous by white banditry as passage between the Scioto and the Great Miami had been because of Indian attack after St. Clair's defeat. There he began to practice his peculiar art.

Apparently, the colonel believed it unnecessary and needlessly expensive to support a gang. Stories of his exploits mention only a roustabout or two, the Negro consort who shared his joys and sorrows and of whom he was extremely jealous, and a junior partner identified only as "Nine-Eyes." The colonel was rather admired by the majority of rivermen, probably because he accomplished his ends through clever strategy rather than brute force—although often enough with the same fatal results. Stories of his exploits were always popular.

Since he was so widely known among experienced rivermen and his basket of tricks had so often been exposed in their off-watch yarning, the colonel must have had to depend almost entirely upon green emigrants as victims.

He usually began his work upstream a suitable distance from his abode. Hailing a passing emigrant boat, he might represent himself as a stranded traveler, one who had survived a wreck and nearly died in the wilderness. Once on board he could collapse upon the bottom planking, surreptitiously pick the calking from the often jerry-built vessel, and arrange to have it sink at or near his home, where Nine-Eyes and his mistress could assist with the salvage of the goods.

Sometimes he claimed to be a pilot who wished only to give warning of a dangerous hazard to navigation below and to offer to guide the greenhorn's boat past it as a friendly gesture of the hospitable West. The immigrant, having slid, whirled, and bumped through the devious passage of the Falls at Louisville but a day or two before, was chary of further risk and delighted to hand the steering oar to this friendly and expansive son of the free country, whereupon Colonel Plug ran the boat hell-bent on the largest snag convenient to his headquarters.

The colonel's intense jealousy periodically reached a heat at which he challenged the always-suspect Nine-Eyes to mortal combat, winner apparently to take all. The stories of these battles have many variations but in most of them the duels end with the combatants getting drunk together, passing peacefully out, and awaking friends—until the next time Plug's suspicions of his lady's faithlessness were aroused.

Like Wiley Harpe's, Colonel Plug's end came in a peculiarly appropriate manner. It was not even in the course of one of his more elaborate schemes that he lost his life; nothing more than a routine operation which did not require any part of his great histrionic ability. On this final exploit the colonel had only slipped up to an emigrant boat tied to the shore while its passengers cooked a meal and slept. He had climbed aboard, hidden himself between the short steering deck and the bottom planks—as any ordinary untalented thief might have done—and waited for the emigrants to push off in the morning.

All went well and when he judged it time to promote a sinking suitably near his headquarters, he began to pick the oakum calking from between the planks. Apparently, however, this traveler had been even more gullible than the average and the Pittsburgh boat-builders had not followed what was claimed to be standard practice of allowing planks for emigrant boats to season a whole day. After only a few picks the planks sprang apart and the water rushed in; the boat sank, right enough, but far ahead of the Colonel's usually accurate estimate, and Colonel Plug himself, trapped below the deck, was the only fatality.

Cave-in-Rock—early refuge for passing mound builders, historic Indians, voyaging French, and weather-bound colonial traders— began to shelter another kind of people some time during the last two decades of the eighteenth century. Organized banditry seems to have begun to make headquarters there about the time that the spot was selected by one Wilson, given name and place of upbringing unknown, who recognized in the cave a strategic and ready-built tavern. He eventually achieved the dignity of a sign "Wilson's Liquor Vault & House for Entertainment," but his clientele could never have been advertised as select.

Just what Wilson's original intention may have been remains to be proved—that is, whether he set up his establishment as a haven for outlaws or merely accepted fortune as it came and entertained such gentry because they were the best available paying guests. Uncertain also is the extent to which Wilson himself participated in the nefarious business carried on there. When public indignation at the percentage of passers-by who were robbed and murdered by habitués of the House for Entertainment finally reached the boiling point, no one stopped to inquire whether Wilson was the chief executive of the bandits or only their complacent host.

There was harbored at Cave-in-Rock, one time and another, most of the riffraff on the lower Ohio and middle Mississippi, and this, since it had previously been chased out of the comparatively lax settled country, was riffraff indeed. Few travelers left accounts of the place at its most populous, except by hearsay—possibly because no forewarned traveler went near it and such a high percentage of those who did blunder in survived to go no further.

Cave-in-Rock played host to many less publicized criminals than the Harpes and Justice Mason and his sons. It came to be feared far more by the traders, the early Kentucky tobacco and ham raisers and Monongahela distillers taking their products to the New Orleans market, than either the Falls of the Ohio above or the perilous Hurricane Island passage below.

Probably only one class of passer-by who was fully informed as to the character of the place stopped at the cave by choice: the

professional keelboatmen must have tied up there once in a while just for the hell of it, to see if anything new in the way of a ruckus might be offered.

These keelboat crews were a wondrous race, whose various exploits have grown no less remarkable in the retelling. One supposes that the reading public must have grown surfeited ere now by tales of Mike Fink, "the King of the Keelboatmen": let us therefore pass him by, as a man already far too well publicized, with the remark that there probably was such a person, that he was well endowed with a gift for boasting in the hyperbolic fashion of river and lumber camps ("I'm a Salt River roarer, half horse and half alligator, suckled by a wildcat and a playmate of the snapping turtle") and that to have so recorded his name among the elect he must have had considerable ability in a free-for-all and some elements of leadership. Even gentlemen of the cloth contributed to the fabrication of Fink's fame. A person of no less presumed probity than the Rev. Mr. John Finley devotes some space in his autobiography to the story of how an anonymous divine (unmistakably the Rev. Mr. Peter Cartwright) vanquished Fink in single combat at a camp meeting at Alton, Illinois. The Rev. Peter, however, in *his* autobiography, says the story has no foundation of fact; that he did whip a few roisterers on that occasion but that none of them was Fink. In time one begins to wonder if the Fink of the other stories might not have been a couple of other fellows, also, and eventually one wearies of Fink.

Let us therefore pass Fink by: no need to load the ample color of the keelboatmen all on his—admittedly powerful—frame.

The keelboatmen were tough enough but they were by no means desperadoes. Half the time their work was inconceivably grueling. For keelboats, unlike flats, carried cargo not only downstream, but poled, rowed, and dragged it upstream as well. They grew tough, those keelboatmen, in their terrific exertions and they liked their relaxation to be equally strenuous. The Robin Hoodish aura that surrounds them probably grew because, like many men whose pay comes hardest, they could be wildly generous upon occasion, and because they usually seem to have confined their more murderous pastimes to the circle of their immediate acquaintanceship or at least to their own social stratum. The fact that occasionally in their roistering they burned, tore down, or pushed into the river whole neighborhoods of river towns was usually overlooked by the law on the grounds that those must have been undesirable sections of the community anyway, since boatmen could have had no reason or desire to visit any other sort.

Every river town had such an undesirable district, but from the very beginning of Ohio River traffic some towns were recognized as being more dissolute, more lawless than others, and—perish the thought!—an unprejudiced visitor may still identify a few of them after two hundred years of progress and culture. To the eyes of keelboatmen no stigma attached to a community so marked, far from it! If a town had a really malodorous repute it was to them, then, an excellent place to lie over and relax a bit. Their relaxing, of course, did nothing to quiet the scene of their repose.

Contrary to the impression easily gained from a study of the WPA murals in modern river-town post offices, the keelboat hand was not necessarily handsome, though he was doubtless of appearance rakishly dashing enough. He usually carried upon his person one or more marks of his avocation; most frequently it was a damaged ear, sometimes bitten off, sometimes cauliflowered (a most expressive term, that, as you know if you are acquainted among the less-successful competitors in the prize ring), or most often simply chawed in a scalloped design around its rim. Or he could have lost the fleshy portion of his nose and sometimes, if he was unusually inept as a fighter, he could have lost that decorative feature *and* an ear—or an ear and a half, or even two ears. If, in addition to any or all these misfortunes, he happened to be a veteran of an Indian foray in which he had got himself scalped (it was not at all uncommon to be scalped while unconscious from a wound and to survive the operation), he was likely to present a rather frazzled appearance; one which could not be rendered attractive even by his red-flannel shirt, remaining hair worn in a shoulder-length bob, wool hat set off with a feather, and skin pants—both skin in material and skin tight in cut.

There were some general social distinctions between those who floated or poled the Ohio before steam took over and these methods of propulsion disappeared.

Lowest in the scale were rafters, unloved greenhorns (except later, in lumbering days) who lashed together any number of logs or timbers from six up, built a lean-to on one end, loaded on their scanty household effects if they were emigrants, or trade goods if commerce was their aim, and floated off downstream. They were practically at the mercy of the current and its many random whims sent them butting and nosing whatever came in the way, from shore line to passing boats.

If the owner was an emigrant he beached his craft when he saw a likely spot—current permitting—and broke it up to use the timber in putting up shelter, if that appeared convenient. Or, if it looked easier to fell new trees, he simply "let her set" to be floated off by the first freshet and later to bring anathema on his anonymous head as it menaced navigation on its way to the Gulf. Few of these raftin' settlers had much truck with land purchase, titles, or such business; being of a class whom rafting suited, they were perfectly willing to squat upon a likely plot until the owner drove them off or wearied of trying and let them stay.

If bent upon trade, the rafter swapped his "notions" as he went along—and being a *rafter's* notions they were not likely to be what is called class merchandise. He then disposed of what remained, along with the logs of his raft, at New Orleans, and started back east afoot or stopped along the way as the spirit moved him.

Next above the rafter in the social scale came the emigrant flatboatman. He might never have seen navigable water until he embarked on it but his boat made some pretense of answering her helm, and sometimes, if he was affluent and cautious, he hired a pilot. Just above him came the cargo flatboat, which, while built for a one-way trip downstream, was usually commanded by someone who had made the trip before.

The bulk of emigration westward on the river by more substantial people was by flatboat. Building of such craft at Pittsburgh or Wheeling, or wherever a good road from the settlements struck the upper river, became an important industry. Flatboats could be built to survive almost any amount of banging which even the island-studded Ohio could give them, or they could be thrown together of a combination of green or rotten planks and shoddy workmanship so that they were scarcely able to float' their own weight in a moderate riffle. Emigrants were warned against purchasing the latter article by newspapers, land agents, emigrants' guidebooks, and well-disposed rivermen. But through impatient haste, ignorance, bullheadedness, or the persuasive lies of the builders they often did so anyway. The result was sometimes that the members of an emigrant family were marooned by the river scores of miles from a settlement without gun, ax, or means of making fire and with only what woodcraft they had picked up in the cities of New York, Philadelphia, Baltimore, or even (perish the thought!) Boston.

Cargo flatboats were something else again. They were usually built by, or to the order of, the jobber of produce or the association of farmers who supplied the cargo. No chances were taken in their construction (except by some occasional obstinate cuss who wouldn't ask or accept advice) and the wood employed was carefully selected in order to be salable at New Orleans.

These cargo boats were built wherever the products to be shipped originated. The Monongahela and later the Kentucky River produced whisky and most tributaries that ran through beech or oak tree land in Ohio, Indiana, and Kentucky contributed hogs fattened on "mast," which were shipped in the form of hams and pickled sides and shoulders of pork. The Kentucky River farmers loaded hemp and tobacco, those on the Wabash offered shelled corn, cherry and walnut planks, ginseng, and so on.

The flatboats carrying produce were usually manned by the owners, their neighbors, and their sons, who, once cargo and boat were sold at New Orleans, walked or rode horseback up the Natchez Trace to Kentucky and the Midwest or took passage on upstream keelboats and later steamboats, if the value of their cargo had been great enough to warrant such extravagance.

If owner and crew succeeded in eluding the bandits who lined the Trace, the social charms of New Orleans, Natchez-under-the-Hill, Memphis, Shawneetown, and the other riverside communities which featured gilded sin as a local commodity, and got home with

their money, they could and frequently did pay for great tracts of
rich river-bottom land with the proceeds of a few such trips. If,
far from the deterring eyes of their neighbors, they slipped aside
from grace to sample the charms of these riverside cities of joy,
they were likely to be puzzled as time went on by the unkindness
of a fate which had sent them and their wives unaccountable stiff
joints and skin ailments and had afflicted their children with blind-
ness and idiocy. The abysmal degeneration of some of the river
fronts plus the ignorance, if not the innocence, of the backwoods
farmers frequently added up to a result most unhappy.

But high in the scale above even the most experienced flat-
boaters stood the before-mentioned keelboatmen—at least until they
were supplanted in the esteem of the marveling public by the
gaudy and princely rich steamboat pilots of the forties and fifties,
complete with the doeskin vests and tremendous watch chains that
were the badge of office.

The keelboaters were the aristocracy of the early day; their
gaudy life the preference of every 10-year-old boy, and the secret
envy, often, of his grandfather.

This upper crust of rivermen was composed, by its own admission, of genuine ring-tailed snorters from away back; children of
calamity and bearers of ill tidings who could lick their weight in
catamounts with their bare teeth while handcuffed, who could out-
drink, outfight, and outcuss anything in this world or the next. The
reason for this necessary supremacy in matters of strength and
stamina? Why, solely that *they* poled, rowed, or snaked their boats
*up*stream as well as steered them down! *They* were the official
cargo carriers, the preservers of upriver commerce—until, of course,
an even more fiery monster, the steamboat, took over.

They were tough—there was no question about that—and
though they were not wantonly murderous as were the Harpes, nor
treacherous as the Girtys or Colonel Plug, nor covetuous as Justice
Sam Mason and General Wilkinson, they were boisterous and
reckless enough to be included here. Even after we discount their
prototype and their prophet Mike Fink by fifty per cent.

They were loud in the announcement of their iniquity to every
shore dweller, every raftsman, every flatboater they sighted. Even

their gargantuan consumption of alcoholic beverages and their dallyings among the painted ladies along the river could not subdue them—for if they overindulged in spirits they soon sweated out the excess against the current; if they acquired some unpleasant complaint in the brothels ashore, their rigorous life either destroyed the frail spirochetes and gonococci or the victims quickly died of their debilitating effects and were replaced by hardy new recruits. As has been said, they were probably not handsome as a class, since even their simple internecine warfare involved the biting of ears and noses, the gouging out of eyes, and another form of mayhem, now presumably extinct, which resulted from the inserting of one's finger inside one's opponent's mouth, between teeth and cheek, and the application of a quick, stout, jerking motion.

The keelers were feared ashore (although their free-spent money was always welcome) because they were possessed of a strong fraternal spirit which prompted them, when one of their number had been inordinately mulcted in either a gambling establishment or a house of dubious repute to (*a*) push the building into the river if it happened to be located at the river's edge, as many were, or (*b*) if that was impractical either to pull it board from board or to set fire to it.

The shore dwellers were not viewed as mortal enemies by the keelers, however, nor were even the rafters whose erratic progress presented such real danger to their lives. As is usual it was the next below themselves in the social scale, the flatboatmen, for whom their special venom was reserved. By report the battles between these two classes, sometimes held upon keelboats hooked to flatboats midstream, sometimes in riverside towns, sometimes on islands selected for the purpose by challenge and acceptance, were fearful to behold.

Tough and practiced in the manly art of self-defense as they were, the keelers often fought on only even terms with Hoosier, Red-hoss Kaintuck, Buckeye, and Sucker farm boys who, though uninformed in the finer points of the art of mayhem, had still been somewhat hardened by stump pullin', loggin', and, as was claimed by many of them, "pushin' a plow and two ox too" through prairie sod or bottom land choked with pea vines and bear grass.

Those were colorful days on the Ohio and its tributaries and on the Mississippi; considerably more attractive in the usual retelling, probably, than they were in actual fact. Perhaps it was as well that, by 1830 or so, the keelers were obviously going—to reappear later as bodacious loggers—the rafts were thinning out and the worst menaces to peaceful travel on the Ohio had become card sharps and steamboat-boiler explosions.

Chapter Nine

1805-1830:
WESTWARD THE TIDE...

It HAD all started before 1805, of course. The claim of the Iroquois to the south bank of the Ohio was extinguished by the Treaty of Fort Stanwix in October, 1768. Loss of the territory could have been of no great importance to them; their claim had always been of a dog-in-the-manger sort, designed chiefly to remind their enemies of who was boss below the eastern lakes.

White emigrants immediately swept in to settle the south bank between Fort Pitt and Limestone; they were considered a mighty host at the time—dozens of them in all. By 1770 the Zanes had founded Wheeling and three years later young farmer-surveyor George Rogers Clark, comfortably ensconced in the pole lean-to he had erected as bachelor quarters on his clearing near the mouth of Fish Creek, was able to write to his brother that "the people are settling as low as ye Sioto river, 366 below Fort Pitt."

In 1773 Thomas Bullitt and James Harrod left Fort Pitt for the Kentucky country, under orders from Governor Dunmore to survey lands intended to be awarded to veterans of the French and Indian War. Harrod made an extensive project of it, surveying tracts from present West Virginia to middle Kentucky and, as already reported, beginning the settlement of Harrodsburg, the first permanent town in that region.

In 1775 Daniel Boone founded Boonesborough and by June, 1776, there were Kentucky settlements enough to warrant calling a convention of their representative citizens.

Defense against the Indians was the problem during the next four years but Clark, Boone, and the others solved it as best they could and the population continued to grow steadily, even during the darkest days.

Refugees, sometimes families who had lost a father to the British or whose houses and barns had been sacrificed to the necessities of either British or American armies, left the war-ravaged eastern seaboard and poured into Kentucky, western Virginia, and what would soon become eastern Tennessee. New families arrived by hundreds instead of dozens on the frontier each spring after 1777, and 1787 saw the claimants beginning to move on the Virginia Military Lands north of the Ohio. In 1788 those first settlers of the Ohio Company landed at the mouth of the Muskingum.

Veterans of the Revolution continued to move west and settlements sprang up along the entire length of the river. Shortly after the beginning of the nineteenth century the Indian title to lands well back from the north bank of the Ohio was extinguished and the troubles over jurisdiction of the commerce on the Mississippi were ended forever by the Louisiana Purchase.

Christian Schultz took a long look at the river in the winter of 1807-8. According to his statement, he made the trip down the Ohio to look into the utility of lands he had purchased as a speculation. For his own amusement he kept a careful journal and, having read the book which Thomas Ashe had written about the Ohio, he decided to publish his own notes as a correction of the fabrications Ashe had foisted upon the British and American public.

The result was Schultz's *Travels on An Inland Voyage,* one of the best of the early journals of travel in the Midwest. He made some enlightening remarks upon the development of the Ohio Valley:

> From the rapid increase of population in this country within the last twelve years, and the immense tracts of the finest lands in the world which yet remain to be settled, you may form some idea of the vast quantities of Indian, European, and West-India goods already consumed, but likewise to what an astonishing amount it must arise in the short period of fifty years.

Schultz demonstrated the saving in time and money to be made in supplying the Ohio by a route from New York by way of Albany, overland to Schenectady, by boat and ship to Presque Isle, then by road to Le Bœuf, and by boat to Pittsburgh and the Ohio. Philadelphia, Baltimore, Alexandria and many other ports had hitherto shared in the western commerce but Schultz rightly assumed that the port of New York could, and soon would, monopolize the bulk of it. He could not foresee the War of 1812 but, once its hostilities ceased and left travel free to follow his proposed route, overland hauling from Philadelphia and Maryland to Ohio ports began to dwindle: Christian Schultz was proved to be eminently right.

> The inhabitants of this country, however, labour under one great inconvenience, which arises from the scarcity of cash. . . . The merchants . . . receive their goods from New-York, by way of Alexandria, Philadelphia and Baltimore . . . Payments are made to them chiefly in the bulky produce of the country, the only market for which is at New-Orleans. The consequence is, that they are constantly occupied in collecting all the specie they can, in order to make their remittances over land. Several plans . . . have been projected for obtaining the money in the country but none I believe have operated so effectually as that in practice in the States of Kentucky and Tennessee. Here a round dollar no sooner makes its appearance than it is divided from the centre into eight equal pieces; and some, I am told, carry their in-

genuity so far as to make *nine* and even *ten* eighths out of a dollar;
securing at once a profit of twenty-five per cent for their labour . . .

From the practice of cutting a dollar into eight wedge-shaped
fractions called "bits" originated the name of the popular though
not officially sanctioned monetary unit "two bits"—equaling one
quarter or two-eighths of a dollar! Schultz continued:

The inhabitants of the Ohio country in general have very little
of that unmeaning politeness, which we so much praise and admire
in the Atlantic States. They are as yet the mere children of nature and
neither their virtues nor their vices are calculated to please refined
tastes. They are brave, generous, and humane, and in proportion to their
population are able to produce the most effective military force of any
in our country. This preeminence may chiefly be attributed to their
exposed situation on an Indian frontier, where they . . . found it
necessary to teach their sons and daughters, as soon as they were big
enough to raise a gun, to load and level the rifle. On more than one
occasion I have seen these Spartan females, while engaged at the
spinning-wheel . . . snatch up the loaded rifle and fell the bounding
deer as he incautiously passed within shot of the cabin . . .   .

The country above Cincinnati is healthy, and free from all kinds
of bilious complaints, although the shores of the river are generally
one continuous and impassable bed of mud and slime. On the contrary,
the shores below [Cincinnati] are dry and gravelly, and frequently
present a clean beach to the eye; yet they are very subject to complaints
of the bilious kind. . . . I think I discovered the cause in the great
difference of the face of the two divisions. The upper being a more hilly
and mountainous country, affords an easier descent for the waters;
while the lower is gradually subsiding into a plain and level country,
where the lands back from the river . . . retain large portions of water
from every rain and inundation; these, for want of a passage to the
river, soon become stagnant and putrid, and give rise to complaints . . .

This country being as yet quite new, it cannot be expected that
any extraordinary advances should have been made in the propagation
of fruit trees . . . Apples, pears, and cherries are therefore not . . .
common on the Ohio, although in some of the old interior settlements,
they have them in abundance. The peach-tree, however, may be said
to cover the banks of the river, as there is scarcely a settlement between
Pittsburgh and the Falls . . . that has not one or more orchards . . .

The following are the natural fruit and forest-trees . . . mandrakes, currants, grapes . . . plums, mulberries, wild cherries, black and white haws, buck-eyes, papaws, and cucumbers . . . black walnuts, white walnut, butter-nut, pines, sassafras, white-oak, black-oak, dog-wood, locust, beech, ash, elm, sycamore, sugar-maple, soft-maple, tulip (magnolia,) black thorn, Jerusalem oak, or spice oak, cotton or poplar, and of willows an endless variety . . .

The War of 1812—not taken as seriously by all Americans as it should have been—slowed commerce and investments but not emigration, and after the terrible winter of 1811-12 there was new and wonderfully fast transportation as the steamboat was introduced on the Ohio. Whatever deterring effect the war had exercised was ended on Christmas Eve, 1814, with the signing of the Treaty of Ghent.

The day was over in which annual emigration to the West could be reckoned by dozens and hundreds; the movement had now really begun—and the Ohio was its highroad.

In 1810 the combined population of Kentucky, Ohio, Indiana, and Illinois was 674,073; in 1830 it was 2,131,296. Indiana's population was multiplied by more than twenty-eight in that twenty years and even that of Ohio, second oldest of the group, quadrupled. All this takes no regard of the hundreds of thousands who came down the Ohio to pass north and south up its tributaries, or west through the Mississippi and its connecting rivers, to swell the population of Georgia, Mississippi, Louisiana, Alabama, Tennessee, Missouri, and Arkansas.

True enough, there were other feasible routes. There were the trails over the western Virginia and Carolina mountains and through the various gaps into Kentucky and eastern Tennessee—that trail through Cumberland Gap which Daniel Boone had popularized, for example—and Americans had also begun western emigration by way of Lake Erie before the end of the eighteenth century. These alternate routes made their contribution, but the Great Lakes as yet gave access to little settled country, only northern Ohio, Indiana, and Illinois, still partly Indian land, and the mountain trails could not compete with the Ohio River when household goods and supplies were to be transported.

Besides the superior ease of floating goods downriver over either sailing or rowing them through the lakes or hauling them over the mountains, the Ohio had acquired a romantic appeal. Some of the same glamour began to attach to boating down the Ohio to set up a home on its banks as would, in 1849, become identified with crossing the plains to California to pan gold—except that the latter was a movement of bachelors while the Ohio fever claimed victims from both sexes.

Newspapers were full of reports, books and pamphlets were written and ballads were composed celebrating the glories of the Ohio country.

Some of the ballads were weak enough in structure and literary style to be worthy of Tin-Pan Alley a century later; they voiced a nostalgia for the pleasures of land clearing and Indian fighting as sharp as the Alley's has been, one time and another, for the South Seas, the Cotton Belt, and the Cow Country. An example of the production of the 1812-1818 period, titled "Banks of Ohio," makes its first appeal to the wanderlust of youth, turns, through a gracefully subtle transition to stir the homemaking instinct, and ends with a sure-fire appeal to simple primordial sex:

Come all young men who have a mind for to range,
Into the western country your station for to change,
For seeking some new pleasures we'll altogether go,
And we'll settle on the banks of the pleasant Ohio.

The land it is good my boys you need not to fear
'Tis a garden of Eden in North America:
Come along my lads and we'll altogether go
And we'll settle on the banks of the pleasant Ohio.

There's all kind of fish in that river for our use,
Besides the lofty sugar tree that yields us their juice,
There's all kinds of game besides the buck and doe,
And we'll range through the wild woods and hunt the buffalo.

This river as it murmurs it runs from the main,
It brings us good tidings quite down from New Spain:
There's all kinds of grain there and plenty it doth grow,
And we'll draw the Spanish gold right from Mexico!

Those blood thirsty Indians you need not fear,
We will all united be and we will be free from care,
We'll march into their towns and give them their deadly blow,
And we'll fold you in our arms in the pleasant Ohio.

Come all you fair maidens wherever you may be,
Come all join in with us and rewarded you shall be.
Girls if you'll card knit and spin, we'll plow reap and sow,
And will settle on the banks of the pleasant Ohio;
Girls if you'll card knit and spin, we'll plow reap and sow,
And we'll fold you in our arms while the stormy wind
    doth blow.

And it was not only illiterate balladmakers, safely resident in New York or Philadelphia, who were moved to verse. Educated young men who had paid heed to the call of the West found themselves even more impressed by the sight of it, more overwhelmed by its vast solitude and its enormous virility, than they had anticipated. Twenty-one-year-old Frederick W. Thomas first attracted wide attention to his literary talent by publishing in 1833 a poem in heroic measure which recounted the impressions the Ohio had made upon him as he floated down it to seek his fortune in the West:

We both are pilgrims, wild and winding river!
Both wandering onward to the boundless West—
'But thou art given by the good All-giver,
Blessing a land to be in turn most blest:'
While, like a leaf-bourne insect floating by,
Chanceful and changeful is my destiny;
I needs must follow where thy currents lave—
Perchance to find a home, or else, perchance a grave.

Happily Thomas seems to have found the welcome that his previous discouragements—financial troubles, his inability to gain a foothold as a writer, and the handicap of a badly crippled leg—had led him to fear was none too sure, even in the West. He continued:

There is a welcome in this Western Land
Like the old welcomes, which were said to give
The friendly heart where'er they gave the hand;

Within this soil the social virtues live,
Like its own forest trees, unprun'd and free—
At least there is one welcome here for me . . .

It is not difficult to trace the principal sources from which
people had come to the valley before the War of 1812. They were
an interesting folk, representing every variety from the absolute
dregs of depravity to the best of the good and godly—and every
shade and nuance of character between.

In 1812—if we may select an arbitrary date for a survey of the
people of the river—the oldest stock represented in the valley by
people whose doings had been newsworthy was the mixture of
French and Indian blood, usually either of the Lakes tribes or those
of the Miami Confederacy. Individuals of such mixed blood were
prominent in the Ohio Valley, especially upon the northern tributaries, as traders and agents for the Indians; Francis Godfroy, trader
near Detroit and later on the Wabash, for instance, and earlier
those remarkable descendants of no-less remarkable Madame
Madeleine Montour—French Margaret (who in 1754 traveled
through Bethlehem, Pennsylvania, on her way to New York "in a
semi-barbaric state, with an Irish groom and six relay and pack-
horses"), Queen Esther Montour, Queen Catharine Montour and
Captain Andrew Montour, associate of Washington and Croghan
before the French and Indian War.

The French of pure blood made somewhat less impression
upon the Ohio Valley than might have been expected. Their tenure
as traders immediately upon the north side of the river after
Céloron's expedition and again after Braddock's defeat, added to
but a few years in all and few of the French at Vincennes, nearest
of the old French posts to the Ohio, have been distinguished in the
affairs of the valley, even to the present. The French made little
contribution to culture except through the work of their early
priests.

One element resident in the two French settlements upon the
Wabash and transacting business on the Ohio, minute as it was,
played an important part in both the commercial and the political
development of the lower Ohio Valley: it was composed of the

Spanish merchants, Vigo, Bazadone, and a few others, only men of any considerable property north of the Ohio in their day, who largely financed the campaigns of George Rogers Clark. Colonel Francis Vigo, formerly of the Spanish army and later American merchant, furnished cash, supplies, and influence to the extent of his resources, freely and willingly—and his estate recovered his investment forty-odd years after he had died in poverty. Bazadone, the other Spanish merchant, was owner of some $20,000 in goods impressed to maintain Clark's force at the close of the Revolution—and he raised a screaming protest which, heard in Virginia, was admitted as evidence to back the charge of irregularity used to relieve Clark of the command that had already cost him so much. These two Spaniards, almost solely, represented capital north of the Ohio to 1785!

Next in priority in the valley (even a bit before the Spaniards) came the Scotch-Irish (actually Scots who had gone to Ireland in the reign of James I) and men from southern Ireland (must we say Irish-Irish?) who settled in Pennsylvania. These, like other settlers later upon the Ohio, ran to opposite extremes—the very steady and enterprising and the very tough, quarrelsome, and wild. It has been suggested that the latter may have been those favored with a larger infusion of the Celt than the former, in whom the sober Scot prevailed.

It was the brawling Irish who became the traders' men upon the Ohio. Lone wolves, they seldom supported families although the paths of their wanderings were certainly marked by increasingly long upper lips and blue-cast eyes among the Indian children—while the steadier Scotch-Irish element introduced cattle raising, set up Pennsylvania farms, developed the distillery business, and contributed the name to America's national drink. All the men from Ireland co-operated in helping to make politics a lively art.

Peculiarly, and fortunately enough for the future of the lower Ohio Valley, some of the Scotch-Irish wandered down the lateral valleys far south of Pennsylvania to settle in the Virginia and Carolina back country. They too were nonetheless ready, when the proper time arrived, to move west and north to meet their Pennsylvania cousins in opening Kentucky, the Tennessee and Cumberland

River Valleys, and from those bases, to open also the north bank of the Ohio.

From the days of the first Ohio Company, before the French and Indian War, the younger or impecunious sons of patrician Virginia and Maryland families saw in the south bank of the Ohio a country in which to build up fortunes of their own. Young George Washington was one of these, and by the time Governor Dunmore's soaring ambition had turned toward seizing the river's head, there were many ready to make the move—especially to the Bluegrass of Kentucky.

New England colonial stock came in with the second (and successful) Ohio Company to settle on the river near the mouth of the Muskingum. New York State stock was represented first by the large body of emigrant New Netherlanders who had never become reconciled to British rule in what was still to them "New Amsterdam." They called themselves the Dutch Company and bought the equivalent of two present counties of land on the south bank of the Ohio near the mouth of the Kentucky River. New Yorkers also were among the purchasers of lands from John Cleves Symmes between the two Miamis, at present Cincinnati.

Others notable in the valley were such of the French emigrant artisans as survived the first winter at Gallipolis, an infusion of the so-called "Pennsylvania Dutch" (who were of mainly German and Swiss ancestry), what would then have been termed a "right smart sprinklin'" of descendants of families already resident on the eastern coast of America generations enough to forget their European origin altogether and many individuals who, because of unhappy circumstance or plain expediency, were willing to have not only their antecedents forgotten, but even their trans-Allegheny names!

From such varied sources sprang the population of the Ohio in 1812.

After 1815 emigrants from any eastern state—or directly from Europe, for that matter, since emigration from Europe to the Midwest was as attractive after the wearisome years of the Napoleonic Wars as it would be today—came naturally to the headwaters of the Ohio. The tributaries, Allegheny, Monongahela, Kanawha, made boating or rafting possible from the north or south, and

Braddock's Road and the main western route over the Forbes, or "Glades," Pennsylvania Road to Pittsburgh fed foot, wagon, or horseback traffic in from the east.

All lines met along the upper Ohio and most further travel continued on the river in whatever elegance of style or absence thereof fitted the emigrant's purse. From various downstream points the newcomers spread north and south from the river into Ohio, Indiana, Illinois, Kentucky, up the Cumberland and Tennessee rivers to Tennessee and North Carolina, or to the Mississippi and the boundless and trackless "Far West."

Lands on the Ohio could be acquired in several ways, some of them legal:

The easiest way to secure land, short of paying hard-earned cash for it, was to enlist for a military campaign which offered a grant of wild land as a bounty. Money for paying a bonus to the citizen who enlisted was scarce in colonial treasuries but land was plentiful, so that the practice of granting a stipulated number of acres to the volunteer developed early. Sometimes he was given additional acres if he brought his gun along and more still for the use of his horse.

By the end of the French and Indian War this system was well established, with Virginia making the most liberal use of it. It was a party sent to the Kentucky shore of the Ohio to survey bounty lands for Virginia veterans of that conflict which, in 1773, brought some of the men who were to be leaders in the conquest and settlement of the entire lower valley. Thomas Bullitt, James Harrod, John Smith, Isaac Hite, James Sodousky, Abraham Haponstall, Ebenezer Severns, and John Fitzpatrick were of the official group and on the way down the Ohio they were joined by two other independent exploring parties which included the McAfee brothers, Hancock Taylor, Matthew Bracken, Jacob Drennon, and others.

After the American Revolution the state of Virginia set aside a tract on the Ohio and Tennessee rivers in western Kentucky for Clark's veterans, the federal government granted a tract adjoining the Virginia Military Reserve in Ohio to veterans of the Revolution and the War of 1812, and Clark's men who had participated in the capture of Kaskaskia, Cahokia, and Vincennes received 150,000 acres in the present Indiana counties of Clark, Floyd, and Scott immediately opposite Louisville. These were the main bodies of land early granted to veterans on the Ohio River: they drew important settlers but the careless manner of their surveys still causes title trouble in their neighborhoods.

It is interesting to note that, characteristically, while other veterans received 200 to 1,000 acres each for privates and 5,000 and even 10,000 acres each for officers, Clark's enlisted men were awarded a parsimonious 108 acres each, his commissioned officers from 2,156 to 4,851 acres, and Clark himself received only 8,049 acres for the Illinois campaign—far from enough, in its value during his lifetime, to equal the money for which he had given his personal bond to purchase supplies and pay expenses!

Military bounty lands shared one general characteristic: from those awarded for service in the Indian skirmishes of early colonial days to those warranted to soldiers of the Mexican War they were not often occupied by the veterans themselves. Of necessity such grants were located in districts not previously settled. Since comparatively few soldiers except the Ohio Valley volunteers had experience in frontier living, most of them considered their wilderness real estate to be worth only what it would bring in cash. This was usually but a few dollars on the best market and, in case the veteran happened to be far removed from his allotment and much in need of a drink, it might go down even to a few cents per acre.

The purchase of soldiers' land warrants became a thriving speculative business for eastern investors or for provident and forward-looking ex-officers of the armies involved. One land company of the former class, composed mostly of Bostonians and New Yorkers, acquired 900,000 acres of the Military Reserve of Illinois, which was located between the Illinois and Mississippi rivers. Veteran officers Nathanial Massie, Duncan McArthur, and Thomas Worthington bought up 1,035,000 acres of the Virginia Military Reserve in Ohio—and, incidentally, made use of their holdings to the public good as well as their own benefit.

No disrepute should attach to either private speculators or ex-officers who bought land warrants: the veterans had little interest in the lands and were anxious to sell; the purchasers took a long-odds gamble in buying.

The actual settler need not go through any elaborate channel such as proving his claim to land through his military service: in the early day, unless he had a preference for legal doings, he could simply wander about until he found a tract which suited his fancy, move in, throw together a shelter of some sort, and depend upon the authority of strong language and his musket to defend his homestead. Frequently this maneuver worked quite effectively, and when squatters became numerous their titles were usually cleared by a Congress equally conscious of the welfare and the votes of its constituents.

If he was not a veteran of the wars and chose to become a free-holder through proceedings more conventional than seizing his land, the emigrant could buy it from a promoter who had acquired a large tract and who would guarantee—if he chose to keep his word—some degree of community improvement and organization.

Many of these promoters were interesting individuals. Vision-ary men, often, who were prompted as much by a desire to improve the condition of their less fortunate fellows and to immortalize their own names as benevolent patrons in the new country as by any yearning for profit, they dreamed strange dreams and did strange things.

Each land promotion rode its own hobby—and they were often stout and useful animals. The Ohio Company, for instance, cer-tainly brought a workable plan for free education to the river. John Filson, through his minor interest in the city of Losantiville (for which, incidentally, he never paid his partners), tried to en-sure cultural and recreational facilities for its future citizens. George Rogers Clark, in his own plan for Louisville, contemplated a system of public parks which, had they been carried out, would have made that pleasant city one of the world's most beautiful, as well. George Rapp planned earthly comfort (though a celibate one) leading to spiritual salvation for the Rappites at Harmonie in Pennsylvania, Harmony on the Wabash, and Economy in Penn-sylvania. Robert Owen, in his *New* Harmony revision of Rapp's Harmony, mentioned the hereafter only to deride but fought gallantly if ineffectually for earthly ease and cultural and civil perfection. The Englishmen Birkbeck and Flower used their settlement as a weapon to fight the threat of slavery in Illinois and to promote free education.

Many of these ventures also had the services of another interesting class besides their actual promoters and stockholders.

Coming before the individuals or companies who received grants or purchased lands for resale, and often working for or with them, came a type of promoter to whom the very idea of the drudgery of homesteading was repugnant. He was ready and willing to promote settlement by hunting, scouting, and soldiering for the domestically inclined; he sometimes had a family, with whom he visited occasionally, and he often acquired vast holdings of land more or less unintentionally—but he almost always lost them through neglect of his own interests in favor of exploring new lands or protecting new settlements for others.

Daniel Boone was one of these, as was Simon Kenton, and even Christopher Gist exhibited some of the characteristics. Many of similar tastes and talents failed to gain the fame of these three but still played their parts in opening the new land.

After the War of 1812 the work of these individualists on the Ohio was pretty well completed. They had no more Indian raids to repel and a townsite could be selected with chief regard for its transportation and drainage facilities and its water supply rather than its eligibility for defense. Three years after the war, Morris Birkbeck (whose analysis might have been colored by the fact that he had some lands to sell in Illinois) said "All America is moving west"—and if that was a slight exaggeration it must have been, from the looks of traffic on the Ohio River at the time, a pardonable one.

After 1796 the prospective settler could, if he so desired, buy lands in the Ohio Valley direct from the federal government, and it was through this agency that by far the best farming lands in the valley were eventually sold, for it is unquestionably true that the early Virginians, New Yorkers, and New Englanders did not recognize the best of the western lands when they saw them: that black loam was, as they believed from experience in the East, "too rich!"

Virtually no public lands were for sale by the federal government in the states of Pennsylvania, Virginia, or Kentucky. Most of the lands on the Ohio shore of the upper river were also either transferred to private ownership by military bounty grants or by sales in huge tracts—such as those to Symmes and the second Ohio Company—but some Ohio lands were sold through the federal land office direct to individual settlers and direct sale by the government accounted for all but relatively small sections in Indiana and Illinois.

Considerable debate and experiment was carried on before a satisfactory system was finally developed for selling lands in the new states being carved from the public domain. The young Republic needed whatever revenue it could get but at the same time a considerable number of statesmen, led by Thomas Jefferson, saw the desirability of putting the land as rapidly as possible into the hands of people who would render it productive.

The first plan tried by the federal government, under the Land Act of 1796, provided that the lands be offered for sale at auction. Half the townships were put up in 5,120 acre tracts and no quantity less than a square mile—640 acres—could be sold. These sales were held in the national capital, so that the buyer had either to make an expensive trip or risk his interests to an agent. The Land Act of 1796 set the price at $2 an acre minimum with a discount of 10 per cent for cash in advance but the full amount was due within a year of purchase when cash was not paid. Few were the young men with farming experience and equipment necessary for clearing and working 640 acres of land who also had $1,152 in cash to invest or who could raise $1,280 within twelve months. Another difficulty was that, with eighteenth-century methods, it took two full military squads of men to farm 640 acres and this new land, offered in such large parcels, could not be worked by slaves because of its location in free territory.

On May 10, 1800, a second bill was enacted for enabling the sale of public lands. Called the Harrison Land Act, it had been drawn up by the Committee on Public Lands of which William

Henry Harrison, a general in the War of 1812, first delegate to Congress from the Territory North-west of the River Ohio, and former governor of Indiana Territory, was chairman. William Henry Harrison, "Old Tippecanoe," had fought over these lands now to be sold and had commanded and governed the kind of men most likely to buy and operate them. He knew intimately the problems of the West, which could not even be conceived by members of Congress from either the rockbound New England states or the slave economy of the South.

The Harrison Land Act provided that, west of the Muskingum River, tracts of as little as 320 acres could be sold; payments could be made over a four-year period and, most important innovation of all, it located four land offices on the Ohio—where a Westerner could reach them without undue trouble or expense. Auctions were to be held in the offices at Cincinnati, Chillicothe, Marietta, and Steubenville at regular intervals, but a man could also come in and buy his land at any time. The sale picked up but the credit system encouraged wildcat speculation.

As a matter of fact even the Harrison Land Act represented a compromise between what Harrison's knowledge of the West indicated was feasible and the grandiose theory of the seaboard statesmen; not until 1820 were regulations liberalized to produce a really successful system. At that time, under the Land Act of 1820, the credit system was abolished (no one had ever been able to suggest a method by which defaulting payments could be profitably collected from settlers sometimes twenty-five miles apart and with nothing upon which to levy more valuable than a brindle ox and a couple of razor-back shoats), the unit of purchase was reduced to 80 acres and the price was set at a minimum of $1.25 per acre, cash on the rail.

Here at last was a plan which worked and, though slowed at its beginning by the depression of 1820-1825, land sales under its provisions brought settlers to the West and, eventually, millions of dollars to the federal treasury where, under the first overgrasping plan, only hundreds had come, and under the second little but promises. Any man with enterprise enough to farm 80 acres could get together a hundred dollars, some way, and could set up to farm without taking a long-shot gamble on four good crop years in a row.

The West began another boom after 1825.

And, as is usual, with the boom came trouble. To the eastern emigrant who had sold a farm, a house, or a shop in the thickly populated country $1.25 an acre for land looked like an open-handed gift, too liberal to pass up. Even the lands for sale by small speculators who had bought at the government price—or sometimes at a discount from disillusioned original buyers—looked like a bargain. Soon everyone was land-crazy, nearly everyone bought more land than he could possibly clear, improve, or operate, and only the constantly increasing stream of immigrants postponed the inevitable bust.

The new arrivals on the Ohio were no less varied in character and background than had been their predecessors except that now, possibly, the majority of fugitives from justice were crossing the Mississippi.

Occasional men of some wealth and considerable enterprise were coming west to invest their capital. They were usually alone on the first trip or two but after 1820 they began to bring their families and to set up in Cincinnati, Louisville, Pittsburgh, Madison, Marietta, and Wheeling the sort of establishments they had admired in the older cities but could not quite afford.

Farming families came, with experience dating back through generations in the dairy country of Pennsylvania and New York to roots in Germany and England; there were Southerners wearied of tobacco raising in the depleted lands of Maryland, Virginia, and North Carolina, anxious to try it again upon the new soil of Kentucky with slave labor or in southern Indiana without. New England farmers also came—men who had never seen an unobstructed furrow a hundred yards long and who were, as a result, more than competent to make every new level acre pay in grain, hay, fruit, honey, nuts, maple sugar, and garden sass.

Artisans arrived, settling in the towns and villages; there were blacksmiths, gunsmiths, shoemakers, harnessmakers (mantuamakers had to await another generation except in the largest towns), coopers, cabinetmakers (who also made the caskets, for the re-

mains of the more elegant among the deceased), and masons, carpenters, millers, distillers, and any other skilled workers who were
willing to take a chance on earning a living in a short-of-money
society in which, if necessary, any head of a family could perform
these functions after a fashion for himself.

The professional men were mostly young and, surprisingly
often, either competent in their own right or quickly rendered so
by experience and necessity. Lawyers, doctors, preachers, teachers
were not questioned too closely as to their formal preparation for
their callings but either they soon overcame any shortages they
might have had in the way of training or they found some other
way in which to earn a living—perhaps as the dancing masters and
singing masters whose advertisements began to appear.

Merchants found their own salvation, or failed to, in short
order. Many a young eastern clerk left his apprenticeship either
with or without the blessings of his master, came to Pittsburgh
with a sack full of notions, secured a boat or a share in one, and
floated downstream swapping by the way until he heard of a likely
town in which to settle. If he had informed himself as to the difference between a beaver skin and a muskrat (poor incompetent
John Filson had not, when he tried trading) and if he knew what
degree of moisture content made a ginseng root salable and what
it was worth on the east coast when it was prime, and if some
wholesale merchant in New York or Philadelphia remembered
him as having a face honest enough to trust with a small consign-
ment of goods, he was as good as rich. Traveling merchants, with
boats, wagons, saddlebags, or only packs filled with buttons, ribbon,
and fancy notions in general, sometimes kept to the road all their
lives. It was hard work but it was fun to be first to carry news from
one settlement to another, and there was always the excitement of
haggling, for no one, least of all the peddler, ever expected a sale
to be made at first asking price.

Later in the nineteenth century some of these peddlers were
young Jews, brought from Germany and Poland by eastern relatives and started out overland with packs of goods for the joint
purpose of earning something to pay back their passage money and

learning, in the quickest possible way, the manner of American life.

Not many of these keen youngsters, sad in their experiences in the Old World, failed to appreciate the virtues of this newest part of the new one. Few indeed were the county seat towns in the Ohio Valley between 1870 and 1920 whose public squares were not dominated '(next to the courthouse) by a thriving dry goods store owned by a benevolent old Jewish gentleman whose greatest joy lay in standing at its door and greeting the sons and grandsons of the people who had first bought trinkets from his pack half a century before and had taught him that there were really places where most folks didn't care where he came from, if his word was good and his goods were sound.

Then there were the others; those "others" who must make up the bulk of the population of any land. Some of them were down on their luck and came to the West by working their passage on boats, by poling rafts, by driving stock, by walking; sometimes they came in families, afoot and dependent upon the generosity of the backwoods which (often to the sorrow of the country) had already become proverbial. Next above them, in precariousness of future, were the "pore relations," grubstaked by relatives or charitable neighbors to a new start. These two luckless classes of emigrants soon proved their metal, one way or another, in the West. There was more than enough work to be done and work alone would ensure a living more readily in the West than anywhere else in the world of that day. If the poor and unfortunate from the East continued to be so in the new country, then the real cause was shiftlessness; if they prospered, then the trouble back home *had* been hard luck, after all.

That term "shiftless" introduces another class of emigrant which has attracted perhaps a disproportionate attention; the class that was, by general agreement wherever it paused, both *shiftless* and *ornery*. Plain shiftlessness could be borne with, but no self-respecting neighborhood need be expected to have truck with a family which was both shiftless *and* ornery, for that implied a graceless state of being from which there was no reform.

It has been hopefully suggested by observers with unbounded

faith in the virtue of the peoples who originally left the mother country for America that the "pore whites" were not really anomalies; that they never had the pioneering spirit at all because they were descended from southern colonial bond servants who had not *come* but had been *brought* to America; that such people thus must ever lack enterprise and independence!

That explanation seems pat enough, until it is recalled that, through English debtor's laws, poor laws, and apprenticing customs, the children of the sturdiest British yeoman (claimed by the same observers to have been the backbone of that nation and ours) were usually bound out as servants or apprenticed to learn a trade. Those brought to America must have averaged about as good blood as those who came overseas after serving out their time in bond or apprenticeship.

As a further argument against loading the shoulders of the English bond servant with all of such a degrading burden it must be recorded that various specimens of the "pore whites" were contemporarily described as round-headed and tow-haired, which would hint at a North European genesis; as gaunt, black-haired, and beady-eyed, which could indicate Spanish blood from down Florida way or an admixture of Indian upon English or French. Also, as we have noted, many of those unquestionably of pure English descent bore and still bear (for they are still about) the names of some of the most distinguished families who came to America. Unlike Negro slaves who took their masters' names for lack of better ones, these people are not infrequently able to recite a memorized genealogy running back—through perhaps a black sheep or two but still straight enough—to a noted governor, divine, landlord, military figure, or even nobleman of the early day!

One fears, in view of these facts, that many of these were nothing more than people who were ornery because they preferred so to be.

Fortunately, perhaps, all of this class which arrived did not stay put on the Ohio. Life in the fertile valley soon threatened to become too strenuous: people as a whole did not seem to be sensibly satisfied that putting in a patch of corn and sweet potatoes occasionally, when no more attractive pastime offered, served man's whole duty; unreasonable legislators passed laws requiring dwellers to pay taxes and help maintain roads. Authorities even insisted that their inoffensive children be exposed to the dangers of book learning and suggested that they wash, even in winter. Such a state of affairs, which interfered with gigging catfish and possum hunting, soon became intolerable to the ornery.

Many of the "pore whites" moved west from Ohio into Indiana, in time, and from Indiana to Illinois. Toward the middle of the nineteenth century they appeared to congregate in Pike County, Missouri, at least for time enough to cause those of them who later finished their hegira in California to be called "Pikes" in Bret Harte's unflattering accounts of them.

Of course not all of these interesting people followed the same pattern; they must have left substantial groups of founding fathers —and necessarily mothers—along their westering route, for even today pockets of their descendants can be found settled upon the poorest lands of southern Ohio and Indiana. Still greater numbers of them exist upon the less-eligible hilltops of Kentucky and Tennessee and many indeed are entrenched in that strongest stronghold of them all, the gullies of the Ozarks of Arkansas and Missouri. They are easily recognized: jealous of their traditions, of their lineage, and most of all of their leisure, they are a people who wish only to live their lives according to their own standards, which, perhaps, may not constitute an aim quite so unworthy as this driving modern world makes it out to be.

So they came to the Ohio in the first great wave—the rich, the poor, the practical, the dreamers, the no-accounts. By the very variety of their places of origin, their characters, their backgrounds, their attainments, and their ambitions they combined in a virile new population for the new country—frequently turbulent but always interesting.

Chapter Ten

JOHN FITCH HEARS THE
MUSIC OF THE STEAM ENGINE

BY THE TIME steam first pulled a payload upon the Ohio that river already carried a water-borne commerce second in the Western Hemisphere only to the lower Mississippi and the populous Hudson. Practically all the Ohio's boats (and most of those used upon the Mississippi as well) were built in Ohio ports.

Arks, flatboats, giant bateaux, and rafts had been constructed by the earliest arrivals and once shipwrights and hands began to emigrate from eastern seaports the abundance of prime timber and easily available iron suggested operations on a larger scale.

Shipbuilding was tried early; it was a hazardous undertaking because of the upper river rocks and shoals which were likely to damage or wreck deep-draft keeled vessels before they passed Louisville. But cheap materials made the gamble appear worth while to some adventurous souls. The first oceangoing vessel of Ohio River make seems to have been a brig of 120 tons built by Commodore Preble at Marietta in 1798 or 1799. The commodore sailed her down the Ohio and Mississippi, across the Gulf of Mexico to Cuba and thence to Philadelphia, where she was sold. Building vessels for world-wide commerce in the backwoods towns on the Ohio caused an occasional unforeseen difficulty: Zadok Cramer tells of a ship built in Pittsburgh and cleared for Leghorn, Italy. Upon her arrival there the port authorities threatened to confiscate her because they assumed her papers were defective and issued at a nonexistent port. Her captain spent some time gathering maps upon which he could show the Italians that such a port as Pittsburgh existed and in convincing them that a stretch of navigable river over two thousand miles in length connected it with sea water.

The construction of keel- and flatboats upon the upper river soon proved to be the safer and more profitable business and at the beginning of the nineteenth century the demands of commerce and travel kept pace with the production. Flatboats were the favorites at first and they served many purposes: as commercial carriers capable of passing from Pittsburgh to the Mississippi in fifteen days; as emigrant boats, for sale at Pittsburgh during the War of 1812 at $1 to $1.25 per foot—$35 for the average—or for special purposes such as store boats or floating mills.

The obvious risks that the merchant or manufacturer took in shipping his goods in his own flatboats, often manned by inexperienced hands, created the demand that brought about the extensive use of the keelboat and its professional crew.

Zadok Cramer explained the situation in advice to shippers included in the 1814 edition of his pilot's guide:

Merchants are beginning to prefer this method [keelboats] for safety and expedition; and instead of purchasing boats and taking charge of them themselves, they get their goods freighted down from Pittsburgh in keel boats by the persons who make them, and who make it their business to be prepared, with good boats and experienced hands, for such engagements.

This method is the safest, if not the cheapest, for this special reason: the cargo is consigned to the care of an experienced and careful man, who perhaps descends and ascends the river twice or thrice in the course of one season, and of course must be well acquainted with all the difficulties of navigating it . . .

As a natural result of this development, companies were soon formed solely to handle transportation, with keelboats—called also barges—run on regular schedules. Apparently the first such company of which the well-informed Cramer knew had headquarters in Cincinnati before 1814:

What adds to the commerce of Cincinnati is the line of barges running regularly from that place to New Orleans, descending loaded with the produce of the country, and returning with cargoes of sugar, coffee, rice, hides, wines, rums &c. and dry goods of various kinds, and cotton from Natchez. Messrs. Baum and Perry and Mr. Riddle, have a line of barges constantly engaged and some others are also employed . . . These barges carry about 700 barrels, and are long in proportion to their breadth, 9 men conduct them down, from 24 to 32 up stream, oars and poles are the principal dependence, they have sails also that are frequently useful. Cordelling, where the water is too swift to be stemmed by the force of the oars, or too deep for the poles, is the only alternative and a bad one it is. They descend from Cincinnati to New Orleans in about five weeks, unless they run day and night. . . . They ascend in about 80 or 90 days . . .

Unquestionably such transportation was big business by 1814, for Cramer had already reported that as early as the winter of 1810-11 the pilot records at the Falls of the Ohio had shown the following merchandise passing downstream through a period *only sixty days in length:*

197 flat, and 14 keel boats descended the Falls carrying

18,611	bbls.	flour	1,484	lbs.	thread
520	"	pork	59	lbs.	soap
2,373	"	whiskey	300	"	feathers
3,759	"	apples	400	"	hemp
1,085	"	cider	154,000	"	rope yarn

721	"	royal		681,900	"	pork in bulk
43	"	wine		20,784	"	bale of rope
323	"	peach brandy		27,700	yds.	bagging
46	"	cherry bounce		4,619	"	tow cloth
17	"	vinegar		479	coils	tarred rope
143	"	porter		500	bu.	oats
62	"	beans		1,700	"	corn
67	"	onions		216	"	potatoes
20	"	ginseng		817	hams	of venison
200	gross	bottled porter		4,609	"	of bacon
360	gal.	Seneca oil		14,390	tame	fowls
15,216	lbs.	butter		155	horses	
180	"	tallow		286	slaves	
64,750	"	lard		18,000	feet	cherry plank
6,300	"	beef		279,300	"	pine plank
4,433	"	cheese				

Also, a large quantity of potter's ware, iron mongery, cabinet work, shoes, boots, and saddlery . . .

Only a year after the shipping reported, a steamboat passed downstream and by 1814 Cramer foresaw the future passing of the keelboat *if* certain monopolies could be voided:

It would be well if . . . the barging business propelled by manual forces could be got rid of and steam power substituted. And this would be done soon no doubt, but for the overwhelming patent of Fulton and Livingston which secures to them *all* the navigable rivers in the United States *for fourteen years* . . .

So let us now look at that patent of Fulton and Livingston, at exclusive charters to navigate, at steam power in general—and at poor John Fitch.

The inventor and the builder of the first American steamboat to operate on schedule with payloads was unquestionably not Robert Fulton but John Fitch, into whose fifty-five years was crowded enough misfortune to have supplied the entire Ohio Valley for a generation. In spite of the fact that his association with the Ohio and the West dated from 1780, in spite of the fact that he had built a steamboat upon tested plans and had secured a charter to operate it upon the Ohio, it was not given to him to first navigate that river under power. That honor went to Captain Nicholas Roosevelt (with the proverbial luck of his clan) in a boat constructed upon the plans of and licensed by Fulton and Livingston.

John Fitch was born in East Windsor, Connecticut, on January 21, 1743. After a few years of schooling he learned the clockmaker's trade, served for a time as a journeyman and suffered the first of his major misfortunes when he married during the latter period.

His wife was, at least to him, a woman of unbearably "turbulent" disposition and conduct. (Late in life he wrote: "I know of nothing so perplexing and vexatious to a man of feelings as a turbulent wife and steamboat building. I experienced the former, and quit in season and had I been in my right senses, I should undoubtedly have treated the latter in a like manner.") He apparently left the lady about 1769.

Settled in Trenton, New Jersey, he followed his trade and at the beginning of the American Revolution began the manufacture of guns. Shop and stock—and Trenton itself—were taken by the British in December, 1776, and Fitch joined the New Jersey troops and spent that dark winter with General George Washington's ragged men at Valley Forge.

After the war, in 1780, he was made a deputy surveyor of the lands back from the Ohio and between the Kentucky and Green rivers. He took a great fancy to the Kentucky country and returned to Philadelphia after the survey was completed, bought a stock of goods for trading with the white settlers, and joined a party which was traveling west in the winter of 1781-82. They were attacked by Indians, Fitch's goods were taken, and he and nine of his companions were captured.

He remained with the Indians north of the Ohio for two years before he was able to escape and make his way to Pennsylvania, where he resumed his old trade. Life as a captive of the Indians could not have provided facility for research but there was probably, in their unexacting schedule, time enough to dream:

John Fitch returned to civilization with a plan for powering a vessel with steam.

At Warminster, Pennsylvania, in April, 1785, he completed a model of a steamboat to be driven by side wheels but dissatisfied (unfortunately for his future) he changed his plans and in 1786 constructed a skiff moved by galley oars at the sides instead of paddle wheels. The oars were operated by a steam engine with a 3-inch cylinder, double action, which transmitted its power to the oars by means of cranks. He could get neither state, federal, nor private assistance to continue experiments on a larger scale.

With characteristic ingenuity Fitch raised $800—by selling a map of the country north of the Ohio which he himself engraved and printed on a cider press! With the proceeds of this sale in hand he began work early in 1787 on a boat, 45 feet long with a 12-foot beam, in which he installed an engine of 12-inch bore to drive the same type of oars he had employed in his skiff model. He tested it on the Delaware River on August 22, 1787, with the delegates to the Constitutional Convention as witnesses.

That fall, his invention having been demonstrated before the unimpeachable scientific authority of the delegates, Fitch succeeded in organizing a company to engage in steamboat construction and was granted exclusive right to navigate the waters of New Jersey, New York, Delaware, Pennsylvania, and Virginia for a period of fourteen years. The company was formed in February, 1787.

This system of monopoly of navigation rights—which was apparently his own idea—would later, when the monopolies were in the hands of others, bring about *his* failure. Such was the fortune of John Fitch.

He began the construction of another boat in October, 1788, and a third in April, 1790. The second was the *Perseverance;* first steamboat in America to carry payloads on regular schedule, it operated between Philadelphia and Burlington, New Jersey, at a phenomenal average speed of eight miles per hour.

Encouraged by the success of the *Perseverance* and prompted both by the attachment he had conceived for the Ohio and by the fact that the Ohio Valley was much in the public mind, Fitch now directed the steamboat company in applying to the state of Virginia for the exclusive right to carry freight and passengers on the Mississippi, the Ohio, and its tributaries. It was granted.

There was, however, a limiting time clause in this charter; in addition, had the matter come to a test, there was a rather serious question as to whether or not Virginia, though it governed the Ohio to its north bank, could grant exclusive rights upon the Mississippi, of which the Old Dominion held only the short stretch of present western Kentucky bank. It is in no way surprising that Virginia undertook to assign these rights—bless her confident, independent old heart!—but John Fitch and his associates might well have scanned her jurisdiction more carefully before they based their hopes upon the charter.

As it happened, the question of Virginia's jurisdiction had small bearing; the congenital ill luck of John Fitch saw to that. The boat he designed to navigate the western waters was so damaged in a storm that repairs would obviously extend beyond the time limit set for beginning operations and the company, the enthusiasm of the shareholders obviously having cooled in the breeze of public opinion against the probable success of steam navigation, decided to make no effort to renew the charter. In 1791, however, Fitch did receive United States patents covering his inventions.

The directors of the company sent Fitch to France in 1793 for the purpose of building a steamboat there. He found that France, torn by revolution, was no laboratory for radical experiments in transportation. The evil genius still active, Fitch took a trip to England, leaving his drawings in France with either Aaron Vail or Joel Barlow. Thus he made possible the strange coincidence by which the plans were examined by artist and civil-engineer Robert Fulton.

Fulton was another American who had been long preoccupied with dreams of sailing without sail or driving upstream against a head wind without manpower. Aaron Vail was American consul at L'Orient. Joel Barlow, American minister of the gospel, poet, parlor intellectual, and pseudo uplifter (remember his sale of Ohio land to the French artisans?), was at the time making himself a minor revolutionary nuisance in the pinkish salons of England and France.

Whoever was the immediate custodian of the plans belonging to John Fitch (some sources say Vail, some Barlow), the drawings were certainly shown to Fulton and were in his possession for some time. Fulton had experimented with paddle wheels unconnected with steam power—Fitch had considered their possibilities but abandoned them in favor of oars; now, aided by the refinements in the matter of power transmission that Fitch had developed, Fulton began to experiment with paddle-wheel driven steamboats. Meanwhile he secured a patron in the person of Robert R. Livingston, United States ambassador to France. Livingston, it should be noted, was not new to the sponsorship of research in steamboat designing: John Stevens, of Hoboken, New Jersey, had begun to experiment in 1791, perhaps inspired by the early efforts of John Fitch, and in 1799 he, with Nicholas J. Roosevelt and Livingston, had secured from the legislatures of New York and New Jersey exclusive rights to navigate the waters of those states, Fitch having meanwhile failed in his effort to put steamboats on their rivers.

In France Fulton began his new experiments in 1802. There he made a working model and built a boat 60 feet in length with an 8-foot beam powered by a steam engine and driven by a single paddle wheel at the stern. Livingston then recommended that the experiments be continued in the United States.

Before leaving Europe Fulton designed an engine and, without divulging the use to which it was to be put, submitted its plan to the British master, James Watt, for criticism and for eventual construction by the firm of Watt and Bolton. That engine was delivered in New York in 1806, installed on the deck of the *Clermont,* and on August 11, 1807, the maiden voyage from New York to Albany was made—150 miles in 32 hours.

John Fitch failed to attract interest to his work in England or France (except in so far as the very unprofitable interest of Vail, Barlow, Fulton, and Livingston was concerned) and, destitute, he shipped as a sailor on an America-bound vessel in 1794.

Arriving in the United States, probably fed up with civilization and its frustrations, he went immediately to the farm he had long since acquired near Bardstown, Kentucky: he found it in the possession of strangers. Possibly, in his preoccupation with

steamboats, Fitch had neglected such formalities as proper registration—even tax payment—but in any case the farm was gone and
he went back east, locating himself in Sharon, Connecticut. But
there was no peace for John Fitch; still troubled by his dreams, he
went to New York in 1796, where he constructed a steam-powered
boat consisting of a ship's yawl powered by a small high-pressure
engine and driven by a screw-type propeller; it was successfully
demonstrated but attracted no particular attention.

Two years later he returned again to the vicinity of Bardstown,
Kentucky, taking up residence in a small neighborhood tavern.
There, his spirit still undaunted, he built a small model of still
another steamboat and tested it on a neighboring stream in 1798.
But this was the end; the boat functioned satisfactorily but so had
those which resulted from his earlier efforts and he was still friend-
less and without recognition: John Fitch poisoned himself in his
garret room on the night of July 1, 1798—four years before Robert
Fulton built his first American boat!

Fitch was reported to have made the request that he be buried
on the banks of the Ohio "where the songs of the boatmen would
enliven the stillness . . . and the music of the steam engine soothe
his spirit." In this, as in everything else, he was disappointed. His
dust still lies in Bardstown, well out of hearing of boatsmen's
songs.

In the journal he left in his room he had prophesied that

*The day will come when some more powerful man will get fame
and riches from my invention; but nobody will believe that poor John
Fitch can do anything worthy of attention.*

In none of this need Robert Fulton be considered the villain.
Most frequently this character was given him and his associates,
especially in the antimonopolistic Midwest, because of the grant
of sole steam navigation rights they held on the Hudson and later
secured on the Ohio and other rivers: actually it had been the
West's posthumous hero Fitch who had first petitioned for such
a monopoly and received it—though the Fitch nemesis had prevented its being profitably used. Fulton had undoubtedly, through
his studies with Watt in England, approached the power problem

in the more scientific spirit, and it was Fulton who developed the paddle wheel successfully after Fitch cast it aside in favor of the far less efficient galley-type oars.

The truth of the matter seems to be that Fitch was an inspired but untutored genius and the victim of a lifetime of ill luck, while Fulton, possessor of a methodical mind and powerful associates, reaped the benefit of these attributes. Whatever the case, steam navigation soon revolutionized travel and commerce in the West and opened an empire to settlement and trade.

Fulton and his associates sent Nicholas J. Roosevelt to the Ohio and Mississippi to explore the possibilities for steamboat operation. Roosevelt had himself been an early experimenter with John Stevens; had, in fact, designed the paddle wheel that was eventually used on the *Clermont,* and had become associated with the Fulton-Livingston interests after the successful completion of that vessel. Roosevelt reported favorably upon the prospects in the West, the Ohio Steamboat Navigation Company was incorporated in December, 1810, and through the political connections of Livingston and another distinguished shareholder, DeWitt Clinton, exclusive rights of operation in the Louisiana Territory were also secured for a term of fourteen years.

The Ohio's first steamboat—indeed the first on western waters —was completed at Pittsburgh by Nicholas J. Roosevelt under a Fulton-Livingston license in 1811. It was the *New Orleans,* a vessel of over 300 tons, and on October 10, 1811, it left the yards on its "shakedown" voyage. It carried its builder and his wife and baby, an engineer named Baker, one Andrew Jack as pilot, six hands, a few servants, and a dog. Roosevelt had made a survey of the Ohio and planned to load coal below Pittsburgh and from an exposed vein he had seen below Louisville, using it as fuel. Wood-yards, those sinks of human misery upon which every traveler later made a comment, were not yet dreamed of.

The *New Orleans* reached Louisville late at night, seventy hours out of Pittsburgh. That Kentucky community awoke to an excitement well-nigh worthy of the implications of the occasion, opinion being divided as to whether the sparks and noise originated in a comet which had hit the river or a fiery raid by Tecumseh's Indians.

Water was low at the Falls and, during the three weeks before rains brought a rise, the *New Orleans* made several round trips to Cincinnati, presumably carrying a selection of local citizens on the first passenger excursions by steamboat in the West. During the last week in November the vessel ran the Falls and headed for Natchez, where she docked on January 12, 1812—having passed through the heart of the country most affected by the earthquake of December, 1811, and having been almost swamped by the turbulent waters it caused. She had certainly undergone more of a test than her owners either contemplated or imagined possible.

Although the dangers of navigation in the unimproved Ohio—snags, sawyers, stumps, sand bars—were far greater to a boat proceeding either up- or downstream under steam power than they were to slow-moving craft which were rowed or poled or floated, steam had come west to stay.

According to Zadok Cramer, the *New Orleans* picked up its first payload of freight and passengers at Natchez for the haul on down to the delta and presently began a regular run between that city and New Orleans. Her operation was profitable from the first:

Her accommodations are good and her passengers are generally numerous; seldom less from Natchez than from 10 to 20, at 18 dollars per head, and when she starts from New Orleans, generally from 30 to 50, and sometimes as many as 80 passengers, at 25 dollars each to Natchez. According to the observations of Captain Morris, of New Orleans, who attended her as pilot several trips, the boat's receipts for freight upwards, has averaged the last year 700 dollars, passage money $900—downwards $300 freight $500 for passengers—That she performs 13 trips in the year . . . amounts to $31,200. Her expenses are, 12 hands at 30 dollars per month, $4,320, Captain, one thousand dollars; 70 cord of wood each trip, at $1.75, which amounts to $1,586, in all $6,906 . . . a nett gain of $24,294 for the first year . . .

Cramer's fear that the Fulton-Livingston monopoly would long delay the general use of steam upon the Ohio was ill-founded.

Monopolies and patents did not hold in the West, the populace not believing in such privilege, and soon everyone who had the urge and could find the capital was engaged in steamboat building. Even during the strenuous days of the War of 1812 boatbuilding continued upon new plans and combinations of old ones freely borrowed until, by 1816, there must have been nearly a dozen steam-powered vessels on western waters. Up to that year, however, steamboats had had one feature in common; all hulls were deepkeeled, in the traditional form of oceangoing ships, with the engines below decks. In 1816 Captain Henry M. Shreve, who had run the 75-ton *Enterprise* to New Orleans in support of General Andrew Jackson's famous defense of that port against the British in 1814, followed a period of experiment by completing his new boat, the *G. Washington*. This, in its several new departures, set the pattern for future construction and completed the progress in design that established the steamboat on the western rivers.

By substituting fixed horizontal cylinders for the oscillating vertical type, Shreve was able to place his engines on the deck rather than in the hold, and thus to substitute a shallow, flatboat-like hull for the deep-keeled type used previously. He also made improvements in engine design but it was this new principle of hull that was of greatest significance. Shreve's boat ran *on* the water; the predecessor of those later boats used in the small tributaries and in the shallow Missouri which were exuberantly described by their owners and passengers as having "no draft at all!"

The *G. Washington*—too obviously something new, something old, and something "borrowed" to escape the attention and interest of the Fulton-Livingston holders of the technical rights—was seized on their complaint when she arrived in New Orleans. There had already begun some litigation in unenthusiastic western courts upon the matter of the protection of the monopoly and of the various patents. Finally when the *G. Washington* case reached the Supreme Court in 1824 (by which time the original charters were being violated by more than a hundred steamboats on the Ohio and Mississippi alone) the court threw American waters open to free competition and nullified, in effect, many of the patents of Fitch, Fulton, and others.

By the end of 1835 some 684 steamboats had been built in the West—including 304 in the Pittsburgh neighborhood, 221 near Cincinnati, and 102 at Louisville and its Indiana suburbs. Improvements in design had been rapid and twenty-four years had already seen the beginning of those efforts of the passenger boats to excel each other in the elegance that eventually resulted in the side-wheeled gingerbread palaces of the fifties.

Steamboats on the Ohio presently—by the thirties, at least—began to be built by organized transportation companies under orthodox business management, by individual merchants and manufacturers, or by private capitalists. Probably, however, most early steamboats were financed, constructed, and operated by more or less amateur groups of investors living in small river towns. In his unpublished memoirs one John Hewitt Jones, citizen of Rising Sun, Indiana, in the clear and cloudless morning of that community's prosperity, gives an account of such a venture which is probably typical of hundreds of others in both procedure and final reckoning. Says Jones:

> Moses Turner, who built the flouring mill and run it several years, now went into the dry goods business and offered me better wages so I set in with him, during which time there was a Steam Boat Company formed.
>
> It was in December, 1834, (a book with the articles of agreement &c I keep as a relic) I had saved every year from my wages and now having a *surplus* thought it a fine chance to invest in a Steam Boat and was urged in it by older men, supposing it would pay. I took a share—$100.00 in stock—but the company departed from the first calculation, which was to build a little flat boat like for $1800.00. It was enlarged on, from time to time, until it cost about $5000.00. Finally in the windup it took all the money I had made, say $500.00, to pay up.
>
> I run on the boat, which was called the *Alpha,* as clerk. First in the Cincinnati and Rising Sun trade and then in the Cincinnati and Portsmouth trade . . . for then goods from New York was sent by canal to Buffalo and then across the Lake [Erie], by canal to Portsmouth and by Steam Boat to Cincinnati &c.

Jones describes a charter trip, a common enough procedure no doubt, but in this case with a cargo of peculiar interest:

We left Rising Sun on Sunday the 13th of December, 1835, on board the Steam Boat *Alpha* for Florence, Alabama. The river full of ice when we left. Saturday, the 19th, found us about 100 miles up the Tennessee River . . .

December 20th, '35, we saw the wreck of the old "Rising Sun" Steam Boat converted into a house 150 miles up from the mouth of the river.

The morning of the 22nd found us at Waterloo, a town without houses at the foot of Muscle Shoals, a sorry set of steam boat men. We had to reship freight in Keels to Florence, above, which delayed us some days.

While there Beatty and Ingersoll, Govt. Agents, came in with about 511 Creek Indians to be removed to the Indian Territory up the Arkansas River to Fort Gibson. We bought two Keel boats and took the Indians aboard, a Keel boat on each side, and started on our trip. Was getting $2200.00 for the trip and were to stop and lay up nights for the Indians to camp out and do their cooking &c . . .

We happened to make Memphis just in the evening and so had to land on the opposite side of the river. Next morning the party who went by land to take the ponies thru were crossing the river so it was impossible to get away for the Indians must see their ponies and would bring them into camp and make a terrible fuss over them and were very loth to part with them. Some offered to sell a nice little pony for 5$ for fear they would not go through the trip . . .

We entered the mouth of White River and went through a cut-off into the Arkansas River on the morning of the 2nd of January 1836.

We anchored out in the river at Little Rock—the agents and the Boat having some business to transact. We had to use this precaution to keep the Indians from getting in to town and getting whiskey, for when they did there was a tear round among the Indians. The women (squaws) would down a fellow and tie his legs and tie his arms and let him lay till he got sober. Little Rock was of about 1000 inhabitants, a very pretty place.

We had lost time and the agents agreed we might run some of the nights to make up. The first night, the river being very snaggy, we stove one of the Keel boats and made a terrible rumpus among the Indians, again having to lay up of nights and leave the sunken Keel boat at Lewisville. It was a very dark night, the stove Keel boat sinking fast with about 250 Indians on board, causing great confusion and such a time to get them and their baggage on the Steam Boat. The yelling of the yellow skins, big and little, old and young was not easily forgotten . . .

It was a fine sight to see the camping of the Indians on the trip. As soon as the Boat was tied to the shore and a plank out the first to leave was the squaws, who gathered up their kit, which was usually tied up by the corners in a blanket in which was their tents, blankets, cook articles &c. They would throw it over their backs and let the tie come across their foreheads, resting on their backs and in one hand take an axe and in the other and under their arm a little papoose and run ashore and up the bank.

They would chop trees and make a fire and prepare supper. I often used to walk through the camp of pleasant evenings. It looked like a little village. They parched corn in a kettle and then would pound it in a mortar or deep cut trough in a log and then boil it up and make a very fine dish which they called "sophka" and would broil their meat stuck on a stick before the fire.

This was played on the violin by a half-breed Indian (Creek) on the Steam Boat *Alpha* which was removing them from their old home in Georgia, in the winter of 1835, to their new home in the Indian Territory:

> Alas! for them—their day is o'er
> Their fires are out from hill and shore:
> No more for them the wild deer bounds;
> The pale man's axe rings through their woods,
> Their pleasant springs are dry.
> Their children—look! by power oppressed,
> Beyond the mountains of the west,
> Their children go—to die,
> By foes alone their death song must be sung.

We only got a few miles above Fort Smith, the water being too low, and had to give up the trip and return. We bought 25 Bbls. Pecans at Fort Smith for 5$ per Bbl. and started back. We had to check up near Van Buren and stay there two weeks, when the river rose and we left and got back to Rising Sun in March, all safe.

It was not long till we sold the *Alpha*. It proved to be poor stock —my last dollar was gone—so I had to start again. . . .

So much for a minor but doubtless typical experience in steamboating: such investments could be remarkably efficient at absorbing the savings of venturesome young dry goods clerks. Of course steamboat operation was profitable under the right management—as witness the Greene family of Cincinnati, which operated its first boat on the Ohio soon after the War of 1812 and whose Greene Line boats still ply the river.

No matter what improvements in operation and comfort followed those early ones by Captain Shreve, steamboats on the Ohio could be, and were, of every shape, size, and design, depending upon what cargoes were expected and which tributaries were to be penetrated. Little more than a flatboat with an engine and high slat sides would serve quite well for carrying slaves and other cattle, while the boat that could attract the new-rich gentry of Pittsburgh, Cincinnati, Louisville, and the prospering back country had better be well supplied with paneled walls, silver service, and gilded lounge furniture.

A boat designed to run from Shippingport, at the foot of the Falls of the Ohio, to New Orleans might be of any reasonable length and beam but if it made side trips up the Wabash it must have an extra-strong hull to meet hardwood stumps; if up the Kentucky it need be short and narrow at the waist to negotiate the bends; if up the Tennessee it must be well planked at bottom to slide over rocks and, should a run on the Missouri be contemplated, the boat must, traditionally, be light enough of draft to "run on a heavy dew."

Mortimer M. Thompson (as a humorist signing himself "Doesticks" he was a popular writer in the middle of the nineteenth century) describes in the light vein a boat which plied the Kentucky, with side trips on the Ohio:

. . . the boats on this stream differ from any others in the world . . . *Steamboat Blue Wing* . . . very much the shape of the Michigan country-made sausage, and is built with a hinge in the middle to go around the sharp bends in the river, and is manned by two captains, four mates, sixteen darkies, two stewards, a small boy, a big dog, an opossum, two pair of grey squirrels, and a cream-colored chambermaid . . .

Night so dark the clerk has two men on each side of him with pitch-pine torches, to enable him to see his spectacles . . . pilot so drunk the boys have painted his face with charcoal and coke berries, till he looks like a rag carpet in the last stages of dilapidation . . . fast asleep with his legs tied to the capstan, his whiskers full of coaldust and cinders . . .

Boat fast aground, with her symmetrical nose six feet deep in Kentucky mud . . .waiting for the mail boat to come along and pull her out . . .

Some steamboats were operated, even in the twenties, in a manner perhaps not elegant but certainly genteel. Captain Donald Macdonald, who always had an eye for the finer things, reported on a trip he made in December, 1824, from Pittsburgh to the Wabash in company with Robert Owen upon the latter's first visit to the site of his New Harmony experiment:

A little before ten o'clock we left Pittsburg in the *Pennsylvania* steam boat, drawing 3½ feet of water. . . . The owner of the boat, Mr. Hart, was on board. The Captain's name was Cunningham. We had from 20 to 30 passengers, and 10 or a dozen deck ones. As the boat was of small size we were very much crowded. . . . The table was cleanly served and amply provided . . .

In the course of the evening several of the deck passengers were joined by some of the cabin ones, and spent three or four hours singing merry songs. All the females sang in turn, and though no elegance of manner was shewn & no charming melody heard, yet natural good fellowship and a friendly desire to please, made the time pass away in a very social & agreeable manner . . .

Before steam had touched the Ohio, that interesting stream had been responsible for a publishing venture which, as traffic increased, became quite profitable. In 1801 Zadok Cramer, of Pittsburgh—already largely quoted here—brought out the first edition of the *Navigator,* destined, through its frequent revisions, to serve as the unique bible of the pilot, the captain, and the apprentice until Samuel Cummings set up as a competitor with the *Western Pilot.* Later there were many such publications, one of the most successful being the various editions of *Conclin's River Guide,* which entered the field from Cincinnati.

. The publishing of river guides was good business; no pilot who valued his life and his job failed to buy each new edition, for, while the Ohio did not jump about over the landscape as did the Missouri, yet the annual Ohio flood was, and is, a fearful and wonderful thing to see. There were always cave-ins and cutoffs enough then, even by Missouri standards, and, in addition, the often high and rockbound Ohio banks forced the flood current fast and deep. What was clear channel in October often became sand bar by March and an island in August; a chain of rocks buried ten feet below the lowest of low water might accumulate stumps, logs, and driftwood which would yield only to blasting.

Yet the worst hazard of life and property was, by general agreement, steamboat racing. A prideful captain with a conviction that his boat was the fastest on the river, a bet on the side, his deck hands sitting on the safety valves, and very frequently too much Monongahela rye under his belt, spelled disaster. (It is a commentary upon the undeveloped taste of our forefathers that it *was* generally Monongahela rye until post-Civil War days, when that delectable product of lower tributaries of the Ohio, sour mash bourbon, finally came into its rightful own.)

The number of boiler explosions reported upon American waters during the first three or four decades of steamboating appears incredible until one recalls that few engines in the early boats had *tubular* boilers, in which steam is generated in pipes coiled within a metal shell. The early boilers were little more than metal tanks filled with water, under which heat was applied to generate steam. The steam was in turn carried off through a pipe to the engine, where it operated to push the pistons in the cylinders and the movement of the pistons was transmitted to turn the paddle wheels. Such boilers were in effect little more than giant teakettles with the lids riveted on. The pipe that carried the steam to the cylinders corresponded to the spout of the hypothetical kettle but with the difference that the escaping steam had the pressure of the pistons in the cylinders to hold it back. Of course there were safety valves on the boilers but they were sometimes deliberately held down to produce extra power and they frequently stuck when needed most.

When excessive steam pressure caused a boiler such as this to

explode the effect was very different from that which would have
resulted from a tubular boiler within a jacket: this "teakettle" type
was often ripped into thousands of jagged pieces which were hurled
unbelievable distances through the air, maiming or killing those
unfortunate human beings who happened to be in their paths.
Sometimes the explosion blasted the entire superstructure high off
the keel, as in the case of the *Whippoorwill* as described in the
ballad tribute to the mighty Steamboat Bill:

> Oh down the Mississippi steamed the *Whippoorwill,*
> Commanded by the pilot, Mr. Steamboat Bill.
> The owners give him orders on the strict q.t.
> To try and beat the record of the *Robert E. Lee.*
>> Steamboat Bill, sailing down the Mississippi,
>> Steamboat Bill, a mighty man was he;
>> Steamboat Bill, sailing down the Mississippi
>> Tryin' to beat the record of the *Robert E. Lee.*
>
> Up there stepped a gambling man from Louisville;
> He tried to get a bet against the *Whippoorwill*
> Billy pulled his roll and it was a bear—
> Just then the boiler busted, blew them up in the air.
>
> The gambler says to Billy as they left the wreck,
> "I don't know where we're goin' but we're neck-and-neck!"
> Says Billy to the gambler, "Tell you what I'll do;
> I'll bet another thousand I go higher than you!"
>
> The river's all in mournin' now for Steamboat Bill;
> No more you'll hear the whistle of the *Whippoorwill!*
> There's crepe on every steamboat now that plies the stream
> From Memphis right through Natchez down to New Orleans.
>> Steamboat Bill's not sailin' on the Mississippi;
>> Steamboat Bill, he was a mighty man:
>> Steamboat Bill's not sailin' on the Mississip'—he's
>> Pilot of the ferry to the promised land.

If the passenger or hand on the steamboat whose boiler ex-
ploded was not struck by flying iron or woodwork or blown into

space, he had an excellent chance of being scalded by the escaping steam or burned in the fire that usually completed such a catastrophe. Failing these, if he had a weak heart, there was every reason why he should die of fright; if he could not swim, the odds were long in favor of his being drowned before rescue on the still sparsely peopled river.

Racing resulted in unbelievable hundreds of deaths from bursting boilers and resulting fire; from steamboats broken in collision or run ashore. Sooner or later almost every boat, in the first century of navigation upon the Ohio, snagged, sank, or blew up—racing or not—but racing was the surest nemesis.

The danger was widely enough recognized. Passengers used every subterfuge, including pre-emption and subsequent defense by fist or bootjack, to secure the berths farthest from the boilers. Everybody decried racing—until the race began. Then it was a timid soul indeed who failed to thrill and to develop a partisanship, conceal it though he might, for the vessel on which he had paid his fare.

For instance, there is the story still repeated in Kentucky, of the staid and prosperous Bluegrass widow: *she* had a mortal fear of steamboat racing but she also had horse-country blood in her veins!

Being a prudent manager the lady chose to accompany her annual shipment of hams, bacon, and lard from her farm to the New Orleans market. In arranging the passage she swore the captain of the boat she patronized to refrain from racing. Competition for cargo was stiff that season and the lady's shipment was large enough to give weight to her desire.

The trip began. The weather was fine, the stage of the water exactly right, the boat smooth-running and well appointed. Soon the widow had her sea, or rather river, legs and took her ease, daytimes, on the hurricane deck. It was here she was sitting when a rival boat, of comparable speed and trim, bore down in the wake of the vessel she rode and held alongside while its captain called a challenge. The captain of the widow's boat ignored it.

The challenger called again—adding a note of derision and offering odds on his bet. Again the captain ignored and after a third unanswered hail the challenger gave up and began to pull ahead.

"Are you going to let that scan'lous-actin' man get away with that so't of talk?" queried the lady of the captain.

"I'm bound by my promise to you, madame," replied the captain ruefully.

"Suh," said the widow, "yo' ah released in view of his impudence. He sounded to me lahk a New Yawk Yankee—You go 'head and beat him, captain, an' teach him bettah manne's an' if Ah may, Suh, Ah'd lahk to take half of that bet!"

The captain dived immediately into the pilothouse, informed the pilot that a race was on, rang the necessary bells, and shouted the required orders; the widow's boat pulled up on the challenger.

It is reported to have been neck and neck until the two boats were within a mile or so of the finish point which the challenger had named; it was then that the widow's boat leaped ahead by lengths.

The race now safe for him beyond a doubt the captain left the pilothouse and entered the engine room, curious as to what gave his boat the final burst of speed that he had not known she had. He found the Kentucky widow in command of the firemen, busily supervising the stoking of the boilers with the last of the fat pork products of her season's farming, broken out of their barrels to give that extra ounce of power necessary to put the impudent Yankee challenger in his place.

Steamboat racing and racing challengers were by no means always so friendly. Sometimes a drunken or simply bullying captain attempted to force a race at all cost. John Randolph Thomas— Poe's friend—tells a story of an encounter of the latter type. It could well be true; may have been so.

Thomas's story opens as the steamboat *Fort Adams* picks up the pilot Samson and his squirrel rifle. Samson is known on the ·river as "Old Kaintuck."

Samson had been hailed, earlier in the day, by the pilot of the boat *Shelby,* who had asked him to come aboard as a relief pilot. One pilot, he had been told, was down with fever; the other and the captain had demonstrated to Samson's satisfaction that they were both drunk. Samson had refused the berth and had come aboard the *Fort Adams* as a passenger.

The *Shelby* had stopped for minor repairs and had been passed by the *Fort Adams,* but the former was now back in the channel and rapidly overtaking the *Fort Adams* as officers, passengers, and Samson watched its progress. The author tells the story as a passenger:

By this time all was excitement on board the *Fort Adams.* The *Shelby* was a larger and faster boat, and she was pressing us hard. I could hear the barkeeper calling out to the steward for more ice—and, as I glanced back towards the bar, I discovered a crowd of persons in excited talk, drinking; among them was the captain.

"Come, let's go up on the hurricane deck," said Kentuck, "and see how matters look."

As we entered the cabin to go forward and ascend . . . a number of ladies rushed from the cabin toward us, exclaiming—

"Gentlemen, they are racing; they'll blow us all up, Gentlemen!"

"Ladies, don't be frightened," said Old Kentucky, in a manner of exceeding courtesy, at the same time taking off his fox-skin.

"Oh! sir," exclaimed a beautiful, delicate looking lady to him in an agony of terror, "don't let them race; I had a brother and sister lost on the *Mozelle.*"

"Don't be frightened, my good lady, don't be frightened," rejoined the Kentuckian; and, shaking her hand, proceeded to the hurricane deck.

The *Shelby* was "barking" after us like a bloodhound from the slip. There was quite an expanse of water in this place, but, as I learned from the Kentuckian . . . the channel here was very devious and dangerous. The captain came to the Kentuckian's side with a flushed cheek, and asked,

"What do you think of it Samson?"

"If I had the strength of my namesake," replied the Kentuckian, "I'd swim out and chuck that boat, cargo passengers and all ashore; as it is, she's too fast for us, and I always knew it. I told you Bob Albert, the pilot there, has been on a bust this week past . . . Beattie's sick; and I saw . . . Albert was tight; he swore you shouldn't beat them if they blew everything up. I tell you, capting, it's my opinion they'll be into us; the channel's too narrow for them to pass us and they've got such a head of steam on and they are so much bigger than we are that if they come agin us we are gone!"

"Kentuck," called Rogers from the wheelhouse, "just step here a moment. You know the channel better than I do. I wonder what those rascals mean?"

The meaning seemed to be to my eye a resolve to run us down; the smoke ascended black and sulphery from her chimneys, with occasional flashes of volcanic fire. She gained on us . . . while the excited crowd on her hurricane deck and guards . . . hurrahed as, by the orders of the mate, they stepped to the center of the boat to keep her righted.

The noise they made . . . with the fearful trembling of our boat, for we had all steam on, too, so alarmed the ladies that, following impulse rather than reason for they would have been safest in the cabin, they hurried on to the hurricane deck, and the one I have spoken of rushed to Samson, who was now at the wheel, and begged him not to race any more.

"Kentuck," said Rogers, "they'll be into us—they mean to run us down—they must all be drunk there."

"Pretty much so," replied Kentuck, "Bob Albert was in for it early this morning; he's the only pilot on board; that is, Beattie's down with a fever mighty low—Bob hates your capting here, and when he's tight he's perfectly crazy."

"We shall all be lost—all be lost," exclaimed the young lady, "Oh! Mr. Old Kentucky save us!"

"Old Kentuck will do that, my dear, if he has to shoot that rascal at the wheel; they're bent on runnin' us down—self preservation's the first law of nature—if these fellows ram us it'll be a clear case of murder and they're hardly six lengths off. Hang it all . . . here Rogers take the wheel a minute and hand me my rifle—you see it's a necessity!"

"Don't kill him!" exclaimed Rogers, nevertheless complying with the request.

"Kill him? No I'll just break that right arm of his between his wrist and elbow, first time he shows it fairly."

So saying the Kentuckian lifted his rifle to his shoulder. We all felt our danger too much to interfere . . . the sharp report of the rifle was heard, all eyes fixed upon the pilot of the *Shelby*. In an instant his arm fell lifeless to his side . . . the *Shelby*, uncontrolled, rushed on to a shallow bar just beside her and in another moment was fast aground.

But strong drink and high-pressure steam were not the only explosives upon the Ohio River boats in the first half of the nineteenth century; politics was a danger which, though it did not often kill, was still likely enough to maim the spirit.

From the beginning of regular steamboat service the Ohio fed western and southern congressmen, big with hopes and schemes, into Pittsburgh for transshipment to Washington at the opening of congressional sessions and fed them back to their constituents, at adjournment, deflated or triumphant according to the degree of their failure or success in the capital.

It was probably to this powerful clientele that the Ohio first owed major improvements in clearing of channels and marking of menaces to navigation. Certainly it received a large share of the early "stump-pulling" appropriations, even in proportion to its importance.

The member of Congress whose boat was snagged and the senator whose vessel stuck on a sand bar were certainly willing to vote appropriations to prevent such assaults upon statesmanly dignity in the future—even when constituents were not immediate inhabitants of the river's banks.

In the fifties the Ohio connected through the Mississippi and its own tributary rivers and canals, including only negligible stagecoach jumps, with the capitals of Ohio, Indiana, Illinois, Wisconsin, Minnesota, Iowa, Missouri, Arkansas, Louisiana, Mississippi, Georgia, Tennessee, Kentucky—and even Michigan and Texas. In theory at least it was possible to convene a caucus of United States senators which might have guided the destiny of the nation aboard a hypothetical and particularly blesséd Ohio River steamboat.

In practice such a caucus was unlikely: senators often carried their feuds with them out of the capital and there must always have been the temptation to secure passage upon a boat uncluttered by other senators who might compete for the limelight. Indiana's monumentally dissolute Edward Allen Hannegan was probably the only specimen of his breed aboard the boat that carried him and the impressionable Mrs. Sarah Mytton Maury (whom he had already won at an "ice cream party" in Washington) upon an occasion concerning which she palpitated in print. She had already described Hannegan as "a true son of the West, ardent, impulsive" and neither term was an exaggeration, in any connotation. She continued:

A devoted lover of the country and of its independence, he so pined at last in Washington, that he was compelled to go home for a fortnight to refresh his spirits and recruit his health.

I met him on the Ohio on his way. "Come home with me," said he to the Doctor and myself, "come home with me, and I will show you the lovely valley of the Wabash. I can endure those hot and crowded halls no longer. I must have free air and space to roam in; I like to hunt when I please, and to shoot when I please, and to fish when I please, and to read when I please. Come home with me, and see how I live in Indiana." But we were bound to the Mammoth Cave, in Kentucky, and could not.

Although Mrs. Maury's quotation is florid enough to have been his very words, it does not ring quite true—the voice is the voice of Hannegan but the words, some of them anyway, sound suspiciously like the words of the British-born authoress. It is to be feared that she hastily assumed, in her effort to paint a sufficiently glowing picture of the chivalry and hospitality of a "true son of the West," that there was a difference, as in England, between "hunting" and "shooting." There was no such distinction in the Ohio Valley and it is not likely that the two phrases would have been used by even Senator Ned Hannegan: the fox was hunted afoot, in the vicinity, and usually with the assistance of a lantern.

In view of such romantic attentions it is unlikely that United States legislators traveled in groups, when single passage was available.

Nevertheless, they probably met in Pittsburgh, en route to the capital, where the Monongahela House dispensed hospitality which met the approval of even the exacting Charles Dickens, and proceeded thence up Laurel Hill and over the jouncing right of way of the stage road to Washington. What legislative intrigues may have been inspired by the select rye whisky, venison steaks, and wild turkey giblet gravy of the Monongahela House may only be surmised in this drab and degenerate age.

Steamboat travel as a pleasure in itself shall be separately treated; that phase of river transportation, except for the German-band-polka-and-bock-beer-picnic variety, was to die with the beginning of the eighteen-seventies in order to give space to a new and frenzied commerce in the transportation of basic commodities.

Shortly after the Civil War newly constructed steamboats began to feature the practical accommodation of payload over elegance; with the peak development of the railroads at the close of the nineteenth century only the most efficiently loaded boat could expect to pay its way. Then, between 1900 and 1915, river commerce almost died completely. When the revival came—but that, also, will be discussed elsewhere—it was with a new sort of boat which owed little to famed Robert Fulton but much to Captain Shreve and poor John Fitch and the screw propeller that had been one of the early inventions he demonstrated in public.

Chapter Eleven

THE TRAVELERS

AN OCCASIONAL specimen had early invaded the land but not until
the second decade of the nineteenth century did the great plague
of journal-keeping travelers descend upon the Ohio.

And "plague," in its Old Testament sense, is the word. Because of the reports of the country that appeared in the books they
inevitably produced, the descent of a fresh batch of these visitors
came to appear to the average resident of the Ohio Valley to have

all the unpleasant features of those disasters which had afflicted Job and Pharaoh's people.

The writers were of all kinds—from Thomas Ashe, the Irish swindler, and William Faux, "the English Farmer," to Karl Bernhard, Duke of Saxe-Weimar and Prince Maximilian of Wied—with Charles Dickens, Mrs. Trollope, Thomas Hamilton, Timothy Flint, and Captain Basil Hall in between—and they all traveled more or less on the Ohio River.

No other American stream was so unanimously favored. No matter what was the eventual destination of the traveler—whether he or she intended ultimately to go south to criticize slaveowners, north to visit British Canada, or west to see the red man in the flesh, the Ohio was generally utilized in some way. This is not strange when it is recalled that 90 per cent of the settled United States was east of the Mississippi up to 1850 and that, after a short overland jump from the principal Atlantic ports, the Ohio led directly to the south by way of the Mississippi or Tennessee, to the west by way of the Missouri or Mississippi and Arkansas, to the north by the upper Mississippi, Illinois, and Wabash rivers or by trails, roads, and eventually canals and railroads to points along the entire northern border on the St. Lawrence and the Great Lakes.

As taken by Europeans, the tour usually followed a popular pattern: A New York debarkation, overland to Philadelphia, Washington, and possibly Baltimore, by ship from Norfolk to New Orleans, by steamboat up the Mississippi and Ohio to Cincinnati or Pittsburgh, overland to the Great Lakes and down the St. Lawrence to the sea or down the Hudson to New York—with, of course, stops by the way. The tour could begin at any seaport along the route, could be traveled in either direction, and had, of course, endless variations of length and side trips. As is obvious, however, some reach of the Ohio formed a part of the way in any case.

The published journals were frequently curious documents. The majority expressed a highly critical view of the Ohio River towns, people, and economy, which made no allowance whatever for the newness of the country or the peculiarities of terrain and climate in which it differed from the authors' particular small patches of familiar homeland. Generally ignored was the fact that here, on the Ohio, a new democratic principle was being tried out in a virgin land set apart from the eastern states and lower Mississippi country, where democracy had been superimposed upon the foundation of a century or two of monarchical European rule. The Ohio country, for all practical purposes, had never been other than free soil, as far as whites were concerned; and if its new citizens found democracy and the break from tradition a bit heady, no thinking person should have been surprised.

The chief thing foreigners failed to appreciate was the vast difficulty of opening a wilderness and the marvelous speed with which it was being accomplished in the Ohio Valley. They heard stories of the country as it had been a generation before, but they had no comparison, then, by which to measure the term "wilderness"; no comprehension of a future development as illimitable as here promised. The nearest thing to the development of the Ohio Valley in European history had probably been the boom in ancient Gaul after Roman conquest—and that had been so many hundred years before that to the nineteenth-century tourist it was represented only by a few dates in history, some changes on maps.

Most of these travelers misunderstood the people as they did the past and future of the country they saw. Many of them—especially the British—evidenced a provincialism in their relations with the citizens of the river which would have shamed an illiterate native wood-lot operator. After harping upon "lack of taste; of feeling" which they discovered among the Midwesterners along the route, they almost invariably manifested an intolerance of the well-meant but sometimes awkwardly proffered hospitality of the settlers that itself set an all-time low in "taste" and seemed entirely devoid of "feeling." Just what the term "feeling" signified to the Englishmen of 1830 to 1840 is not clear anyway, but by their account Midwesterners always lacked it.

As an example of the typical traveler and his host, let us consider the fictitious—but by the record entirely plausible—relations of Squire Smith of Smithport, Ohio, and Ponsonby Ricketts, the British journalist:

Squire Smith, we'll say, hailed from Westmoreland County, Pennsylvania. His father mended shoes for a living and preached on Sundays; he farmed forty acres and taught his children to read, cipher, and write a fancy hand, with illuminations of birds in flight and scrollwork by way of decoration.

The elder Smith had books—*Pilgrim's Progress, Robinson Crusoe,* Fox's *Martyrs, The Patriot's Monitor,* and several others— and he always made it a point to give each of his young ones a book on his or her birthday. He couldn't get quite the kind of books he wanted and knew must exist, because books cost cash money and the stocks of the Westmoreland County merchants and the peddlers were somewhat limited; but he always told the children that better books were to be had, which they might sometime buy for themselves.

Well, Daniel (that was the boy who would become Squire Smith later) grew up and went to Pittsburgh, where he worked his way through two terms at the grammar school. Then he went home, sold the heifer he'd raised, and got a job as a hand on a flatboat in order to work his way down the Ohio.

Daniel got to Marietta and found work right off. That fall he took the ten dollars he had saved, walked up the Muskingum a piece, and found a school which needed teaching.

Three years of that, with farm-hand work between terms, not only put together a nest egg of more than two hundred dollars but gave him plenty of time, evenings, to "read law" surrounded by the overwhelming majesty of the local lawyer's library—more than fifty books, there were, in all. He walked down to Marietta, passed the bar examinations, and was admitted to the full-fledged practice of law.

Dan Smith opened an office in Marietta in the other room of his house—he needed a house now, having a wife who had come out from Pittsburgh and married him during the second year he taught school. He began to get some law cases right off.

They were mostly land cases—for people's land was pretty much mixed up as to title, what with purchases from the Ohio Company, or buying soldiers' land warrants in the Virginia Military Reserve and swapping with squatters and one thing and

another—but most of the litigants paid something in money, produce, or land. The last kind of payment accounts for the way
Smithport came to be laid out.

Lawyer Smith straightened out one client's claim, an old fellow who had fought in the Revolution. He'd been trying for
thirty years to get a clear title to the land that Virginia had given
him on the north bank of the Ohio on account of his having
fought with General Washington and lost a leg doing it and
when Lawyer Smith got it fixed up the old fellow was mighty
grateful. He told Smith he'd pay him in cash or give him a third
of the land, right back from the river. Smith took the land.

Pretty soon Lawyer Smith went down to the tract, found it to
his liking, and decided to lay out a town and sell lots. Why not,
it was a good, high, healthy location and everybody else was doing
it? He laid out the town and people bought lots: wasn't long
until it was a right promising place, big enough to need a lawyer,
even. He sold his house in Marietta and moved down.

Once he had sat as a special judge in Marietta, and besides he
still owned quite a few of the lots in Smithport and had an
interest in a steamboat, so naturally he became *Squire* Smith and
began to buy all the books he wanted, even modern ones, like
novels, that no one *had* to have.

He read his books, too, and enjoyed them and passed them
around to his friends. When, in 1828, the newspapers announced
that the immortal Ponsonby Ricketts, one of the squire's favorite
authors, was planning a trip to America where he intended to
penetrate the interior and see how people really lived, the squire
wrote a respectful letter to him in London and told him that if he
came down the Ohio and had time, he and Mrs. Smith would be
delighted to have him stop in Smithport as their guest for as long
as he liked.

The squire was pleased and flattered when Mr. Ricketts wrote
from New York that he would stop. The squire, with pardonable
pride and excitement, told a good many people about that letter.

At receptions in his honor in New York and Boston and Washington, Ponsonby Ricketts met a great many people who had long
admired his works, who said frequently that he was the only

author with whom they found themselves completely en rapport and who thought he had such a feeling for the finer things. These were mostly ladies, but there were some gentlemen with their whiskers parted in the middle who felt pretty much the same way, and said so. Ponsonby did not think much of his admirers of either sex, and admitted it in the letters he wrote back to London for publication in the *Weekly Tuppence Mirror of What Is Proper.*

Presently, after the more lucrative lecture dates had been filled, Ponsonby announced that he would now leave these metropolitan scenes of imitation Old World glitter and go to The West, where men were even then said to be men and where democracy flourished. He was weary of the counterfeit intelligentsia, he said, and surfeited with its ignorant flattery.

So Ponsonby took a stage from Philadelphia to Pittsburgh, over the mountains and through other hazards. It was a harrowing journey: there was a *swineherd* on top of the coach (called a "stage" through some unaccountable American foible) and an innkeeper inside! Riding with the gentry! Imagine!

He boarded a steamboat at Pittsburgh. It moved downstream quite rapidly; but it had only two classes of accommodation, cabin and deck, and everyone ate at one table and with disgusting speed, as if they were hungry! There was venison, bear, pork, beef, squirrel, chicken, turkey, fish, and ham, and it was well enough, one supposed, but some of the bread, a serving man informed Ponsonby, was made of Indian corn! Fancy!

Lecture stops were made at Wheeling and Marietta. People crowded round and shook hands, endlessly. They said they had read Ponsonby's books! One man, particularly, said he had read some of them twice. Ponsonby was later informed that the man was a *blacksmith;* an artisan! *Read!* There were lesser stops too, but at Jonesville (a dismal place with only one inn) there were only 342 paid admissions while *Darby's Gazetteer* gave the population of the place as 388; and there was a pewter fi'penny bit in the receipts at Brown's City! The Americans have no regard for the finer things of life—only money, money, money!

Presently Ponsonby arrived at Smithport. Squire Smith's

"carriage" met him at the landing—and such a carriage! Upholstered in cowhide and actually driven by Squire Smith! Ponsonby
could not help fancying what second cousin twice removed Lady
Euphemia Ricketts would have said, had that equipage wheeled up
to Ricketts Towers! Clever woman, cousin Euphemia, of the
branch of the Ricketts family that sprang from the night the king
visited Ricketts Manor in 1678.

Not only did the squire himself hold the lines but upon the
seat by him sat *Mrs.* Smith! Of course the poor soul meant well,
but she stepped from the "carriage" and extended her hand and
shook! Well, Ponsonby could scarcely restrain his mirth. She
shook!

But there was no place else in Smithport to stop, where there
was no charge, at least, so Ponsonby allowed the squire to load his
boxes into the carriage, except those which wouldn't go in and
had to be transported by a carter. (Called the squire "Dan," the
carter did! and the squire did not even summon a constable!)

The squire said that the house was Ponsonby's but obviously he
exaggerated, as did all Westerners, because when Ponsonby suggested that the Berkheimer girl, whom Squire Smith referred to as
"the help," be sent out for a pot of porter, the squire said he would
rather she didn't go because she was a nice girl and had never
been in a tavern. Fancy! The squire walked down and got it
himself!

Ponsonby stayed a week and the Smiths did their best to make
him happy; even got the Berkheimer girl to boil the leg of lamb
she'd started to roast—though she said she considered boiling good
lamb unchristian and went home that night and did not reappear.
Ponsonby seemed a little queer about some things but the squire
said to his wife, "Other lands, other ways." Mrs. Smith's colored
woman couldn't do Ponsonby's shirts to suit, and they weren't
quite right after Mrs. Smith had done them over herself, but on the
whole she and the squire thought they had entertained him pretty
well; and when he finally left he said he was "really quite
obliged!" That convinced them; the visit had evidently pleased
him!

That's what they thought until a year later when the squire

went to Boston on business and bought a copy of Ponsonby Ricketts's new book, *Democracy in the Backwoods; or Seven Days in Smithport.* After he had read it the squire showed his true colors as an ignorant backwoodsman with the characteristic American hatred of the British that dated back to the sound trouncings his Majesty's army had administered to the colonies in 1776 and 1812, before the British cast them off completely! The squire threw the book in the fire and aimed a kick at the cat, which had rather taken to Ponsonby during his visit.

Now of course neither the squire nor Ponsonby nor Mrs. Smith nor the Berkheimer girl actually lived and breathed under those names but they will certainly serve as composite figures; and for every little peculiarity exhibited, *actual* characters among the travelers and their hosts displayed one ten times more startling!

As has been said, the backgrounds of the travelers—European *and* American—were highly miscellaneous, their real reasons for traveling sometimes curiously at odds with the ones they alleged in the reports they published, and their motives in making the tour about evenly divided between the philanthropic and the highly materialistic. Much of what is pertinent of the things they said of the river and its people is reported elsewhere in its proper chronological place and the full text of the works of the more celebrated is readily available, but who some of the more intriguing of the personalities among them were, whence they came and why, seems worth a little attention in itself.

For instance, there was Captain Gilbert Imlay, born in New Jersey about 1755, a British officer in the Revolution, who came to Kentucky as a commissioner connected with the survey of new lands after the war. In London by 1792 he published his *Topographical Description of the Western Territory of North America* and followed it with a novel attacking "present practices of matrimonial engagements." He went to France and, evidently in keeping with the views expressed in his novel, lived for a time with that extremely liberal intellectual, Mary Wollstonecraft, and left her and their daughter, Fanny, for good in 1796. Mary Wollstonecraft's letters to Imlay are published in her works; she tried to drown herself at the time of their separation and, by one evidence and another, it is apparent that he made quite an impression upon the blue-stocking lady.

Imlay disappeared soon after the separation, but Mary took up residence with William Godwin, married him just before the birth of another daughter, Mary, and died shortly after that event. Had Imlay's affections been more constant, he could as well have fathered this daughter, also, and thus have become father-in-law to Percy Bysshe Shelley.

Thomas Ashe, who called himself Arvil in America, was the possessor of a background entirely appropriate to his picaresque career as world traveler, pseudo scientist, and defaulting agent: "Third son of a half-pay officer . . . received a commission in the 83rd regiment which was almost immediately afterwards disbanded . . . sent to a counting house at Bordeaux . . . suffered a short imprisonment for wounding in a duel a gentleman whose sister he had seduced. . . . Returned to Dublin . . . appointed secretary to the Diocesan and Endowed Schools Commission, but, getting into debt, resigned his office and retired to Switzerland . . ." (One may wonder if T. F. H., writing in the *Dictionary of National Biography* did not mean "fled" rather than "retired.") The best known work of Ashe, *Travels in America,* published in London in 1808, gives a highly colored account of adventures he may or may not have had on a grand tour of America from the Mississippi east.

It is a smaller work, *Memoirs of Mammoth and Other Bones Found in the Vicinity of the Ohio,* published in 1806, that connects him with one of the shabbiest of those many little frauds perpetrated on the inhabitants by travelers in the West.

The Big Bone Lick of Kentucky had long been known. It was considered one of the wonders of the western country and Dr. William Goforth, a Cincinnati physician with an inquiring mind, had made a careful survey of the place and assembled a fine collection of bones including, by report, the skeleton and tusks of a mammoth nearly complete enough to be articulated easily. Thomas Ashe called on Dr. Goforth during his travels; and the doctor, probably flattered by the attention of such a distinguished scientist as Ashe represented himself to be, and eager to present his find to the scientific world, "loaned" the bones to him for study. Upon his return to England Ashe sold the bones profitably to the Royal College of Surgeons in London and to private collectors throughout the British Isles—without profit, needless to say, to Dr. Goforth.

Morris Birkbeck was a traveler—or rather an emigrant and investor—of a very different stamp. Born to a prominent Quaker family in Settle, England, he became impatient of his lack of privilege as a British citizen and came to America in 1817. He and his old friend, George Flower, secured a large tract of land in Illinois (on the Wabash above New Harmony, Indiana), which they sold to English farmers and tradesmen, chiefly through the agency of Birkbeck's two books, *Notes on A Journey in America From the Coast of Virginia to the Territory of Illinois,* 1817 (which went through eleven editions in English and one in German in eight years) and his *Letters From the Illinois,* 1818, almost as successful.

Unlike that of many of his fellow promoters, Birkbeck's land was worth the selling price and the emigrants he attracted were mainly substantial citizens. They prospered and left an indelible mark upon the lower Wabash, being chiefly responsible, incidentally, for keeping slavery out of Illinois.

Birkbeck's patent enthusiasm for the new country made him the butt of many a slighting remark in the works of the British journal keepers who conformed to the usual pattern; and the reports on immigrant successes in the West incorporated in his *Letters From the Illinois* brought forth a rash of volunteer writers of books purporting to give accounts of immigrants who had *not* succeeded and of the crudities of America that had brought about their failures.

Captain Donald Macdonald was another "half-pay" officer in the British army. (The annual stipends of only those half-pay British officers who toured America in the twenties and thirties must have been a serious drain on the royal treasury; and it may be presumed that *some* half-pay officers toured the Orient or stayed at home during those two decades.) Macdonald came to the United States as a sort of secretarial assistant to Robert Owen in 1824.

It is his minutely detailed diary that gives the only day-to-day account of the activities of the indefatigable Owen upon his first

visit to the Ohio and Wabash rivers for the purpose of buying the
Harmony Community from Father George Rapp and setting up
New Harmony as the hopeful nucleus of the New Moral World.
Macdonald was an uninspired but agreeable young man whose
most remarkable characteristic was his attention to detail.

Unfortunately his accounts of towns, people, and traveling
conditions in the West are probably almost as untrustworthy as
are some of the deliberately derogatory reports. Macdonald was as
starry-eyed as the visionary Owen; to him all he saw was bright—
he even found that, with all the visible defects of Shawneetown, the
innkeeper "was very attentive" and had all the latest papers.

While the fact is not stated in so many words, Captain Basil
Hall must also have been a "half-pay" officer—although of the
British navy rather than the army.

Born in 1788, second son of Sir James Hall, Bart., he joined the
navy in 1802 and his travels as a private citizen—they covered Asia
and North and South America—resulted from an interest aroused
while on duty in foreign waters.

He was anxious to avoid "mentioning circumstances or even
making allusions, calculated to give pain," according to his statement in *Travels in North America in the Years 1827 and 1828,* but
the book is as full of picayunish complaints as any pre-Trollope
writing. Hall, unlike many of his fellow travelers, was a scientific
observer of some distinction and was above distorting facts, though
he could and did present them in their most unfavorable light.

Captain Hall's great contribution is not to be found in his
writing, but in his drawings of scenes and people in the American
West and South. He had with him on his American tour an ap-
paratus called a "camera lucida," which, through an arrangement
of mirrors and lenses, cast a reflection of any scene on which it was
focused upon a sheet of paper, where its outlines could be accurately
sketched in pencil by anyone with moderate talent for drawing.

He made hundreds of such sketches, all photographically accurate, of towns, buildings, boats; whites, blacks and Indians; rivers,
hills, and forests. A few were reproduced and published in 1827
under the title *Forty Etchings, From Sketches Made With the
Camera Lucida, In North America* and the entire collection of

several hundred drawings embodying the American frontier scene in most minute detail is now preserved in the Library of Indiana University.

Captain Hall lost his mind in 1842 and died in an asylum two years later.

Most cordially hated—and most widely read—of all the British visitors was Mrs. Trollope, mother of Anthony. While she had in the person of her husband, Thomas, an agent sufficient to turn any matron's blood to vinegar, even her British contemporaries agreed that her criticism of the citizens on the Ohio went a bit on the acid side.

Thomas Trollope had been educated for the law, but had failed at it because of "his irritable temper"; he had tried farming and failed at that, cause unknown but suspected. In 1827 he and his wife (he fifty-three; she forty-seven) listened to the glowing reports brought back to England by spirited, erratic Fanny Wright, herself a long-time Ohio River traveler in the interests of her various campaigns for social reform. Apparently these conversations convinced the Trollopes that a complete lack of experience was sufficient qualification to ensure success in the mercantile line on the Ohio. The venture was to take the form of "a bazaar for the sale of fancy goods," according to Mrs. Trollope's statement. She was to operate the business while Thomas followed other pursuits.

She came to Cincinnati, bringing with her the numerous Trollope children and a young British artist, and since no building in the town could possibly do justice to the scope or spirit of the business about to be established, she erected a special edifice to house it. This building was generally agreed to be the most loathsome architectural hodgepodge on the North American continent of its day; and judging from pictures of it which are preserved, it probably would have held its own even in competition with the architectural wonders of the President Arthur period, had it withstood the ravages of time on shoddy workmanship and unsuitable materials.

Trollope's Bazaar was opened for the convenience and improvement of the people of Cincinnati; and, allowing little time to accomplish these worthy purposes, it failed. The Trollopes returned to England.

No failure was the book Mrs. Trollope later wrote about the nature and foibles of her American neighbors: her *Domestic Manners of the Americans* must have outsold most of its contemporaries two to one. The British bought it because it bore out their conceptions of American character, and Americans bought it to see if it was really as libelous as it was reported to be.

Meanwhile Thomas tried real estate investments in London and—surprise!—he failed again. His final failure came in 1835 with his death "from premature decay, partly induced by an injudicious course of medicine," according to his sober biographer.

Though she might have been justified in so doing, it was probably not Mrs. Frances Trollope who administered the medicine; in her writing she exhibits only the most profound respect, even veneration, for Thomas.

After the death of Mr. Trollope the family seems to have begun to prosper almost immediately. Mrs. Trollope began to produce novels which enjoyed a wide sale in spite of their almost universal lack of merit and she lived to see her son Anthony father a literary output of stupendous volume, which is now enjoying a limited return to favor.

Frances Trollope's *Domestic Manners of the Americans* infuriated the dwellers on the Ohio as nothing had before nor has since, and with considerable cause. As residents of a new and raw land they were hypersensitive, but that phenomenon alone could not account for the bile aroused. Only a few years later "Doesticks" (Mortimer M. Thompson) was a popular humorist on the river because he could write such pieces on Indiana and Kentucky as:

Have got over on the Indiana side, principal difference to be noticed in the inhabitants is in the hogs; on the Kentucky side they are big, fat and as broad as they are long: on this side they are shaped like a North river steamboat, long and lean.

I just saw two of 'em sharpen their noses on the pavement, and engage in mortal combat; one rushed at his neighbor, struck him between the eyes, split him from end to end . . . This is decidedly a rich country; the staple productions are big hogs, ragged niggers, and the best horses in the United States. The people live principally on bread made of corn, whiskey ditto; and hog prepared in various barbarous ways. They give away whiskey and sell cold water . . .

But Mrs. Trollope's day, her nationality, and her vitriol made her remarks something else again!

The Ohio country of the Trollopiad was wild enough and crude enough, in places, but it made quite a different impression upon the traveler who could realize that the wildness was a comparative matter and that the crudity would pass. The Rev. Mr. John A. Clark, rector of St. Andrew's Church in Philadelphia, for instance, made the passage from Pittsburgh to Cairo sometime in the thirties. While he was appalled at the drinking, swearing, and gambling, he did not let these objectionable features shade the wonder of the river's development and the beauty of the scene—wherein he differed from the Trollopian school, which preferred to see first the shadow.

In his little-known book, *Gleanings By the Way,* published in 1842, he gives one of the best accounts of the physical appearance of the river in the thirties. Leaving Pittsburgh—

Our backs turned upon the clouds of smoke that hung in dense masses over what has been called the Birmingham of America. As we stood on the deck, we seemed at the moment of starting enclosed by a forest of dark funnels peering up from countless steamers lying along the shore. More than forty of these were clustered together in the same group . . . It is said there cannot be less than seven hundred steamboats moving on these western and southwestern rivers.

We were fully in the stream!—we began to feel that we were borne on the flowing bosom of the Ohio! . . . We thought of the vast territory it watered—its majestic length—the scenes of Indian warfare that had been enacted upon its shores and on its surface, long before the axe of the white man had felled a single tree in those vast and unbroken forests that stood upon its banks, and were reflected from its mirrored surface. It was even then *the beautiful river,* as the name *Ohio* denotes. It is said that "the line of beauty" is not a straight but waving line. If so, this river is richly entitled to its name. From first to last it moves in "the line of beauty." So winding is its course that we usually do not see, as we are passing along upon it, more than half or a quarter of a mile

in advance . . . Thus we see it in distinct sections, each section resembling a beautiful little lake, surrounded by its own sweet and particular
scenery—shut in by its verdant and variegated banks and wood-covered
hills, and ornamented by one or two, and often several little green
islets, around which the parted waters wind romantically.

Below Wheeling the good parson found himself growing still
more expansive in the Ohio sunlight:

I know of nothing more delightful than to sit at one's ease, and be
wafted down such a beautiful stream as this, winding its graceful and
circuitous way through groves and grass covered fields, and beauteous
woodland scenes. Occasionally we see the banks surmounted with lofty
bluffs that lift their proud summits up towards the clouds—and then
succeeded by bottomland studded with trees that bend over to dip their
pendent boughs in the glassy surface that sweetly reflects them . . . One
acquires as he proceeds westward, largeness and expansion to his ideas:
his mind is carried out of its former habits of thought, and swells away
into the vast dimensions of the majestic rivers, and boundless tracts of
country, over which the eye expatiates . . .

Different, this, from that western journey of the Rev. David
Jones sixty-odd years before! Yet there were still great tracts but
thinly settled, even in Clark's day. His steamboat laid up on the
Kentucky shore in the rocky passage below Maysville for the funeral of its owner, brother-in-law of the captain:

Most of our passengers having landed, the coffin was brought out
from the boat and conveyed toward a cottage . . . The first thing that
attracted our attention was the number of horses fastened to the fences,
and equipped most of them with ladies' riding saddles.—Around and
within the house we found a large company assembled. I was sorry to
see so many rotund and rubicund faces among the men, bearing un-
erring indications of intemperance. The fair daughters of Kentucky
were certainly on this occasion more happily represented than the
stronger sex. They were, however, very peculiarly dressed. They generally wore a sun-bonnet, which had a long frill or flounce that hung
like a shawl over their shoulders, and carried in their hands little
riding whips . . .

No local minister was available, so the Rev. Mr. Clark performed the service when that lack became evident. He found the manner of conducting the obsequies of interest:

The funeral began to move off in the following order or rather disorder. First, the four bearers took the lead, carrying the coffin on two rudely hewn sticks, prepared for the occasion. Then followed four or five of the near relatives, all abreast. Then came the bereaved widow, riding on horseback, and after her all the assembled crowd, male and females, hurrying on twelve or fifteen abreast of each other . . .

Below Louisville Clark again found time to relax and enjoy the passing scene:

When the sun began to decline and we again found ourselves gliding as by enchantment over the surface, and sweeping through the midst of the beautiful scenery of the Ohio, I felt that I had passed into a new world. As I traversed the deck of the boat, and saw reflected from the smooth and mirror-like bosom of the river, the luxuriant foliage, rich and dark by its own deep verdure,—the smooth green bank that sloped down to the water's edge, as though to kiss the smiling surface that slept so quietly below—the abrupt precipitous bluff, starting up like a mound of earth or a wall of solid masonry— and the head-land sweeping off into sloping woods that towered in majesty above the stream, I could not but feel, and could scarcely refrain from exclaiming aloud, how beautiful and surpassingly lovely are the works of God!

While we were leisurely sailing along today, the weather being oppressively warm, and the heavens very bright and sunny, and not a breath of air stirring, pyramids of snow-white clouds began to pile up in the northern and western sky. These masses of cloud seemed heaped together in every fantastic form. They towered aloft like huge mountains of snow . . . Through these masses . . . there occasionally appeared large interstices, like deep caverns, opening into the blue profound!—long vistas through which we could seem to catch a view of the inmost heaven. Suddenly a tremendous gale struck us; the waters of the calm Ohio were thrown into the utmost commotion, and the wind came down upon us with a power that threatened to shiver the steamer into a thousand atoms. The heavens gathered blackness, and the whole dark firmament presented a surface every now and then lit up with a sheet of the most vivid fire. The waters ran very high, the wind roared, and

the thunder was awful. . . . Then the rain fell in torrents, as though the
waters of the river itself were scooped up and poured upon us . . .

We stopped toward evening to take in wood on the Kentucky
shore. We there saw for the first time the native cane-brake. A wood-
cutter's hut was near. A little ragged boy came out followed by two
large dogs, and a little pet fawn. The dogs seemed to be very fond of
this innocent little thing . . . It seemed as it skipped along, and played
around the footsteps of the child, very affectionate and confiding . . .

Just at nightfall we passed the steamer Louisiana in distress. She
had run upon a reef of rocks and was in a sinking state . . .

Clark's boat passed Paducah; and the mouths of the Cumberland and Tennessee rivers both appeared, as they do to travelers
today, large far beyond the expectation:

The Ohio here, having received its last large tributaries, had
become very deep and broad. Its banks were covered with tangled
underwood, and dense forest trees—presenting a scene of unbroken
wilderness. Now and then a woodsman's hut was visible on the shore,
and a little boat fastened to the bank . . .

Clark noted the difference in color between the comparatively
clear Ohio and the very muddy Mississippi, but mainly, since
Louisville, he had been preoccupied with dirt of another kind: that
in the conduct of crew and passengers.

There was no question about it: steamboat life below Louisville
was more sinful, even though more elegant, than upon the upper
river, and the quick eccleciastic eye of the reverend gentleman was
caught and fascinated by the goings-on:

Almost all the hands on board of steamboats, down even to the
little boys, utter an oath almost every other word. *Profane swearing* is
one of the crying sins of this western world. Oaths the most horrid are
awfully common among all sorts of people. Amid these scenes of varied
beauty, where creation appears so lovely we may truly say,

> . . . Every prospect pleases
> And only man is vile.
> In vain with lavish kindness
> The gifts of God are strewn.

I have already spoken of the annoyance . . . from the profanity of those we encountered. And I may now add that gambling is another of the vices that are rife here. On our way from Louisville to St. Louis there has been one incessant scene of gambling night and day. We have evidently had three professed gamblers on board. I am told that there are men who do nothing else but pass up and down these waters, to rob in this way every unsuspecting individual they can induce to play with them of his money. . . . Another crying sin, which abounds on board the western steamboats, and is fearfully prevalent through every portion of the western region, is *the free and unrestrained use of ardent spirits as a drink;* usually on board these western steamboats whiskey is used just as freely as water. All drink. The pilot—the engineer—the fireman—all drink. The whiskey bottle is passed around several times a day, and then the dinner table is loaded with decanters. I am satisfied that more than two-thirds of the disasters that occur on board these steamboats, are attributable to this free use of ardent spirits . . .

Charles Dickens came to the Ohio in 1842, the same year that the Rev. Mr. Clark published his book.

Dickens's object in visiting the United States was partly rest and a change of scene but largely to see what could be done toward establishing an international copyright agreement which would protect writers, particularly British and American, from the piracy of booksellers and publishers.

Dickens had suffered particularly from this, since he was more widely read than any other author to his day. Cooper and Irving were the Americans most popular in England; and they, too, suffered, as did every writer including Samuel Langhorne Clemens. Forty years later Clemens practically hung on Queen Victoria's skirts to further Dickens's aim—and with much better success. A reasonable beginning toward international protection was finally made by Congress in 1891 and was put in practical form in 1909.

On his tour in the forties, Dickens fell an immediate victim on the eastern seaboard to the American tendency to worship visiting celebrities. This American passion for overstrenuous welcome must have been an unpleasant trait, no matter how well meant, and it continued down to, and perhaps reached its height, with the public asininity that marked the triumphant tour of the party of minor

royalty led by Queen Marie of Rumania in the nineteen-twenties.
There is ground for hope today that it is wearing itself out on
domestic movie stars and reigning crooners and that it is beginning
to confine itself more exclusively to the very young and the otherwise unemployed.

Dickens's reaction to this worship, together with his disappointment as a young liberal that *all* social abuses had not been
immediately checked by American democracy, is said to have been
the cause that resulted in a disgust with the United States and its
institutions which continued throughout his life. However his oft-
repeated comment in *American Notes* upon the "dullness" of his
fellow passengers on the two steamboats that carried him from
Pittsburgh to Cairo, his inference that the only one of them who
evinced any intelligent interest or familiarity with British writers
was a Choctaw chief who invited him west to write of the tribe,
suggests a possible additional cause for his disappointment; could
it be that the great Dickens was a bit disappointed by the failure
of valley travelers to recognize his famous person?

Travel on the Ohio by foreign diarists either fell off after
Dickens or, in the glorious steamboating days of the forties and
fifties, diarists found so little of which to complain that their writing
was colorless and unsalable. For those Midwesterners whose dander
rose at the very thought of a journal-keeping traveler, there was,
after Dickens, little cause for irritation.

Foreign—and what to dwellers on the Ohio was as bad, New
England—travelers still used the Ohio; but their comments were
limited mainly to the cities where they stopped, to the vulgar
display of wealth by their fellow passengers, to the gustatorial feats
accomplished at table and bar, and to the gambling on the boats.
The chief circumstance that saved the Ohio from further caustic
comment was probably the fact that it was no longer the West.
The West, now, was across the Mississippi; and travelers who
passed down the Ohio toward it felt bound to save their derogatory
adjectives for the new marvels of crudity they expected to find
ahead.

Chapter Twelve

PLEASURIN' THE BACKWOODS

INFORMAL homemade entertainment by whites certainly came to the Ohio with the earliest explorers—especially if they happened to be French.

One may be certain that Robert Cavelier, Sieur de La Salle, had in his little party (if, indeed, La Salle *had* such a party) which visited the river (if it *did* visit the river) at least one half-breed Canuck boatman who could sing a ballad or two for his fellows as they squatted around the campfire after supper had been eaten and the pot rinsed.

Certainly, if La Salle failed, the first French party that did reach the Ohio had such an entertainer, and the traders and boatmen who followed from New France made sure that at least one such artiste was included in the crew. Soon such popular fellows began to carry with them crude flageolets when they signed for a long voyage in the wilderness (that was about the only musical instrument light enough to make no difference in the delicately balanced bateaux) which could be played to inspire those leaping, heel-cracking dances for which French woodsmen were to be famous for generations to come.

Presently the gourd fiddle was introduced and continued to serve isolated or impoverished virtuosos well into the nineteenth century. Constructed upon the general principle of the violin, it substituted a large dried gourd for the conventional body and was played by a bow which was exactly that—a bow made of hickory and strung with a single thong or length of deer gut. Whether the frontiersmen ever found an entirely satisfactory substitute for professionally made strings is to be doubted, since much was made of the capture of a boatload of French traders who had violin strings among their goods which were appropriated to enable George Rogers Clark's slave, Pompey, to string his fiddle and play a Christmas program for the dog-tired soldiers of Clark's western army.

The dulcimers of the southern mountains—mainly women's instruments—may be supposed to have arrived with the first immigrants to Kentucky from the Virginia and Carolina back country and doubtless they were played in the long, locked-in evenings in the forts at Boonesboro, Harrodsburg, and Ruddle's Station. These instruments had come to the south Atlantic coast with early colonists (one would scarcely look for dulcimers to the north, among the Puritans) and are still played today by the people who were left marooned in the mountains when their cousins moved on through the Cumberland Gap.

Vocal music always followed the explorers, though it was more common to the people of some cultural backgrounds than to others. The French, again, were probably the first upon the Ohio to make much of their singing. Earliest travelers who wrote of New France—

after the days of Hennepin, Joutel, and those others whose interest was solely in the wonder of discovery of this new vast land—mention the songs of the boatmen. Those they sang while at work during the day had a dual purpose; to synchronize the paddling or poling and to pass the time and lighten the terrific labor. A good leader of songs in each bateau or canoe was fully as important as an adequate supply of the scant rations served up twice a day; and a lively and inventive singer—even though he might shirk his paddling a little—was always in demand.

The boating songs mostly followed the same pattern; a few lines, preferably topical and relating to the lusty entertainment waiting at the end of the voyage, to the foibles or peculiarities of individual crew members or to men and women they knew at home, were followed by either a chorus of the last line or two repeated or a nonsense pattern. Such nonsense choruses now survive in children's games and in the play-party of those mountaineers who worship a jealous God complacent of playing but dead set agin dancing.

Such a bateaumen's song, for instance, might be lined out extemporaneously by the leader, with Joe, the second paddler on the left, as the subject—

> *Old* Joe Boisblanc *he's* a mighty *li*ar
> Has a *wart* upon his *u*gly nose!

—paddles dig the water on the accented syllable, make the stroke, recover, and dig again on the next accent. The chorus takes up the refrain:

> Wart upon his ugly nose,
> Wart upon his ugly nose,
> Old Joe Boisblanc's wart
> Upon his nose!

Everyone joins, including Joe, who is meanwhile racking his brain for a verse which will ridicule some peculiarity of the leader or his ancestry.

Other songs of the French were traditional ballads and dance tunes, imported from the homeland but corrupted by word-of-mouth repetition and the admixture of Indian and New World words and interpretations. Some of those which survive can be traced to seventeenth-century European sources, even as can those of the hill and river people of British Isles origin.

If the Irish employees of the Pennsylvania traders on the Ohio sang songs, they were probably old-country ballads—possibly some naughty. If Gist, young Captain Harry Gordon, the Rev. David Jones, or George Washington wrote down the words of the latter it was with private repetition in mind, not publication. Scotch-Irish, on the other hand, ran to psalm tunes.

The Dutch, a people less spontaneously musical than French or even English, left no discernible mark upon the Ohio; and the Spanish, though their songs were extremely popular among the Creoles of Louisiana, did not penetrate to the Ohio in numbers sufficient to exert much influence. The early Germans upon the upper river—westward fringe of the so-called Pennsylvania Dutch—richly endowed the valley with their music, as did the Germans who emigrated in such numbers to Cincinnati, Louisville, and southern Indiana in the middle of the nineteenth century.

Not only songs and styles of singing them were included in the German contribution, but many an elderly Ohio Valleyan must recall as a pertinent feature of the steamboat excursions and river picnics of his youth the blare, blast, thump, and tootle of the German band, as essential to those occasions as the steamboat and the refreshments. And German bands were not even new upon the Ohio then: the Rappites lived to the measure of a brass band at Harmonie and Economy, Pennsylvania, and Harmony, Indiana, in the earliest quarter of the nineteenth century—though it must have been given only to staid and doomful tunes, with few tootles.

The settlers from Virginia and the Carolinas brought ballads as well as dulcimers and theirs are almost the only ones of the early songs which are still sung, here and there, for popular entertainment rather than self-consciously as exhibition pieces by impassioned delvers into folklore.

Among these were the songs of "Barbary Allen" and her woebegone sisters, and many a brave account of the doings of ladies fair and young sprouts of the nobility. Brought to America straight from Jacobean England, these songs sometimes underwent queer manipulations of their wording through word-of-mouth repetition in the new land. There were dance and taproom songs as well, and soon all of them contributed their themes, tunes, and lyrics to the lusty war songs of 1812. Most composers borrowed tunes, as did Francis Scott Key when he wrote "The Star-Spangled Banner" and a great many used whole phrases of older lyrics in these and the campaign songs that grew in a bumper crop when Old Tippecanoe Harrison ran for the Presidency in 1840.

By 1840 the Ohio Valley had also begun to grow its own songs, under the expansive influence of an era of comparative prosperity which left time for such activities: after that came Stephen Foster with his sweet romantic pieces and his lively tunes for the minstrel men; the sad songs of the Civil War and, in the twentieth century, the development of Louisville jazz which in its early days (the 1920's) was as distinctive as New Orleans barrelhouse and tail-gate, or "Chicago style," but not so widely appreciated. All flowered and left their marks on the musical culture of the Western Hemisphere.

The fact that there is little new in the twentieth-century developments in "hot" jazz becomes apparent after a bit of research into the archives of the fifties; the more complicated forms as illustrated in the treatment of Bizet's *Carmen* which reappeared so successfully as *Carmen Jones,* the oratorio and recitative passages in *Show Boat,* the jive talk with which Cab Calloway made his fame, and the semirecitative lyrics which W. C. Handy, the maestro, used in his time-tested "St. Louis Blues"—Mortimer M. Thompson seems to have heard a performance on the Ohio which included the genesis of them all!

In his book *Doesticks, What He Says,* published in 1855, Thompson gave an account of a ride from a point a short way up the Kentucky River to the Ohio and across to the Indiana side on a small steamboat which progressed through vicissitudes which included running aground, stopping while the pilot sobered up, and running aground again. In the course of this rather lengthy voyage —at least long in elapsed time—the author had an opportunity to

hear what he called a "grand oratorio by the nigger firemen." He
reports it—

(Grand opening chorus) "Ahoo—a-hoo—hoo-oooo—a-hooo—a-hoo
—a-hooo—a-hoooo-oo!

The dashes in the following represent the passages where the
superfluity of the harmony prevented the proper appreciation of the
poetry.

"Gwin down de ribber—a-hoo-a-O!
Good-bye—nebber come back—debbil—beans
Grey-haired injun—Ya-a—a—aaaa—Ya-a-a-a-a-a-a-a—
Ga—!" (leader of the orchestra) "dirty shirt massa, got de whisky bottle
in his hat, dis poor ole boy nebber get none—
A-hoo—a-hooo—a-hooooo" (ending in an indescribable howl).
(Pensive darkey on the coal heap)—"Miss Serefiny good-bye—farewell;
nebber get no more red pantaloonses from Miss Serefiny—Oho—Ahooo
—Ahooo—O!"

(Extemporaneous voluntary by an original nigger with two turkey
feathers in his hat, and his hair tied up with yellow strings)—"Corn
cake—'lasses on it—vaphuns—" (meaning waffles) "big ones honey on
'em—Ya-a-a-a-a-a."

(Stern rebuke by leader)—Shut up your mouf, you 'leven hundred
dollar nigger."

(Leader improvises as follows) "Hard work—no matter—git to
hebben bym-bye—don't mind—go it boots—linen hangs out behind—"
(here having achieved a rhyme, he indulges in a frantic horn pipe.) "My
true lub—feather in him boots—yaller gal got another sweetheart—
A-hoo—a-hooooo!—Ahoooooo-oo-O O O O !!!!—Hoe cake done—
nigger can't git any—old horse in de parlor playing de pianny—You-
a-a-a—Ga-Ga-Ga!" Captain here interferes and orders the orchestra to
wood up—and so interrupts the concert.

Where there was music there was often dancing. The Indians
danced—complicated ceremonial patterns and for amusement—
though few of their dance forms seem to have been adopted by the
whites except for those employed after Ernest Thompson Seton in
boys' and girls' camps.

The first whites who indulged in mixed dancing upon the
Ohio—as distinguished from the solo performances of the French
voyageur—were almost certainly those who emigrated from the

southern Atlantic coast. Puritan influence still held some sway in New England when western emigration began, and New Yorkers and Pennsylvanians were not notably dancing people. Strait-laced Presbyterian rule, which early descended upon wild Pittsburgh, frowned on lighter pastimes even while leaders among the Virginia settlers in Kentucky were encouraging dancing classes as necessary to the proper development of the country. As early as 1787 a correspondent to the Pittsburgh *Gazette* wrote to the editor:

To what a height this contagious distemper is arrived . . . is but too visible. Every village which has the least pretension to gentility, has its assembly; every tradesman must have his fine horse, his club parties, and his card parties. Singing, dancing, fiddling and gaming, are no longer mere amusements, they are ranked among the important occupations of the day, among the principal duties of human beings.

In short, gaming and luxury is arrived to such a pitch, and become so universal, that we may, with great propriety, be called a nation of gamesters . . . America is game mad; thousands of my good countrymen seem to be quite out of their senses.

From the earliest historic day upon the Ohio gambling was available to the white man even though he might be separated by hundreds of miles from the nearest of his fellows.

The historic Indian had little feeling for personal property and usually regarded an accumulation of it as a burden; he was thus a willing and enthusiastic gambler, interested chiefly in winning—or even losing—for the fun of it. (Modern Indians sometimes exhibit this same devil-may-care attitude in the manipulation of Texas and Oklahoma oil leases with dazzling results, and there was recently a good part-Micmac college president who could hold a deck of cards in one hand and flip up any card for which his friends might call.) At the time of their first contact with whites every tribe had some variant of the "moccasin game" (which is preserved upon the county fair circuits into the twentieth century as the "shell game" or "thimblerig") and most Indians stood ready to stake their furs, their clothing, their wives, or any other transferable property in a game with either fellow Indian or visiting white.

After the introduction of cards the natives took to them with avidity and the Indian became, by report, quite as expert in the various card games as the African is supposed to have become in handling the cube-cut dominoes, once he encountered them.

Professional gamblers, like professional play-actors, musicians, and athletes, followed long after the amateurs had cleared the way. Endless yarns are told of them, sometimes by themselves, sometimes by others.

The Rev. John A. Clark, whose observations have already been quoted, watched such gentry at work on the steamboat ride between Louisville and St. Louis in the late thirties:

> We saw one victim fall into the clutches of these blacklegs. He was a young merchant, I believe, from Chillicothe, Ohio. He was first induced to play a simple game of cards. A slight sum was then staked to give interest to the game. He was allowed for a while to be successful and to win of his antagonist. He played on till be became perfectly infatuated. He would hardly stop long enough to take his meals. Being fairly within their toils, large sums began to be staked, and this young man did not see the vortex into which he was being borne until he had lost six hundred dollars. In this deep gambling physicians and judges who were present participated . . .

Lotteries were considered perfectly legitimate in the early days upon the Ohio, as they had been in the colonies and continued to be for some time in the Republic. The fancier forms of gambling—roulette, faro, etc., are said to have been imported through New Orleans, whence, also, are supposed to have come the earliest professional gamblers. If this was the case, the Louisianians found many apt pupils in the non-Latin sections of the Mississippi and Ohio, for by 1850 almost all sizable ports were called home by some professionals who apparently sprang from every imaginable background.

With settlement came a greater emphasis on the early and often pretty crude forerunners of "spectator sports," with betting as a chief attraction. Horse racing, shooting matches, bear baiting, cockfighting, and gander pulling all had their enthusiastic fans, especially on the south bank of the Ohio.

Horse racing was probably the favorite, and its requirements were not exacting: the only necessities were two horses, two riders, and a stretch of ground more or less level. Saddles were desirable but not essential, and halters would serve if bridles were not available. Two or three spectators added to the joy of the winner but their presence was by no means mandatory.

At the time of its first settlement by whites of largely British descent, the Ohio River had already been the rough dividing line between the warlike Iroquois League and those of the conquered peoples who had chosen to move south; it had been the line to which New France claimed sovereignty, and it would become the divider between free and slave territory. It is also another barrier seldom mentioned in the history books; the dividing line between runnin' race horses and harness race horses. That is a distinction still of prime importance to a small but resolutely tradition-loving section of American people.

Of course horse races were first introduced to the West in Kentucky, and at a time when it would have been difficult to get two race-minded horse owners together in what would later become Ohio, Indiana, and Illinois. They had to be running races in the earliest instances because there was probably, at the time of the first match race, not a wheeled vehicle west of the Alleghenies except the *calèche,* clumsy solid-wheeled cart of the French settlements.

The "Quarter Race," so called, was the first form of contest; the purse a side bet or a jug of drinking liquor. Matches were sometimes by challenge, sometimes made by a self-constituted handicapping committee selecting what appeared to be the most evenly matched animals in the community.

The distance was a quarter of a mile or thereabouts—hence the Quarter Race—and though it was extended as time went on, these straightaway dashes continue to be Quarter Races today and the horses wont to run them are Quarter Horses.

Very early, still in the eighteenth century, blooded horses began to be brought out from Virginia, Maryland, and the Carolinas. Annapolis had had a jockey club since 1745; Charleston since 1734. These importations came especially to the Bluegrass area around Lexington, where horseflesh seemed to thrive and where interest in horses had apparently already spread its still thriving roots. Soon enough, when the state should have had plenty of legislative business more pressing for the public good, Kentucky passed a set of laws

which governed the penning of unblooded stallions and set up a
code for the improvement of the breed of horses. In 1789 Lexington
laid out its first race course, four miles long and making use of
parts of the city streets. Kentucky horses soon became a notably
important article of export from the state.

North of the Ohio the horse business was taken less seriously,
at least as far as the aspect of speed was concerned. Kentucky horses
were (and are) respected; but it was not until after the middle of
the nineteenth century, when Rhode Island, Massachusetts, and
New York State animals had been crossed with the Kentucky lines,
that horsemen, especially of Ohio and Indiana, developed a new
and valuable property—first the midwestern trotting, later the pacing
horse. They were mostly of the Hambletonian and Morgan lines,
to be driven in harness to, earliest, a cart, and later an especially
designed sulky.

Understand, there exists no statute which forbids the breeding
of pacers in Kentucky or of running horses in Ohio and Indiana;
but they still remain the specialized products of the two sections.
One of the country's famous harness horse meets—the Lexington
Trots—is held in the Bluegrass; and the spectator has a feeling that
in an afternoon there, rather than at the famous Kentucky Derby,
is to be seen the true aristocracy of the dyed-in-the-wool Kentucky
horse lovers—but that is a feeling which is best not expressed, at
least in the presence of publicity-minded Kentuckians to whom the
importance of the Derby is paramount.

As time went on, horse racing became popular enough all along
the river, although part of the time forbidden by law in Pennsylvania, but in Kentucky it was enormously so. Its ancient rules and
customs were there first formally observed in the West, and the
sport enjoyed the patronage of a class of citizens which made it
respectable. Such was not always the case elsewhere. Horse racing
was generally held to be an agency likely to increase the use of
profanity and to encourage idleness; so positive was that feeling
among the clergy and more militant laymen (again excepting
Kentuckians) that usually the three deadly sins on the frontier were
named as "drinking, swearing, and horse racing." In order to
achieve a place of infamy so important as to supplant nine of the

Ten Commandments, these pastimes must have been carried on in a particularly vicious manner which has since been allowed to deteriorate.

Shooting matches were popular as soon as *all* the powder and lead in the West was not desperately needed for warfare and protection. They served two purposes: as militia training and recreation. They, too, were an object of denunciation from the pulpit, perhaps because liquor usually flowed freely and occasionally the contestants left the targets in peace to shoot at each other, or because a certain amount of cussing was necessary to steady the marksman's nerves after a bad shot.

Downright brutal and degenerate frontier sports—brought from ancient England by way of Virginia, according to report—were gander pulling and bear baiting. In the first, a well-greased live gander was hung by its feet from a limb while the competing young bloods rode hell-bent in turn beneath and tried to seize the head and snap the neck. Bear baiting was customarily a contest between dogs and a bear in a pit or pen—a sport common over the world, apparently, to every people who owned fighting dogs and had bears available. Cockfighting was (and surreptitiously still is) conducted in a few localities; but this, as an old hand at the mains has said, is less brutal than its companion sports: "Them roosters is just doing to each other what they'd try to do if they met behind the barn, only these know how to do it artistically," he remarks, perhaps with some justice.

The advent of professional entertainers was delayed past the explorers, past the traders, and past Boone, Putnam, and the seekers for homesites. They must have begun to arrive, however, as soon as the most rudimentary vestige of an audience had gathered at the places where flatboats were launched. The first professionals (any who cared to proclaim themselves such who could sing a song, recite a piece or dance a jig, and pass the hat with the faintest hope of having something more valuable than spittle dropped into it) must have been either the very dregs of the stage-struck, whose complete lack of talent barred them from spear carrying in the sorriest eastern theaters, or adventurers who were prepared to try anything short of work for a few pennies or a drink.

The very earliest of these unfortunates came northwest to
Wheeling from Williamsburg, Richmond, or Charleston, where
the theater had flourished for years before the Ohio country was
opened. To them the performances put on at Wheeling or elsewhere on the Ohio must have been solely as a means of getting
west for other purposes; not even the least informed could have
hoped for a theatrical career on the Ohio in the seventeen-eighties.

Even so, "theater" in its classic sense made its appearance upon
the river, probably, before there were newspapers to report its arrival or criticize its productions. For the earliest issues of western
journals carry announcements of entertainment of one kind or
another and it is inconceivable that a singer, a Grand Dramatic
Company of two or three members, or a juggler could have failed
to secure downstream transportation before the weighty printing
press and type necessary to produce a newspaper.

Cincinnati seems to have had the first theater built or remodeled
especially for the purpose. It opened in 1801 which, incidentally,
was only seventy-five years after the first theater in America, at
Williamsburg, Virginia. Alexander Drake's company, of which his
family formed the nucleus, is generally given credit for bringing
popular drama to the Midwest in general, beginning in 1816. His
contribution is not to be ignored but thespians surely trod the
Cincinnati boards during fifteen years before, including many
amateurs. Amateur theatricals had been popular in America for
a century, though frowned upon by the clergy and outlawed by
many state legislative bodies, and Cincinnati had at least its share.

Theater buildings were few indeed along the river until after
1815. Hastily converted warehouses, churches which had been
abandoned because of splits in congregations over abstruse questions
theological, and buildings vacated by bank failures were frequently
pressed into service, or if only temporary quarters were required for
a one-night stand, the county courtroom, the local assembly hall,
or the tavern dining room would serve very nicely once a calico
curtain had been strung up and a few shaded candles set out for
footlights.

The way of the theatrical manager among the inexperienced
but enthusiastic patrons of the drama in the new West was strewn

with thorns, as is attested by a considerable body of their reminiscence. Most charming among those who wrote of their experiences
in the back country is Sol Smith, comedian-manager who dared the
Ohio River towns before undertaking the cotton states and Missouri.

Sol saw his first play as a boy of fourteen in Albany, New York,
and was immediately stricken with a passion for the theater. The
company was that of the aforementioned Alexander Drake, which
was shortly to go west upon the first of its many tours.

Nine years after they dazzled Sol Smith, members of the Drake
company were fellow passengers of Captain Donald Macdonald and
Robert Owen on the steamboat that carried them on their way to
Harmony, Indiana. Macdonald is said to have fallen in love with
one of the numerous Misses Drake during this short acquaintance;
the romantic note is not introduced in his diary but there is a
view of members of the family and company as they appeared in
December, 1824:

A Mr. & Mrs. Drake & a young man, all comedians, were likewise
cabin passengers . . . This evening Mr. Drake played on a violin &
accompanied his friend who sang several songs. About 11 o'clock at
night we overtook a float, or barge, in which were Mr. Alexander
Drake & his wife & children, Miss Drake, and one or two of their
company. They had tired of waiting for the rise of the river, & had
started 9 days before us in this float. Taking this party on board occasioned quite a theatrical bustle. . . . A little before sunset we arrived
at Maysville . . . Here Mr. Drake's party quitted us, as they were to
proceed thence the next morning to Frankfort through Lexington.—
Mr. Drake, the father, emigrated to this country from England 14
years ago. He now owns four theatres, those of Lexington, Frankfort,
Louisville and Cincinnati . . .

But back in Albany in 1815. Young Sol Smith's infatuation
was so violent that, a few weeks after the Drakes had started west,
he ran away from home to follow them. Arriving in Pittsburgh
he found an established theater (under the management of a man
who existed under the name of Entwhistle) which featured a
stock company headed by a Mr. and Mrs. Legg. Apparently only
the Drakes were for Sol, however; he passed on down the Ohio
as a working passenger on a skiff by which a man with his wife

and baby were moving west. Finally he arrived at Louisville, where the Drakes were appearing.

The company then consisted of Alexander Drake, S. Drake, S. Drake, Jr., James Drake, Julia Drake, Mrs. Mongen, Mrs. J. O. Lewis, and Messrs. Savage, Blisset, Cornele, and J. O. Lewis. (J. O. Lewis must undoubtedly have doubled as scenic artist, since, about ten years later, he conceived the idea of painting and publishing a collection of Indian portraits and scenes of Indian life which was the earliest and is now rarest of this type of work. Valuable as is J. O. Lewis's *Aboriginal Portfolio* historically, one is compelled to conclude, from the evidence of his drawing, that the Drake company played before scenery of no great artistic merit.) Evidently the Drakes had no paying position available to young Sol's embryonic talents, for he went to work in a printing shop.

Louisville's dramatic fare that season included *The Forty Thieves, The Miller and His Men, Honey Moon, Hunter of the Alps,* and *Richard the Third*—all of which one supposes the stage-struck Sol arranged to see, even on the pay of a printer's devil.

Smith walked up to Vincennes—living on apples en route— and was employed on Elihu Stout's Vincennes *Sun,* first newspaper in the state of Indiana. Mrs. Stout, memorialized by others besides Smith as a strong-minded woman, was carrying on a practice said to be common in free territory north of the Ohio. She had brought one of her slaves with her, when she came to Indiana from Kentucky several years before, as an "indentured servant" and, says Smith, "He was to be free at the age of twenty-one and he was now at least thirty-five! Mrs. Stout made him believe he was but fourteen and that he had seven years to serve."

As an indication of the rapid spread of interest in the theater in the western country it may be noted that even in not particularly culture-minded Vincennes, "A Thespian Society was formed— Mr. Stout was appointed one of the managers, and Bradford and myself were allowed to become members." The society produced *John Bull, 'Tis All a Farce,* and other pieces.

After a time Smith went to Cincinnati, studied law, founded a newspaper, and ran a singing school. By then that city had the Columbia Street Theatre, the Globe, located on Main Street, and several usable halls—and Smith finally organized a dramatic company of his own.

Some of the best actors of their day played the West in the late twenties and early thirties—Edwin Forrest, Thomas Cooper, Eliza Riddle, Mrs. Cargill—and original plays were being produced occasionally. In the season of 1822-23 Cincinnati saw Edwin Forrest star in *Modern Fashion,* written by Sol Smith's brother, and also saw Forrest play a Negro comedy part in Sol's own *Tailor in Distress.* The larger places, Cincinnati, Louisville, Pittsburgh, now supplied audiences which appreciated good acting and production, but an occasional "green-un," fresh come to town from the piney woods, often added an unbilled extra note of comedy, and in the small settlements which the companies necessarily played on the road between better engagements (show business was almighty uncertain even in those days) the audience reaction was highly unpredictable.

The great Thomas Abthorpe Cooper was playing Othello in Cincinnati in 1822 when a country girl strayed into a stage box. At that moment Cooper, who happened to be facing her, gave the line—

"Here comes the lady!"

and the maiden, smiling her thanks, stepped on the stage and seated herself in a vacant chair.

Starting up the Ohio by wagon with his company, Smith stopped now and then to play in an empty storeroom but found "The people were indifferent to our indifferent performances." At Wellsburg (now West Virginia) the company was urged to play a Saturday night performance. Smith rented a room and printed bills, but eight o'clock came with not one ticket sold. The price was cut to twenty-five cents, eventually sixty people invested the two bits each, and the show went on; but the sheriff interrupted the afterpiece with a writ requiring the payment of a license fee of $40. Smith—then and later resourceful in handling such emergencies—invited the sheriff to have a seat and then ordered the cast to extend the performance with a series of the specialty numbers of which every actor in those days was required to have a repertoire. Since the first curtain had been long delayed by customer resistance to the original 50-cent price, the hour was already late when the sheriff arrived: when a few of the specialty numbers had carried the performance past midnight, Smith rang down the curtain and called the sheriff's attention to the fact that, it now being Sunday, the writ could not be served. The company gathered its baggage and was far away before the church bells rang.

Traveling between Steubenville and Pittsburgh on the Virginia side of the Ohio River the company put up at a small country inn. At breakfast some of the members discussed a performance they had recently given while the establishment's "hired girl" listened horror-struck, as presently developed, to their remarks. One character had been too slow "after murdering Don Guzman," they decided, and "the combat with Ferdinand was shockingly bad." Another actor had "murdered Dr. Pangloss" very well but had been "too drunk to guard the prisoner," and so on.

Breakfast over, the troupe came out to board their wagons and found themselves in the possession of a constable and a posse:

"Stranger," the constable addressed Smith, "you're in Old Virginia and you mustn't think of getting off! We don't mean to let pirates pass through here, no way, no how!"

One more of Sol Smith's favorite anecdotes illustrates both the conditions under which plays were first seen in backwoods towns and the attitude toward them which the church and its leading members often took.

In 1829 Smith's company was playing in a tavern ballroom. The landlord, a pillar of the church, was anxious to see the show but hesitated to appear out front for fear of his pastor's wrath. Smith seated him on a chair in the wings and the play went on with the landlord one of the most appreciative members of the audience.

Smith, himself, playing Sancho in *Lovers' Quarrels,* had a line advising another character to "save something with which to pay his board." This was too much for the landlord, a liberal man and by now an enthusiast for the drama; sticking his head on stage from behind the wing he shouted reassuringly: "Mr. Smith, don't mind your board; go on with your play . . . if you haven't the money at the end of the week I'll wait."

He received, according to Smith, the greatest hand of the evening.

Since theatrical companies, with their costumes, props, and such scenery as their degree of prosperity allowed, could travel most quickly and easily by river, the showboat was a natural development—first as a carrier of Thespians and equipment only, later as a floating auditorium. On the Ohio the showboat reached its first flowering and existed in every conceivable stage of development, from a few logs lashed together—Huck Finn's unwelcome companions, the Duke and the Lost Dauphin, could have been real as easily as not—to the floating theatrical palaces of the late nineteenth century.

The first example of the showboat as a floating theater that seated an audience as well as transported a troupe seems to have been built at Pittsburgh in 1828. William Chapman, an emigrant English actor-manager, had it constructed, employed his six children as actors, and toured the Ohio and Mississippi. There was soon competition, but the history of showboat development is less easily traced than is that of theaters, which were more an object of local pride and received regular newspaper publicity.

After the Civil War there came a considerable revival of the business, in fact the showboats reached the height of their popularity between 1875 and 1900. Captain A. B. French can be considered the king of the showboatmen and his boat *French's New Sensation* became the most famous if not the finest of the craft. Fourteen of the boats are said to have survived into the nineteen-twenties and several still tour profitably.

Another popular feature of entertainment and instruction made use of boats on the Ohio in the nineteenth century: it was the moving panorama, an Italian invention which was already long popular in London and Paris before it was introduced in America. First of these reported upon the Ohio, Wabash, Mississippi, and other western waters was a juvenile production by George Banvard, who later made a fortune from the business in Europe and the eastern American cities.

The panorama consisted of a picture of, for instance, a battle scene, a river bank, or a view of some historic site, which was

painted upon a long strip of canvas and exhibited to the audience
by being rolled from one turning upright drum to another. The
length of the canvas (Banvard finally achieved a strip over a mile
from end to end), and the speed with which the rollers were
turned, governed the length of the entertainment, which was accompanied by a lecture on the subject presented.

Young Banvard was an unenthusiastic apprentice to a Louisville druggist. He was probably fifteen or sixteen years old at the
time his preference for drawing pictures on the wall of the back
room in which he should have been rolling pills attracted the unfavorable attention of his employer and brought him a severe
reprimand.

He promptly ran off and made for New Harmony, where he
may have known artistic pursuits were encouraged. Probably advised and instructed by the great Charles Alexandre Lesueur—who
had seen a great deal of scenery around the world and loved to
paint it—Banvard headed a group of teen-age impresarios who
painted a panorama of the scenery on the Wabash, built a raft
to carry it, and floated downstream exhibiting its wonders to the
residents on the lower Ohio.

Banvard evidently realized, after this experiment, how appropriate the medium of the panorama was for reproducing river
scenery; the pictured banks appeared to pass by the audience exactly as they would from a moving boat. Years later he completed
his masterpiece—the mile-long canvas that pictured the whole
length of the Missouri and Mississippi rivers—and exhibited it
triumphantly throughout the civilized world.

The minstrel man in blackface apparently originated on the
Ohio and was certainly in his ideal setting there. Credit goes to
Thomas Dartmouth Rice, who, born in New York in 1808, served
an apprenticeship in various eastern theaters before coming west.
He appeared with the ubiquitous Drakes in Louisville in 1828-29
and got the idea for his "Jim Crow" number about that time. He
first tried it out in either Louisville or Pittsburgh and the song
was put in shape for orchestration by a Louisville music teacher.
It was an immediate success and continued enormously popular as
long as Rice appeared on the stage.

Rice probably did not take part in what later became the standard minstrel show form—half circle in blackface, the interlocutor in the center, and the comedians, Tambo and Bones, at the ends during the first part, with specialty numbers following in the second, called the "olio"—that pattern developed much later: Rice did his act as a specialty on regular theatrical programs.

Steamboats must have been carrying dozens of blackface artists within a few years of Rice's invention, either passing between engagements in company or singly, down on their luck and picking their banjos and singing their songs in payment for their passage. The first *reported* minstrel troupe that specialized in blackface exclusively was appearing in New York in 1843. A quartet, it included another famous minstrel man—one who, with Harriet Beecher Stowe and Sir Walter Scott and John Brown, helped to start a great war. He was Dan Emmett, composer of a song called "Dixie."

Christy's Minstrels were organized at least four years before the outbreak of the Civil War and it is this great troupe which finally set the traditional and never-changing pattern for classic blackface minstrelsy.

Blackface characters and minstrel shows were always popular on the Ohio and both Rice and Christy encouraged Stephen Foster to compose suitable songs and bought and popularized his productions.

Perhaps there may be a shadow of sacrilege in the statement, but the early camp meetings and great revivals could be classified as entertainment—for so they were doubtless considered by a great many of those who attended them!

Camp meetings were not new in the period 1797-1801, nor were revivals; but the Midwesterners, taking their reawakened interest in religion hard, promoted· such sessions as had not been seen before in America.

The settlers had brought their various creeds with them across the mountains, but in most localities until the end of the eighteenth century religion was not· often allowed to interfere seriously with the business or pleasure at hand. The same was true of the 1775-1800 period in Europe; and the original settlements in America, while not godless, had become rather complacent: even New England, with a surfeit of sanctimony in its background, enjoyed a relapse.

But by 1800 the pendulum had swung too far; there was danger of atheism—product of French revolutionary thinking—for the intellectuals and a loss of faith through neglect and lack of inspiration for the common citizen. Some thoughtful ministers of the gospel began to bestir themselves, more or less violently according to their temperaments.

Most violent in the Ohio Valley was James McGready, Presbyterian minister of Logan County, Kentucky. McGready's preaching was of the most forceful and, singlehanded, he seems to have aroused the first interest, which resulted in the Great Kentucky Revival and continued its reverberations in the surrounding states for decades following. So successful were his militant sermons that he soon gained support from the formerly inimical Baptists and Methodists, whose ministers joined in the almost unbelievable excitement that marked the camp meeting at Cane Ridge, Bourbon County, Kentucky, during August, 1801—the Great Kentucky Revival.

The attendance at that meeting has been estimated at something in the neighborhood of 20,000 people: a tremendous total in view of the still sparse settlement of the country and the terrific difficulties of overland travel. The meeting is a story in itself, but its great feature was the prodigious manifestations of the converts—the speaking in unknown tongues, cataleptic seizures, dancing, rolling on the earth, and wild singing—which had not been seen since the great revival in the seventeen-forties. These, sad to say, were to set the pattern for possibly less spontaneous demonstrations at future revivals and camp meetings; eventually they would cause such affairs to take on some disrepute in the eyes of thinking Christians.

In spite of the excesses that marked it, the Cane Ridge meeting was highly beneficial in reawakening the Ohio Valley to an interest and a participation in religious activities. The social side of the camp meeting began to develop in the eighteen-twenties. Sessions were held in some pleasant grove, rather than in and around a church building, and lasted four or five days. Families came in wagons and slept in tents. Cooking was usually done by groups clubbed together for the purpose and elaborate fare was the order —naturally every respectable housewife saved her best pickles and preserves for the amazement of others, while her husband pulled down the best hams from his smokehouse. Plenty of jugs were stowed conveniently in the bushes, men sometimes withdrew a short distance between sermons to match the speed of their horses, frenzied courting was the order of the day for young folks, and a general good time was often had by all. Some of the finest fights in western history—excepting the classic battles of the keelboat-men—took place when rowdies invaded the camp meetings intent upon breaking them up. Elder Peter Cartwright, the husky Methodist already mentioned as *not* having whipped Mike Fink, acquired a legendary stature which featured his prowess with his fists on equal terms with his ministerial labors. The hymn singing on these occasions was often something to remember, especially in Kentucky and Virginia where the slaves who had been brought along to mind the children and horses and do the cooking were allowed to add their harmony from the back benches. Such meetings were noble spots in which to campaign in politics and there was no reason why the candidate should not add to his stature by an eloquent testimonial as to the talents he had received from the Almighty—with perhaps an inference as to how they might best be occupied in the public good.

All these attractions made camp meeting revivals fun but they were not, thought the serious-minded, always necessarily conducive to the teaching and living of Christianity. It began to be noted that the percentage of illegitimate births sometimes took a leap three-quarters of a year after camp meetings; that some of the converts who rolled in the straw, frothed at the mouth, and shouted loudest were repeaters who, after they had received comfort and material aid from the innocently pious brothers and sisters present, were likely to backslide promptly into their old ways until the next revival; that too many exhorters appeared to exhort best after a dose of something out of a jug, and that some in attendance seemed to be present more in the interest of selling those same jugs and swapping horses than in exposing their souls to salvation. The leaders of most of the established sects began to take thought, and the camp meeting in its most violent form died out.

People on the Ohio were athirst to better themselves. They took to Culture (with the capital C), isms, and movements almost as soon as they had driven the catamounts from their hog lots.

Lectures and readings were an early attraction, for the lecturer could travel light; need not, in fact, even carry any great amount of information in his head.

Singing schools also were popular, the master requiring only as many copies of the *Sacred Harmony* or other songbook as he hoped to sell to his pupils. Sol Smith (who could act, preach, lecture, teach, print, edit, or practice law as the situation demanded) set up a typical singing school in Newport, Kentucky, in 1823. He had got married and found himself in need of funds so he purchased four dozen songbooks and organized a school. It met in the grand jury room of the county courthouse and received good patronage. Smith says he sold his books "on a credit—that was the way I had purchased them."

Also early in the field of instructive entertainment came the museums of natural wonders located in permanent halls or constructed on flatboats for easy transportation. Exhibits varied, but at least one two-headed calf appears to have been a mandatory item. (Since the calves were always stuffed, it was not impossible to manufacture one if the genuine article was not readily available.) Even the shoddiest of these "museums" and the later medicine shows, which usually embodied some of their features, could take in the unwashed and unpretending elements of society.

The element of the citizenry that considered itself refined patronized the ever-changing fads of phrenology, mesmerism, spiritualism, Owenism, temperance, and—beginning in 1825 with the lectures of the beautiful Fanny Wright—divorce reform, emancipation of the female wearing apparel, and women's rights in general. Phrenology was perhaps the greatest craze of all—even the Smithsonian Institution went so far as to publish a "phrenological chart" of the skull of a long-departed midwestern mound builder, who was found to have been, in his prehistoric life, long on "Destructiveness, Amativeness, Philoprogenitiveness, Cautiousness, Adhesiveness and Approbativeness" and short on "Constructiveness, Benevolence, Reverence, Marvellousness and Initiation"! Not all double-talk originated in Washington under the New Deal.

From a combination of the American passion for being instructed, for sitting through lectures on any subject, and the camp meeting evolved the Chautauqua. Even though it bore the name of a New York State lake and camping ground, this institution had its real flowering in the Ohio Valley during the decade before World War I.

Organized concerts and musical programs (exclusive of the specialty numbers that, as the "concert," followed theatrical productions) began to be offered in the eighteen-twenties, and before many years world-famous artists such as Jenny Lind and Ole Bull toured the Ohio. However, general interest in and wide appreciation of so-called classical music upon the river had to await the day when Cincinnati and Louisville received the great emigration of educated Germans after the European revolt of 1848.

Travel itself became entertainment, first by steamboat in the forties and fifties and then by crude and uncomfortable but novel railroad excursions in the seventies and eighties. Certainly there was no comparison in the comfort of the two modes—the drafty, dusty, smoky, wooden railroad coaches and the elegant steamboats of the golden era. Only the remarkable speed of the railroad could account for its ultimate victory.

Real luxury and high romance rode the river to the Civil War. Those were the days of the pistol and the sword cane, of Spanish sideburns, ruffled shirt bosoms, and doeskin vests; of chignons and spit curls and Spanish combs, of sloping alabaster shoulders and hoop skirts: of adventurous ladies with, perhaps, a touch of the tarbrush, of innocent cotton-rich virgins with designing mamas; of young blades familiar with the code duello, of smooth gambling men, and of coltish farmers with well-filled purses. By tradition, also, that was the remarkable era in which camellias, moonlight, and warm soft breezes prevailed three hundred and sixty-five evenings to the year, rather than seasonably, as before and since.

Whenever steamboat travel is represented on stage and screen it is that of the fifties—and perhaps its romance is not so farfetched at that: there is, even yet, something about evening on a river boat . . .

On the Ohio fun and frolic of a baser and more boisterous sort was available for those who preferred it. There were plenty who did, either by reason of depraved taste, in the bravado that is usually the product of unsureness and a feeling of inferiority, or only as a result of recent escape from a too-oppressive previous restraint at home.

Ladies of lax virtue had existed in the colonies as they had always, everywhere. Through the War of 1812 "camp followers" of both sexes were recognized and regulated by military order; and while some of these were traders, male hangers-on, and wives and children of officers and men, a great many were ladies who dispensed their favors in return for a share of rations and the adventure that offered. They were required to cook and wash for the men—perhaps as a measure to justify their presence and the drain upon the public rations they consumed—but their major professional calling was obvious.

The statement has been made that paid prostitutes did not exist in America before the nineteenth century except in the Latin ports—notably New Orleans. Those who so state cannot have read the Pittsburgh *Gazette* for March 17, 1787, which carries a tragic poem on the subject. The meter is unsure but the scene is definitely Pittsburgh, well into non-Latin territory:

ELIZA: *Or the Cause of Prostitution and Its Fatal Effects.*

> Eliza! once the admir'd of all the swains
> Of Hudson's city and of Hudson's plains,
> Had by some damn'd destroyer been betray'd,
> And like a villain robb'd her of the maid:

Eliza's parents, perhaps no less villainous than the accused party, drove her from home:

Robbed thus of ease, she sought this city's haunts
Where prostitution amplifies its grants:
The Cyprean vot'ries pleased with something new,
From her sweet form unhallowed raptures drew.

Presently Eliza became infected with a disease which "dooms its patients to the worst of fates," was deserted by all and—

She sought this cell her wretched life to end:
Which could no beds of down or covering boast,
To screen its victim from the impending frost. . . .

I stood aghast, and view'd that dropping eye,
That once bright with the diamond's luster vie.
Those lips that once could with the ruby boast,
Gasping for breath and yielding up the ghost.

There is a great deal more in similar tone but these brief lines should prove the case: there was prostitution slightly before 1800— and at the very fountainhead of the Ohio!

That the Ohio had nothing quite comparable to New Orleans or Natchez-under-the-Hill was due, in some unrestrained localities, only to the fact that there was not the concentration of traffic of the lower Mississippi to furnish patronage enough to warrant operations so extensive. Some Ohio River settlements did their best, in the earliest day, to equal the two southern paragons: notably there were Red Bank, Cave-in-Rock and Shawneetown, which once included, within their limits, the very dregs of mongrel humanity, and early Jeffersonville, Indiana, Yellow Bank, and Wheeling appeared to have made every effort, before 1800, to maintain a liberal attitude toward sin.

Some of these communities were later reformed but others, and some of newer origin, bear marks reminiscent of their early open-mindedness to this day—even certain districts of fair and generally high-minded Louisville occasionally receive a Derby Week influx of ladies apparently not painted but literally *enameled* who rent one-story business buildings otherwise unoccupied during the year and seem to enjoy the sunlight that can be absorbed while leaning from doors and windows en déshabillé.

But old Shawneetown is too nearly dead to be evil; Red Bank, as Henderson, is eminently respectable, and Cave-in-Rock is the haunt of school picnics. Wheeling is a prosperous city and Mount Vernon, Indiana, seems to have relapsed into a state of some rectitude while Jeffersonville, of late years, has specialized only in gambling establishments and a wilderness of "marriage parlors," rooming houses, and the highest incidence of establishments featuring malt beverages and slot machines in the Midwest—sinful enough, perhaps, but in no way gaudy.

Chapter Thirteen

THE STARRY-EYED UTOPIANS

BEING the highroad to a new land the Ohio River was naturally in position to catch the imagination of the theoretical Utopians, of the fuzzy-minded do-gooders, in a very early day.

European dreamers began to speculate upon the perfections in government and economy that might be developed in its valley when the first extensive reports of its wonders coincidentally met the wave of reform that swept through the Old World in the latter part of the eighteenth century. Shortly a few of the many plans for setting up earthly heavens in this part of American back country began to take some semblance of shape—especially where, as in the case of William Penn's even earlier project, the entrepeneur possessed ample qualities of sound leadership, a reasonably well-filled purse, and influence at home.

As events finally demonstrated, William Penn was definitely neither starry-eyed nor fuzzy-minded; not that competent Quaker promoter! But his was certainly the first of the reform-seeking movements whose influence reached to the Ohio. Unquestionably the principles upon which his "Holy Experiment" was founded—complete freedom of religion, of enterprise, of immigration; fair dealing for all, including Indians, and the absence of military power even for defense—appealed to his transatlantic contemporaries as being revolutionary enough and his liberal aims, though modified as such must always be, set the pattern for the better part of what we like to consider the American way of life.

Those Moravians—the members of the Unity of the Brethren—also had plans Utopian and not solely aimed toward the salvation of souls, as were those of most other Christian missionaries. They hoped to bring education and earthly economic improvement to the Indians of the Ohio Valley as well as spiritual reform: except for the fact that the site of their early experiments happened to neighbor upon that section of the Ohio which rejoiced in the greatest concentration of fanatical Indian-haters in the entire country their efforts might have left a mark upon the Indians as significant as those of Penn upon the nation.

But some of William Penn's philosophy had been cast aside before the Ohio was much settled; Pennsylvania had a military body long before it ceased to be a royal grant and it came to appear that any hesitation in sending armed men against the Indians in further conflict with the principles of founder Penn was due to political and economic reasons rather than to sentiment: the Moravians were so few and the atrocious fate of their converts so disheartening that the results of their good work were in the end of little significance.

These two practical agencies of reasonable reform having passed from the scene—and that ultraradical experiment which involved the development of the Union itself being already successful—the stage of the Ohio Valley was set for such new productions as growing religious, antireligious, and social ferment in late eighteenth-century Europe might originate. The stage was not long vacant.

None of the social or socioreligious experiments that followed on the Ohio were of lasting importance but several were curious and interesting and most of them had pathetic aspects of one sort or another.

First of the Ohio Utopias to be preached was the profit-motivated and thoroughly disreputable colonial venture of the Scioto Land Company and the Rev. Mr. Joel Barlow at Gallipolis. No planned economy this, its significance lies in the fact that there *was* no plan—only the grossest of deceptions. The promise of salubrious, semitropical climate and cheap new lands drew willing purchasers among the French artisans, farmers, and small merchants who saw their savings endangered in the conflict between the aristocracy, the intellectuals, and the sans-culottes in their homeland. They expected no unearned security but only that, joined in a community of common interest, they could begin a new and better life. Some of those who survived the freezing weather of the first shelterless "subtropical" winter and who withstood the shock when they learned that Joel Barlow and the elders of the Scioto Land Company had taken money for lands to which they had no title did go on to build a fuller life, not in a community of mutual interest but as individual settlers badly handicapped by unfamiliarity with language, farming methods, or customs of the strange land and almost altogether without financial resources. Thus ended that Utopia.

First of the religious community experiments in the Ohio Valley were those of the United Society of Believers in Christ's Second Appearing—called by their neighbors "Shakers" in the interest of brevity of description.

The Shakers were founded by Ann Lee, an Englishwoman who had been converted to Quakerism, had come to believe herself a reincarnation of Jesus Christ, and had thus felt justified in formulating a new creed of her own. "Mother" Lee came to the United States in 1774 with a few of her followers. She made further converts during the unsettled years of the American Revolution and began to set up the communities for which the society became famous. The first of these in the lower Ohio Valley began its organized existence about 1800.

The religious meetings of the Shakers featured singing, marching, and the rather violent group dancing movements from which evolved their popular name. Members of the faith subscribed to a regimen which featured community of labor and property, equality of the sexes, celibacy, and separation from things worldly in their self-sufficient communities. The members supported themselves by farming, gardening, and the manufacture of plain, well-made furniture, leather goods, hats, woven fabrics, and by the packaging and selling of roots, herbs, seeds, and nursery cuttings. Shaker merchandise enjoyed a universal reputation for excellence; their seeds were probably the first marketed in this country under a full guarantee and the term "Shaker-built" came to indicate a high standard of quality in advertised merchandise.

There eventually came to be a total of seven communities in Ohio, Indiana, and Kentucky but several were experimental and were soon abandoned. Most notable of those which flourished in the valley were Union Village (west of present Lebanon, Ohio), Pleasant Hill (near Harrodsburg, Kentucky), and South Union (near Bowling Green, Kentucky).

The former two communities were founded in 1805, the last in 1800. All three, at their most flourishing, included plain, substantial "family buildings"—each community was divided into groups of thirty to fifty men and women who lived as brothers and sisters—an inn, meetinghouse, and various buildings for manufacturing, processing, storage, and stabling of livestock. All were located in the center of gardens, orchards, fields, and pasture.

The Shakers lived well—none better—and while Mother Ann Lee had held that all pleasure was sinful, reliable reports from visitors to their communities would seem to indicate that Shaker life was often far from the solemn existence she had designed.

The sect was doomed by both the requirement of celibacy and the growing enlightenment of Americans which rejected an interpretation of a God jealous enough to make such unnatural demands in the name of worship upon creatures made in His own image. The Shakers have passed from the Ohio Valley and their villages have either been put to some more practical use or set aside as monuments to the futility of their builders' celibate lives.

The second experiment in religious community living reached the Ohio only shortly after the Shakers and enjoyed its greatest prosperity during the same period. Unlike the democratic government of the former, this was a dictatorship—though a reasonably benevolent one. It was that of which Father George Rapp was the leader. No dilettante, no liberal thinker, no intellectual, Father Rapp aimed only for the life hereafter and abundant living on earth meanwhile. His people left little behind them except some now long obsolete improvements in manufacturing and agriculture, and made small contribuation to the progress of the valley while they survived. Even so, at the height of their prosperity their communities were among the wonders of the new West.

George Rapp, Württemberg farmer and vinedresser, was born in 1757. He became an enthusiastic lay worker in the Pietism movement and, in about his thirtieth year, began to preach in his own home to a congregation of his neighbors. The peculiarity of Rapp's interpretation of the Scriptures was the mark of his cult: he advocated celibacy and taught that the second coming of Christ was imminent.

Persecution at the hands of the German Lutheran Church in Europe became more and more stringent and reports of the freedom that was to be enjoyed in the United States appeared to offer an ideal escape. Rapp crossed the Atlantic in 1803, accompanied by his adopted son, Frederick, and other of his followers. They had some $20,000 of combined capital and they proposed to locate on free soil, where they could carry out both their religious and their economic plans without interference.

They eventually bought about 9,000 acres of land near Zelienople, Pennsylvania, about twelve miles northeast of the junction of the Beaver and Ohio rivers. On this tract they planned the town of Harmonie and in the fall of 1804 transported to the new country one hundred and thirty-five families, with their seed, vine cuttings, implements, and goods.

There was an immediate defection of about three hundred of

the immigrants under a lieutenant named Haller, who began a separate colony in Lycoming County, but about forty families followed Rapp to the unimproved lands of the original purchase in November and these were joined by another fifty families in 1805.

Under a rigid and inspired discipline which was lacking in the other mass emigrations of the early nineteenth century, Rapp's colonists began to prosper almost immediately. In the summer of the first year they built forty-six log houses, a large barn, and a gristmill, and they cleared over two hundred acres of land.

By 1810, according to John Bradbury, who reported a visit to Harmonie in that year, the Rappite community comprised

a large inn, a frame barn, 100 feet long, a blue dyer's shop, an oil mill, and a tannery.

There was a saw mill, brewery, brick meeting house, dwellings, stables, a two hundred and twenty foot bridge over Conaquenesing Creek; a fulling mill, hemp-breaking mill, and a combination grist mill and wine cellar. Their annual agricultural production was forty-five hundred bushels of rye, the same of wheat, six thousand bushels of corn, ten thousand of potatoes, five thousand of oats, four thousand pounds of flax and hemp, a hundred bushels of barley brewed into beer and fifty gallons of sweet oil pressed from the white poppy and they had begun the manufacture of broadcloth and the operation of a carding machine, two spinning jennies and a factory with twenty looms.

At the time of Bradbury's visit, he estimated the Rappites' assets as:

	Dollars
9000 acres of land, with improvements	90,000
Stock of provisions for one year for 800 persons	25,000
Stock of goods, spirits, manufactures, leather, implements of husbandry, etc. etc.	50,000
Dwelling houses	18,000
Mills, machinery and public buildings	21,000
Horses, cattle, hogs and poultry	10,000
1000 sheep, one third of them Merinoes, of which one ram costs 1000	6,000
	220,000

Sometime in 1805 the Rappites had bound themselves to an agreement of which the salient points specified:

1. All cash, land and chattels of every member to be a free gift for the use and benefit of the community, and to be at the disposal of the superintendents as if the members had never possessed them; members pledge themselves to submit to the laws of the community, to show a real obedience to the superintendents, to give the labor of their hands for the good of the community, and to hold their children to do the same.

2. George Rapp and his associates to give each member such secular and religious education as would tend to his temporal welfare and eternal felicity, to supply members all the necessities of life, to support them and their women and children alike in sickness and health and old age.

3. In case of withdrawal, a member's money is to be refunded to him without interest; if he had come in without capital such a sum to be awarded to him as his conduct as a member would justify.

Intoxicated by their apparent economic potentialities and the religious freedom of the New World, and falling more and more under the spell of George Rapp's hypnotic personality, the Harmonists developed a confidence in their leader which amounted to worship and which resulted in a testimonial gesture to him which took the form of an endorsement of the policy of celibacy, which he advocated.

This action was taken in 1807. Subscribing married couples separated and lived in the men and women's dormitory and children became the general care of the community. Giving in marriage by the church—which was now wholly and completely embodied in George Rapp—became a thing of the past. The rule of celibacy was not unanimously observed then or later—even so, there were enough endorsers to cause the society to appear strange and unnatural (and therefore to be feared) in the eyes of its backwoods neighbors. It was this custom of the Rappites, fully as much as the jealousy of their material prosperity usually believed to be the reason, that caused the unfriendliness and distrust with which they continued to be regarded until the last of them had passed beyond the reach of human disapproval.

Backwoods gossip spread tales about Rapp and his followers. Perhaps there was no basis for them, but they persisted and the unnatural way of life the Germans chose certainly gave grounds enough for suspicion. From those of sexual irregularities the rumors grew to include every kind of barbarity and degeneracy— including the reports, put into circulation years later, that George Rapp had castrated his own son, in the interest of that young man's purity; that he had caused the assassination of his adopted son, Frederick, because the latter refused to put away his wife; that he had murdered a servant girl—a sort of vestal virgin in his home, the more romantic have it—and sealed her body in a passageway still believed to connect the cellar of his house in the Indiana Harmony community with the Wabash River.

To return to the early Pennsylvania days of the Rappites: In spite of the fact that their wealth had increased from $20,000 to more than $220,000, the people became dissatisfied (or their leader had caused them to become so) by the year 1811.

Reason generally given for this was that the lands were badly located. The community was indeed ten or twelve miles from the waters of the Ohio, which furnished their avenue of export, and a warmer climate would certainly permit further specialization in the vine and fruit culture that most of the members had carried on in Europe. Perhaps, in addition to these matters of interest to the laity, George Rapp was not averse to moving still farther from the influence of the settlements that could eventually bring about the defection of those of his people who might harbor worldly leanings.

Whatever causes were chiefly responsible the sum of them was a decision to move west and south. In 1813 they selected a great meadow on the Wabash, thirty miles by water above the Ohio but only fifteen or so overland. It was secluded and it was fertile, with flatlands for tilling and warm hillsides for vineyards. Besides, while immediately accessible by the then little-traveled Wabash, it was removed from the traffic with which the Ohio River was beginning to be crowded, and it was protected from Indian incursion from the north by Vincennes, the Indiana Territory capital.

The Rappites purchased a tract of land approximately 30,000

acres in extent and a pioneering party went ahead in June, 1814, to prepare shelter and clear land. In the spring of 1815, again with its seeds, tools, household goods—and this time with a year's supply of provisions and its livestock as well—the rest of the community embarked by boat for the new site.

The society sold its property in Pennsylvania, with improvements, for $100,000—not necessarily at a sacrifice, since $129,000 is represented in Bradbury's inventory of 1811 by what may be considered plant investment and an undivided part of even this was in equipment and machinery which was probably carried to the Wabash or sold separately.

Besides the $100,000 from the sale of land, the return from the inventory of "goods, spirits, manufactures, leather, etc., etc.," must have brought enough additional money to total a most substantial capital with which to start the new town in Indiana. They called it by the same name as the old but, where the old had been spelled "Harmonie," they usually employed the spelling "Harmony" in the West.

The advent of these several hundred new and rather strange citizens into Indiana Territory attracted immediate and not altogether favorable attention. They made up a rather substantial percentage of the white population of the southwestern part of the territory and very likely caused the settled population—or at least its less liberal element—to "view with alarm."

There was really very little about the Rappites, except for the rumors that had no doubt preceded them, to alarm even the most timorous. They were plain, quiet, even stupid-appearing people. They went to work immediately clearing fields, setting out vines, building houses and a church, and tending their stock. They laid plans to further their trade, since all the products they had manufactured before, and of which they had at least a small stock on hand, were in great demand on the lower Ohio.

In the third year of the existence of Harmony on the Wabash, a measure was taken which greatly strengthened Rapp's already firm hold on the people. The agreement of 1805 was abrogated so that all previous claims to private property were abolished. The great book in which had been preserved the record of original contributions of individuals was destroyed by consent of the members and the society became irrevocably communistic. The characteristic Württembergian thrift that had hitherto made the society economically successful now bound it certainly together—for it was next to unthinkable that a mature Württemberger would consider withdrawing from a society of which his own property was an indivisible part.

Three years saw much building at Harmony. Log houses, set back from the street, were built first for temporary use and when permanent residences were put up in front of them the former served as storehouses or stables. Some of the new houses were brick and some were of poplar and walnut, insulated with a biscuitlike wadding of clay and straw. A considerable number of both types are still in use in the town.

The years 1817 and 1818, besides (or possibly because of) being the period of the destruction of the original records of investments, were a time of great public works. A handsome brick residence was built for Father George Rapp and the colonists completed the combined fort and granary that still stands, and which—although its chief utility was certainly in the latter field—is nevertheless one of few reminders in the Midwest of the day when a blockhouse was occasionally a very desirable public utility. The building is 40 by 70 feet in area and had, originally, 5 floors. The stone walls are 5 feet thick at the base, with 16 portholes for ventilation or defense, while the upper walls are brick. The roof was originally covered with tile.

The manufacturing that had proved so profitable in Pennsylvania was an even greater asset here, in the midst of a relatively untouched and noncompetitive market. Unfortunately there was no Bradbury to set down a chronological record of their achievements and to take a comparative inventory, but diary-keeping Englishmen soon began to visit the neighborhood and nearly all of them mentioned Harmony.

Among the earliest of these was William Faux. Unfortunately his *Memorable Days in America,* published in 1823 after his visit in 1819, is so full of misinterpretations and misrepresentations that it has been generally discredited as a historical source; it remained for

John Woods to give the first comprehensive eyewitness account of Harmony on the Wabash.

Woods lived for a time at Albion, Illinois, and he visited Harmony in February, 1821, near the peak of its prosperity. He reports that the community's business was diversified, employing carpenters, wheelwrights, smiths, tanners, saddlers, harness makers, shoemakers, weavers of linen and woolen goods, curriers, distillers, malt- and brew-house employees, and millers.

The Rappites had set out orchards and vineyards, "from which they make a small quantity of wine, not of the best quality . . ." and they had a large stock of merchandise on hand. "Their store goods are of very considerable value . . . their business is carried on in the name of . . . Mr. Frederick Rapp; all accounts being made in the name of F. Rapp only . . ."

There was much business and agricultural activity: "manufacturies, distilleries, malt and brew houses, steam, corn, and fulling mills; also a large barn in which was a powerful eight-horse threshing machine, with a winnowing one attached to it . . . many men were employed in putting up wheat . . . fifty women and girls breaking flax in the streets, and all seemed fully employed. They are a most industrious people; but the great part of them are not very enlightened."

In support of Woods's claim for the industry of the Rappites we may well consider a letter written from Indiana and published in *Niles' Register* in the next year. The writer gives the production per day of the Harmony industries as follows:

Hatters and shoemakers,	Value per day			$ 30.00
Distillers and brewers,	"	"	"	30.00
Spinning and carding,	"	"	"	15.00
Blacksmiths and coopers,	"	"	"	15.00
Various cloth (cotton),	"	"	"	25.00
Various cloth (woollen),	"	"	"	70.00
Flannels and linsey,	"	"	"	20.00
The tannery,	"	"	"	15.00
Wagon makers and turners,	"	"	"	12.00
Steam and other mills,	"	"	"	15.00
Saddlers, etc.,	"	"	"	15.00
				$262.00

—which was, of course, in addition to the great agricultural production of the society.

Other travelers sketched in details of the picture. In the years 1822 and 1823, William Blaney paid his visit to Harmony:

> They have indeed proceeded with everything with the greatest order and regularity . . . Everyone belongs to some particular trade or employment, and never interferes with the others, or even indeed knows what they are about. The only occasion on which they are all called out, is in the event of sudden bad weather, when the hay or corn is cut but not carried. In such a case Rapp blows a horn, and the whole community, both men and women, leave their occupations, run out to the fields, and the crop is soon gathered in, or placed in safety . . . Over every one of these trades there is a head man, who acts as overseer, and who, in particular cases, as with the blacksmith, shoemaker, &c., receives payment for any work done for strangers. . . . The head man, or foreman, always gives a receipt for the money he receives, which receipt is signed by Rapp, who knows thus every cent that is taken, and to whom all the money collected is transferred. When any one of their number wants a hat, coat, or anything else, he applies to the head man of his trade or employment, who gives him an order, which is also signed by Rapp, after which he goes to the store and gets what he wants.
>
> They have one large store in which is deposited all the articles they manufacture. The neighboring settlers for miles round, resort to this, not only on account of the excellence, but also the cheapness of the goods . . . The Harmonites have also branch stores in Shawnee town, and elsewhere, which they supply with goods, and which are managed by their agents.

Blaney did not mention the fact, but Rappite mercantile business was by no means confined to purchases from the stores. References are found throughout the literature of midwestern description to the "Rappite trading parties" that set out in season and which called on storekeepers and individuals on the Wabash, Ohio, and Mississippi rivers with merchandise of Rappite manufacture. The trade was extensive and the "rose" trademark, burned upon whisky, wine, oil, or butter cask and stamped upon hats, shoes, and yard goods gained the respect of the western consumer.

Elsewhere Blaney called attention to the growing animosity

against the Rappites and explained the economic reason for it: that upon their arrival on the Wabash, the people in the surrounding country found their presence of benefit, in that they supplied the much-needed goods, but as time went on, due to the fact that they paid no wages and were able to sell below independent merchants, and that they took all cash they received immediately out of circulation, the Rappites were beginning to be considered a drain upon the surrounding territory.

The reasoning of the settlers may be fallacious—but their conclusion is understandable.

In spite of this prosperity and security, in spite of the fact that there was no evidence of serious unrest (except for that of the occasional young couple prompted by those urges which had been moving their ancestors back to Adam and Eve) and in spite of the beauty and comfort of the town that the society had hewed, spaded, and quarried out of the countryside, someone decided, at this time, to move the Rappites on.

Was the decision made by spiritual Father George Rapp or by business manager Frederick? Which held the greater power by this time? Had the increasing importance of manufacturing and trade lifted Frederick to control; did he see the community's present, with sizable settlements growing on all sides, as a favorable time to sell for a quick and sure profit? Or did Father George, as had been suggested through the years, believe he could best hold his flock only when it was working under pressure, only when it was inspired to work to its limit to build itself the comfortable home that was the only desire of most of its members?

Whether George or Frederick Rapp originated the idea, it was decided that the time had come to sell and land was purchased back in Pennsylvania, immediately on the Ohio River this time. As it happened, the means through which the most profitable market for the Indiana property might be reached was readily at hand by 1824.

The town of Albion, in Illinois, had sprung up in what was once called the "English Prairie Settlement"—the tract already described as having been bought by Morris Birkbeck, George Flower and others, in Edwards County, Illinois. That project also had a Utopian tinge, albeit an orderly and staid one. Their sales efforts had been directed toward small farmers in England who were making hard going in the changing times of the industrial revolution. The project under the joint banners of cheap land, low taxes, and educational opportunity had attracted considerable attention.

In the summer of 1824 one of the settlers at Albion, George Flower's father, Richard, decided to take his younger son, Edward, and return to England. The trip was made partly in the interest of real estate promotion and partly because the outspoken anti-slavery views of the three Flowers had brought threats of violence from the ex-Kentuckians and ex-Virginians who made up the bulk of the southern Illinois population outside the English Prairie.

The Flowers had been in fairly close touch with the Harmony community, especially with the business agent, Frederick Rapp, and they were undoubtedly aware of the fact that the Rapps had already purchased 3,000 acres below Pittsburgh, on the Ohio, and that a pioneering party was soon to go there to begin clearing and constructing buildings in preparation for the general removal of the society back to the East. Frederick Rapp evidently had no doubt as to his ability to dispose of such a well-developed community as Harmony, at least in parcels, but Richard Flower had a better idea: he thought that the community might be sold in England as a unit, and he evidently proposed to the Rapps that he be given the agency for such a sale.

The Rapps readily agreed that Flower should receive a commission of $5,000 if he succeeded in making the sale. Whether Flower then thought of Robert Owen as a prospect is not known. He undoubtedly knew of Owen and his theories—some of Flower's people at Albion were fellow travelers of Owen's—but in any case the idea occurred to him as soon as he reached England.

Flower called on Robert Owen and his son, Robert Dale Owen, at New Lanark in September and followed that with other calls and correspondence in which he gave glowing accounts of the virtues of the lower Wabash and the Ohio Valley.

"The offer tempted my father," wrote Robert Dale Owen.

"Here was a village already built, a territory capable of supporting tens of thousands, in a country where the expressions of thought were free, and where the people were unsophisticated . . .

"I listened with delight to Mr. Flower's account of a frontier life, and when one morning my father asked me, 'Well Robert, what say you, New Lanark or Harmony?' I answered without hesitation, 'Harmony'."

Robert Owen seems to have made his decision solely upon the evidence of Flower's report and the few printed accounts of the place that may have come his way. On October 24, 1824, he sailed from Liverpool.

Finding a buyer for the little Rappite empire on the Wabash was as simple as that! Fourteen months and one week after Flower's sailing, Robert Owen had arrived in America, had released a mighty blast of publicity, had gone down the Ohio to the Wabash, and had signed a contract to purchase the town and surrounding lands. That was on January 3, 1825; the purchase price was apparently $100,000.

Already parties of Rappites were moving up the Ohio to "Economy," the new community in Pennsylvania. They were shipped like cattle, except only for the fact that four-legged animals were usually shipped downstream.

Again well-built, well-designed houses were going up and by May 17, 1825, when the last of the Rappites arrived in Pennsylvania on the steamboat *William Penn,* 3,000 acres were ready to cultivate and there was work at hand for all except the small party Father Rapp had agreed to leave in the West to undertake the hopeless task of teaching Robert Owen's followers to operate the mills and manufactories.

Economy, the village that was beginning in 1824, has recently been engulfed by the manufacturing city of Ambridge but six acres of its land and some of its finest buildings are maintained by the Harmony Society Historical Association as a Pennsylvania state shrine.

Back in the years immediately following the return of the Rappites to Pennsylvania none of the lavender glow that now attaches to the historic shrine obtained in Economy, and there was no suggestion that such an end was in store for the place. All was business in this third home of Father Rapp's people in America, even more so than in the preceding two. Their leaders, George and Frederick Rapp, had learned a great deal about agriculture, manufacturing, selling—and about how to get the most out of the labor of the people.

Some little slackening of the early rigorous exclusion of things worldly had taken place even in the Indiana settlement. Some song and sermon books had been printed and a school had been kept for the children of the community (the presence of young children was accounted for by the statement that they were orphans from Europe) and interest in reading, writing, and in the arts increased after the return to Pennsylvania.

As might have been expected, defections of young people increased slightly, here nearer to the eastern fleshpots, but there were new recruits also: prosperity reigned unthreatened until 1832.

Among the recruits were the (spurious) Count Maximilian de Leon and forty followers from Germany. These arrivals' ideas were soon found to be at considerable variance with the Rappite tenets. Among other impieties, the count advocated marriage and he soon gathered around him some two hundred and fifty apostate followers of Father Rapp—about one-third of the whole membership—and proclaimed himself Rapp's successor.

A bitter dispute followed, ending only in a compromise under which de Leon and his party left the community immediately and his two hundred and fifty matrimonial-minded ex-Rappite recruits were permitted to follow him. They took with them something between $100,000 and $150,000 as their share of the Rappite property.

The seceders moved ten miles down the Ohio and began another community (the present town of Monaca, Pennsylvania) but de Leon had neither the iron will of Father George Rapp nor Frederick's talent for commerce. The new society was short-lived.

After the rebellion was over, the Rapps still had five hundred followers—but Economy's eventual doom was no less certain, for the five hundred were the oldsters, to whom dreams of wedded life meant little, and Frederick Rapp had but a few more months to live. He died in 1834.

Manufacturing naturally declined after his passing, the average age of the members increased, and children were not adopted in sufficient numbers to perpetuate the society. In 1847 Father George Rapp went to meet the God with whom he had so long sought intimate contact but his passing made little difference; the ways of the few effective followers who remained were permanently molded to his order of living. There now came bad investments, disastrous loans, suits by ex-members and collateral heirs, but they made little difference—the assets declined, and so did the membership. The last trustees of the property, Dr. and Mrs. John S. Duss, were able, after years of hard labor and persecution by would-be heirs, to see a representative section of the community preserved for posterity.

So passed a Utopia of the God-fearing after a century of existence: how long, then, could a community of professing atheists hold out? Let us return to Robert Owen's *New* Harmony on the Wabash.

Robert Owen did little to improve the Ohio Valley economically or socially through his original experiment but others whose imaginations were touched off by sparks generated between the conflicting elements in his erratic personality left beneficial marks upon the midwestern scene which are still discernible.

Sources from which sprang the inward turmoil that drove the ambitious little dry goods clerk turned industrialist turned humanitarian crusader are evident in the events of his early life.

Robert Owen's childhood and early business life might well have been the model for one of Horatio Alger's standard biographies—although it most certainly was not, Horatio being a God-fearing man who had no literary truck with doubters. Owen was born in Newton, Montgomeryshire, England, in 1771, the sixth of seven children of Robert Owen, a saddler and ironmonger of modest position in life. It is a commentary upon the vagueness of Owen that he does not mention his mother's first name in his autobiography and his biographers have been unsuccessful in learning it.

As far as Robert Owen's statement is concerned he sprang into the world, full blown, as a schoolboy of five or so. Taken at his own evaluation, he must have been a perfect little prig. He states

that it was at this time in his career, shortly after he had begun
school, that his habit of thoughtfulness and consideration was
accidentally formed and his life, as he viewed it, began:

> I requested that this breakfast might be always ready when I
> returned from school, so that I might eat it speedily, in order to be
> first back to school. One morning, when about five years old, I ran
> home as usual . . . found my basin of flummery ready, and as I supposed
> sufficienty cooled for eating . . . but on my hastily taking a spoonful of
> it, I found it was quite scalding hot. . . . The consequence was an instant
> fainting, from the stomach being scalded . . . my parents thought life
> was extinct. However, after a considerable period I revived; but from
> that day my stomach became incapable of digesting food, except the
> most simple. . . . This made me attend to the different qualities of food
> on my changed constitution, and gave me the habit of close observation
> and continual reflection; and I have always thought that this accident
> had a great influence in forming my character.

At seven the dyspeptic young paragon had mastered the rudiments of education as offered by the village schoolmaster, and that
gentleman applied to the senior Owen for permission to make the
younger his assistant, or usher. Young Owen held that post until
his ninth year.

He was much interested in religion, although he says that his
reading of the differing tenets of the various sects soon led him to
doubt them all. According to his autobiography:

> I wrote three sermons, and I was called the little parson. These
> sermons I kept until I met with Sterne's works, in which I found among
> the sermons three so much like them in idea and turn of mind, that it
> occurred to me as I read them that I should be considered a plagiarist,
> and without thought, as I could not bear any such suspicion, I hastily
> threw them into the fire . . .

Robert Owen was between seven and nine years of age and
Sterne was presumably at the height of his powers at the time of
their respective and coincidental authorship. He continues:

> But certain it is that my reading religious works, combined with
> my other readings, compelled me to feel strongly at ten years of age that

there must be something fundamentally wrong in all religions, as they had been taught up to that period.

During my childhood, and for many years afterwards, it never occurred to me that there was anything in my habits, thoughts, and actions different from those of others of my age; but when looking back and comparing my life with many others, I have been induced to attribute my favourable difference to the effects produced at the early period when my life was endangered by the spoonful of scalding flummery . . .

I entered, however, into the amusements of those of my own standing . . . I also attended the dancing school for some time, and in all these games and exercises I excelled not only those of my own age, but those two or three years older, and I was so active that I was the best runner and leaper, both as to height and distance, in the school. . . .

I was too much of a favorite with the whole town for my benefit, and was often pitted against my equals, and sometimes against my superiors in age,—and sometimes for one thing and sometimes for another . . . One instance of this made a deep impression on my mind. Some party bet with another that I could write better than my next eldest brother, John, who was two years older; and upon a formal trial, at which judges were appointed, it was decided that my writing was the better . . . From that day I do not think my brother had as strong an affection for me as he had before this unwise competition.

—should the reader begin to share brother John's feelings, he must remember that Robert Owen was past eighty when he set these memoirs to paper, and that the Messianic delusion had long since taken a firm grip upon him. In middle years he was humorless, usually slipshod in his reasoning, sanctimonious, and prissy; but he was also superlatively kind of heart, generous, and always optimistic of the innate goodness of his fellow man. It is for his always good intentions that he must be remembered.

In 1780, when he was nine years old, he left school and was employed in a small draper and grocer's shop in the town. After a year of service he decided to move to London to join his eldest brother. William Owen had gone to London, found employment, and married the widow of his employer. Through William, Robert was soon established as a draper's apprentice at Stamford, Lincolnshire. Robert Owen was now approaching the discreet age of eleven.

At the end of his three-year apprenticeship, he secured a position
with a London firm, where he stayed for a year before moving on
to a wholesale and retail draper's house in Manchester.

Here he remained until he was eighteen. Manchester's cotton
manufacturing enterprises were beginning their first boom (due to
mechanical improvements in processing, cheaper cotton goods were
beginning to supplant the hitherto pre-eminent woolens in the
British market) and young Owen was also able to pick up, on
the side, some information about manufacturing methods.

He had formed an acquaintance in Manchester with Ernest
Jones, a mechanic of considerable ability, who proposed a partnership under which the two should manufacture machinery for
spinning. They had barely got into production when Jones found
another prospective partner with more money and the two bought
Robert Owen out, giving, for his interest, the machinery they had
already manufactured. This included six spinning "mules," a reel,
and a machine for packaging yarn.

With this newly acquired equipment and in quarters rented
advantageously Robert Owen set himself up as a manufacturer. His
business prospered comfortably and soon an established manufacturer of the place offered him a position as superintendent of his
large and flourishing mill. By 1796 he had become a junior partner
in the then-organizing Chorlton Twist Company, which began
operation in 1798.

Robert Owen had read, thought, and conversed much on the
social and economic injustice that the beginning of the industrial
revolution was already working on the country people who
swarmed into Manchester to man the mills. He was particularly
interested in the lot of the children, many of whom were employed
for full thirteen-hour shifts at the age of only five or six years. His
interest in these mill children, whose lot he was to do most to
ameliorate, was keen from this time on. Probably his sympathy was
first enlisted by bitter memories of his own childhood—some such
memories might also account for the fact that he almost completely
ignores his parents in his autobiography and for his evident exaggeration of his juvenile prowess.

On a trip in the interest of the new company, Owen visited

Glasgow and met Miss Anne Caroline Dale. She was the daughter of David Dale, a wealthy Scotch banker, head of the Old Scotch Independent Branch of the Presbyterian Church and owner of cotton mills at New Lanark.

Owen soon found further occasion to visit Glasgow and, incidentally, Miss Dale. Together they inspected New Lanark—where the Dales had a summer home—and the mills, and Owen was immediately taken with the attractive surroundings of the town. Eventually he proposed marriage to Miss Dale and, according to his report of her rather involved syntax, she replied, "if you can find the means to overcome my father's objections, it would go far to remove any I may now have, to the request you have made."

Owen had not then met David Dale, and thinking to prepare the way for his matrimonial negotiations by a business call, he went to Mr. Dale and, probably with no expectation at all of actual purchase, inquired if the New Lanark mills were for sale? Dale was gruff at first but young Owen succeeded in convincing him of his sincerity, and after some negotiations with his partners in Manchester, he found himself a part owner of the business.

His organization took over the establishment, some diplomatic assistance by friends secured Mr. Dale's consent to the marriage, and it took place on September 30, 1799.

Shortly Owen and his partners found the old management of the New Lanark property unsatisfactory, and Owen himself took over its direction. In January, 1800, he and Anne Caroline removed to New Lanark—which thus became the theater of the performance for which Robert Owen is most justly famous: that of labor reform.

The English and Scotch mills of that period were perfect infernos: places in which those who entered did so without hope of emerging except as blind or maimed or tubercular wreckage. Wages were infinitesimal, hours were desperately long, sanitary conditions were intolerable. In the spinning mills, a large percentage of the laborers were children under twelve years of age.

In New Lanark alone four or five hundred apprenticed pauper children were employed besides the adult workers and their numerous offspring. While David Dale "had solaced his conscience by lodging these children comparatively well, and giving them religious instruction," he had gone little further in the interest of their welfare, moral or economic. Older workers at New Lanark, like the pauper children, were housed in company buildings, one room to a family. Doubtless conditions elsewhere were much worse but even in New Lanark almost any change would certainly be an improvement.

The young manager's first step was to abandon the system of employing pauper children apprentices. He followed this with a rearrangement of the mill machinery to secure better light and ventilation, improvements in the mill buildings and houses, and the installation of sanitary facilities in the town: even those moves attracted considerable attention and comment in the neighborhood.

In 1812 he published his *Statement Concerning the New Lanark Establishment,* in which he described the current state of the local mills and the improvements that had been made during the preceding twelve years.

In the period between 1814 and 1825 Robert Owen introduced the measures upon which his philosophy of paternalistic guidance and co-operative enterprise was founded. He cut working hours drastically, hired no children under ten years of age, introduced schools and even provided classes for the adult workers. He organized co-operative stores and encouraged the workers to undertake gardening and handicrafts of all kinds. To the great horror of some of his pious Scotch neighbors, he also encouraged musical evenings, dancing, drills, and other social pastimes.

He published frequently and mailed thousands of marked copies of pamphlets and newspapers containing his speeches and letters to people who he thought might be helpful in promoting his campaigns for governmental regulations on employment, the control of capital, and the new Owen system of education. He traveled widely, meeting men prominent in the day, and he soon became one of the storm centers of the industrial revolution that was making Britain great.

From his first view of New Lanark he had believed that manufacturing should be carried on in small, self-sufficient communities. As his crusade progressed it was constantly brought home to him that even such a community, if located anywhere on the British

Isles, would still be hampered by criticism and the limitations placed upon it by a reactionary national government and tradition-bound neighbors. Recollections of the obstacles that had slowed his accomplished reforms put him in a mood to be immediately receptive when such a possibility as the purchase of Harmony in Indiana should offer.

The Owen children had been brought up in most pleasant surroundings. At first taught by private tutors, Robert Dale, David Dale, and Richard, at least, had attended the advanced school of Emanuel von Fellenberg at Hofwyl, Switzerland. Fellenberg's plan, with the Pestalozzian system, was later to be incorporated in the educational scheme at New Harmony.

Of Robert Owen's sons it was certainly Robert Dale who was most like his father, although he lacked entirely the business acumen that was so strong a component of the older Owen's character in his early years. It was he who was engaged in helping his father to lay plans for the new co-operative community (its location not yet chosen) when Flower reached England with the Rappite commission, ready to execute, in his pocket. The result of Flower's visit has already been noted.

Robert Owen, accustomed to a less-volatile audience in England, very early made the mistake that proved fatal to even a reasonable trial of his experiment in the United States. When he landed—before he had even called on George Rapp to discuss the purchase—he announced the opening of his community, welcomed recruits, and invited all those who were interested and dissatisfied with their present lot to join him; at his expense! He issued manifestoes at Albany, New York, Washington, and Philadelphia and his plans for the creation of a "New System of Society" in a "New Moral World" were avidly copied by rural newspapers throughout the country and, especially, from one end of the Ohio to the other. Finally he came to Indiana and, Harmony bought and a payment made, returned east and continued his interviews, his addresses, his harangues to any group he could buttonhole. By April, before the last of the Rappites had departed, he began to reap his harvest.

The United States had its full share of malcontents and Owen's invitation to gather around a free board and an open purse was too good to pass up. When he returned to the west for a brief visit to his new property in the spring hundreds of volunteer followers were already on the ground. Most of them had lifetimes of failure behind and nothing except their own enthusiasm to warrant hope for anything else in the future, but delighted with their zeal their host ignored these obvious shortcomings and gave them his blessing. He arranged for their feeding and clothing and—with not the slightest thought of the practical necessity of putting them to work in the fields and mills—dashed off to England to close his affairs and to plead for even more recruits.

During the following year, and indeed until his negotiable assets were entirely expended, Robert Owen fed and clothed and entertained a turbulent mob of the incompetent, the quarrelsome, and the visionary which at one time totaled little less than a thousand men, women, and children!

The honeymoon of the parlor intellectuals on the banks of the Wabash began as might have been anticipated: the recruits venerated Owen's principles but they could not agree upon the exact terms under which they would accept his charity, even while they existed on it! Always they were vocal:

Hearing Mr. Owen I am always in the hills . . . We are confident Mr. Owen will furnish us with every necessity . . . New Harmony is now so filled we scarcely know where to put those who arrive. Hundreds are still on the road. . . . The laurels of Owen of Lanark will be green . . . I regret Mr. Owen's departure but all I fear is the party spirit, a good deal of which was shown in the election . . . I cannot say there appeared much harmony among us . . . industry was manifested in accusing others of doing little. There has been a Reign of Reports from the Committee . . . idle nor industrious are neither of them satisfied. Those who come to look at us seem highly pleased, but they see only the outside . . .

The hopeful scientific economy of the New System finally began to operate: the "practical farmers" organized—and elected a professional bookkeeper president! Resolutions and plans for farming the lands were drawn up but the season was too far along to plant crops. A member later reminisced: "The most inefficient persons pushed to the superintendency . . . and when rejected would sulk and shirk. Factories were mismanaged, cattle and sheep were neglected, fences were being used for firewood . . . the drones became the masters of the hive."

There was a brighter side, of course. Efforts at self-improvement were made during the brief intervals between discussions of principle and bickerings over the quality and quantity of free food which Robert Owen's agents issued. Cheerful, vague, William Pelham of the New Harmony *Gazette* wrote of the weekly program: dancing on Tuesdays, community meetings on Wednesdays, concerts on Thursdays, and "Fridays something else which I do not recollect and Saturday evenings not appropriated . . . Sunday the Rev. Mr. Jennings commonly delivers a lecture without any text . . ." Blessed by New Harmony, branch Owenite communities were formed in the Ohio Valley, at Yellow Springs in Ohio, Albion, Illinois, and in Allegheny County, Pennsylvania—although only the first ever passed the planning stage and even it was of but a few months' blooming.

A school was organized under the direction of the Rev. Mr. Jennings to the relief of the parents but the sorrow of the pupils; "no meat, no eggs, no pies, nothing, in fact, but corn mush and milk. We were marched about with ice beneath us with no shoes, no stockings . . ."

Of course the governing body was at fault; why not, where all was designed to be freedom? One member, typical of many, wrote to a relative that the "Committee announces that all power is invested in them alone . . . our government is an aristocracy, a despotism!"—but of course Robert Owen, now reported as planning to embark at Liverpool in October, would solve everything, reform all shirkers, cause the fields to blossom, and increase the subsistence allowance of the multitude.

Meanwhile the New Harmony *Gazette* undertook a campaign to familarize the public with the happy life in the New Moral World. Even it was forced to admit that "Our manufacturing and mechanical branches may be considered in a state of infancy," but there were always the concerts and lectures! Early issues of the paper containing reports of progress were mailed to Robert Owen's

list of prospective patrons but young William Owen, at the time
the *Gazette's* editor, was inquiring privately by letter of his father
as to "What to do with those who profess to do anything and
everything? . . . they are perfect drones . . . we have got rid of
many such, although we still have a few left."

Robert Owen returned to the United States but not immediately
to New Harmony. He stopped in the seaboard cities to lecture, to
invite recruits, and to advertise for mechanics, masons, carpenters,
and joiners to go immediately to the Wabash to begin a new community building—while William Owen, horrified at the latter
move, wrote to his father: "We were astonished to hear you had advertised for so many hands. . . . It will be impossible to give them
houses or even rooms . . . I repeat and impress upon your mind
that *we have no room for them.* As for building houses . . . we have
no lime, no rocks, no brick, no timber . . ."

Dissatisfaction within the community grew as its members
read the descriptions of their phenomenal success published in the
East, where a whirlwind campaign was waged by Robert Owen,
Robert Dale Owen, Macdonald, and Stedman Whitwell. The four
called on the great and near-great; Whitwell presented a model of
the proposed new community building to the president at Washington; Macdonald and Whitwell called on Thomas Jefferson to
discuss architectural problems and higher education; while Robert
Owen, in brief intervals between speaking engagements, concentrated his attention on the Philadelphia philanthropist and patron
of learning, William Maclure, in hope of getting his support in
the experiment.

Presently Maclure agreed to finance and direct the educational
operations at New Harmony and he carried out his promise.
(Though very shortly after he and his teachers and scholars arrived
there came an end of Robert Owen's ready cash and, concurrently,
an end of Owenism and the New Moral World on the Wabash.)

On January 12, 1826, Owen returned to New Harmony for his
second view. He addressed the assembled inhabitants on the same
evening and, after he had "earnestly recommended unity & brotherly
love," he expressed himself as amazed at the progress made in his
absence! He declared that the "Half-way House" had now been

passed and that the time had come to form the permanent organization to supersede the Preliminary Society. A Committee of Seven was chosen to draw up a new constitution—although no progress had yet been made in commerce or agriculture—and the citizens began a new series of meetings to debate and consider it. The debate became acrimonious, then violent, and Captain Macdonald described the beginning of the next and final stage in a succinct if bald entry in his diary: "a small society of American Backwoodsmen separated from the rest, and next a large one of English Emigrants. The remainder then had a misunderstanding."

The disillusioned were now departing by dozens—with a consequent rise in the average social fitness of the remaining inhabitants—but the time was too late. More seceding communities flourished for a few days, received a grant of land or buildings, and broke into dissension or disappeared from the record. By the end of 1826, Robert Owen found himself unable to meet his coming payment to the Rappites and Maclure purchased buildings for his schools but refused to share Owen's obligations otherwise.

In May, 1827, Robert Owen addressed those of his followers who were still in the town and willing to listen to him for old time's sake, packed up his belongings, and departed—already inspired by dreams of new methods of benefiting his fellow man elsewhere.

Maclure's people stayed on, as did Robert Owen's sons, and with the governing of the New Moral World no longer an issue for debate, there began the golden years of culture in Posey County, Indiana (in which, contrary to tradition, there is no *"Hoop-pole Township"*). They were golden years indeed and *their* post-Rapp, post-Owen influence had not only a profound effect upon culture and learning in the Ohio Valley, but a considerable one upon the entire United States.

Other minor establishments designed as Godless Heavens upon Earth besides New Harmony sprang up on the Ohio or adjacent to it. For instance, there was Utopia (itself), the remains of which still stand upon the Ohio's north bank about forty-eight miles east of Cincinnati.

Judge Wade Loofborough, who was either unfamiliar with

the experience of Robert Owen or fancied that he could succeed where an Englishman might fail, bought a strip of good land fronting the river and organized a communistic settlement of a dozen or so families upon it in 1844. The literature concerning the judge's experiment is less voluminous than that of New Harmony and there seems to be no positive evidence as to whether the buildings erected represented a co-operative investment, or whether he had, like Robert Owen, footed the bill himself in the interest of mankind. Whichever the case, the history was the same: trouble arose over liberty or the lack of it, over division of labor and goods. The members bickered through a couple of years and dissolved their association. As in the case of the Rappites' Harmony on the Wabash, however, a buyer was waiting. The community was purchased by a congregation of spiritualists under the earthly guidance of John Wattles.

Those people—around a hundred in number—added to the buildings but in doing so they gave evidence that they had either proceeded without ghostly sanction or had depended upon an impractical spirit for guidance: they built their new dormitory too near the river and it collapsed during the flood of 1847, killing several of their members. This community also dispersed, leaving only a few neat brick buildings upon the higher ground to mark the site.

Earthly paradises founded upon a community of property and effort simply did not prosper for long upon the Ohio. It is an interesting fact that by far the majority of subscribers to Rapp's movement and the leaders in that of Owen were foreign born, fresh from the confused and unpromising European scene of the nineteenth century. The native American could easily co-operate with his neighbors in the exchange of labor, in making loans, in giving charity—he demonstrated this every day upon the frontier—but even when his laziness or cupidity had led him to join such a movement as Owen's, he soon found the idea of communal property repugnant; in a society such as Rapp's he resented arbitrary leadership, in a Shaker community he came to despise even comfort and good living when it had to be purchased at the expense of independence.

Even the transplanted European was likely to abhor a life of regimentation, once he had seen what miracles of independence were being worked around him. Westerners, even by adoption, treasured their personal liberty, though many of them could not forbear the occasional opportunity to get something for nothing for a time: that was, and is, a hopeful sign.

Chapter Fourteen

LEARNING ON THE RIVER

THERE need be nothing surprising about the fact that the first
educated men who appeared upon the Ohio were priests and military engineers, surveyors, and the emissaries—or as frequently
clerks to the emissaries—of the various governments who came at
one time or another to explore or claim the territory.

Whoever may have been the first French officer who visited the
stretch of the Ohio between the Wabash and the Mississippi the
chances are that he had with him a priest of the Roman Catholic
Church. The priest perforce was a learned man; in secular as well

as spiritual matters he exercised his learning as chaplain to the expedition and frequently as its clerk and journalist as well. He also exerted himself in saving Indian souls as opportunity offered along the route and in locating favorable spots for permanent missions.

Pierre Joseph Céloron, for the French government, and Christopher Gist, for the Ohio Company, seem to have made the first comprehensive written reports upon the river in 1749 and 1750. Gist also had the status of a representative of the government of Virginia, since he worked under secret instructions from the governor as well as for the stockholders of the original Ohio Company: besides, he was a surveyor, by some accounts, and thus further maintained the pattern.

The first accurate geographical and topographical reconnaissance of the Ohio River resulted from an expedition of 1766. That report upon the country and the river's course was the work of Lieutenant Thomas Hutchins—whose civil and military activities on the river have been mentioned heretofore. He was accompanied on the assignment by Captain Harry Gordon, draftsman of an accurate and rather beautifully executed map evidently designed to accompany Hutchins's notes. Something must have gone amiss in the project, for, while Hutchins's report and a map of his surveys were published as *A Topographical Description of Virginia, Pennsylvania, Maryland and North Carolina . . .* in London in 1778 and brought him a fame which survives, Captain Harry Gordon soon disappeared from the American scene and his detailed map of the river itself existed until recent years only in the original preserved in the Library of Congress.

Hutchins's work should in no way be minimized. His was the first book published which pictured the Ohio and its valley accurately and in correct geographical proportion. Both Hutchins and Gordon, at the time of these operations, were officers in the British colonial forces stationed at Fort Pitt.

Of course some men of education saw and even resided upon the Ohio in the early years of settlement but made no contribution other than their presence—one of sometimes dubious value. Such was Harman Blennerhassett, for instance. Unquestionably he brought the first extensive library and the first collection of scientific instruments and paraphernalia to the river and, a graduate of Trinity College, Dublin, with the B.A. and LL.B. degrees, he was probably the first dweller upon the Ohio educated in an institution of then world-wide reputation. Since, however, he used his experimental materials and his library solely in his own amateur dabblings in the realms of learning—since, far from educating or improving his neighbors, he even cut himself off from intercourse with the common herd in a stronghold of which the Ohio formed the protective moat and welcomed only such representatives of the American aristocracy as erratic Aaron Burr—he need scarcely be considered here. Better devote attention to men who, even though educated in institutions of academic standing inferior to Trinity College, made popular contribution to the culture and the well-being of the valley: the scientists, the doctors of medicine, the lawyers, and the teachers.

Perhaps to Pennsylvania-born John Bartram, called "The Father of American Botany," belongs the distinction of having been the first student of the natural sciences to visit the waters of the Ohio. He visited the Forks and the results are published in his *Observations on the Inhabitants, Climate, Soil, Rivers, Productions, Animals, and Other Matters Worthy of Notice, Made by Mr. John Bartram in His Travels from Pennsylvania to Onondaga, Oswego, and The Lake Ontario in Canada,* which was brought out in London in 1751.

Other men of science essayed an examination of the lower river. In the little-known account of his captivity by Indians in 1788 Thomas Ridout reports that he was taken to an Indian town on the lower Wabash in which he saw "several rich suits of clothes . . . belonging to some French gentlemen, taken about the same part of the Ohio in which I had been captured. As they made resistance all were killed. They proved to be three gentlemen— agriculturist, botantist and mineralogist—about to explore the country . . ."

No clue exists as to who these Frenchmen may have been but, had they known it, there remained a consolation for them; at least countrymen of theirs would successfully begin the work they had evidently hoped to carry out. Apparently the first scientist to make a complete survey of the river was François André Michaux. Mainly interested in the botany of the valley he made a downstream trip in 1802, collected specimens, took notes, accurate if brief, and described the flora of the river banks. The results of his observations were published and republished in both France and America in his *Travels to the Westward of the Allegheny Mountains in the States of Ohio, Kentucky, and Tennessee*, the first edition of which appeared in Paris in 1804.

Born in France, François André Michaux was the son of the distinguished botanist André Michaux, who brought him to the United States in 1785. François, fifteen years old at the time of their arrival, accompanied and assisted his father in the latter's comprehensive botanical survey of the country around Charleston, South Carolina. André Michaux made a remarkable collection of the plant life in that neighborhood, prepared and preserved his specimens with infinite pains, and in 1790 transported his collection and his son back to France—just in time to make the former available for wanton destruction by the revolutionary mobs and the latter for active participation in the Revolution. These misfortunes did not prevent André from publishing his monumental folio, *Histoire des Chênes de L'Amerique,* in Paris in 1801.

Young François seems to have made a considerable impression upon the new leaders who came into power with the revolutionary movement, for he was returned to the United States as an agent for the Republic of France.

It was in 1802 that he took time from his governmental agency to make the trip on the Ohio that furnished material for the book already mentioned. He returned to France but later he spent three years in this country while completing and publishing (1810-1813) the monumental and significant *The North American Sylva,* based upon the work he and his father had begun in 1785. On this occasion he also distinguished himself by being one of the two passengers carried on the trial trip of Robert Fulton's steamboat, the *Clermont.*

In 1792 another Frenchman sailed from Havre with the purpose of traveling in America to see what a democracy, in a slightly more mature state than was that of his own country, might have to offer. He was Constantine F. Volney (recently he had been *Le Comte de Volney*) and the French Revolution had brought him imprisonment for ten months, the loss of his title and most of his property.

Volney's interest, besides his first in observing the working of democracy upon the inhabitants, was chiefly in geology, soil, and climate; his book was entitled, in its English translation, first published in London in 1804, *View of the Climate and Soil of the United States of America . . .*

There were other writers, of course, but these—Bartram, Michaux the younger, Volney, and Hutchins—seem to have been most widely circulated. Certainly their books, taken together, provide a fairly comprehensive view of the natural history of the Ohio River in the light of scientific method of the day.

Settlement increased after the nineteenth century opened and travel and residence achieved a degree of safety which encouraged more comprehensive scientific inventories. Threats of Indian attack, of losing oneself hopelessly and permanently in the wilderness, of succumbing to either savage men, reptiles, beasts, or equally fatal starvation were lessened to a point where a man of inquiring scientific mind might actually think of carrying on his studies as a resident of even the wildest stretches of the river. The first to make such a decision (although he refrained from mentioning to his family or friends that scientific study was his real object lest he be suspected of mental instability) was John James Fougère Audubon. In 1808 he floated his family downstream from Pittsburgh to Louisville.

Writing of the changes on the Ohio two decades later, he recalled the river as he first knew it:

Nature, in her varied arrangements, seems to have felt a partiality towards this portion of our country. As the traveler ascends or descends the Ohio he cannot help remarking that, alternately, nearly the whole length of the river the margin on one side is bounded by lofty hills and a rolling surface; while on the other extensive plains of the richest alluvial land are seen as far as the eye can command the view. Islands of varied size and form rise here and there from the bosom of the water, and the winding course of the stream frequently brings you to places where the idea of being on a river of great length changes to that of floating on a lake of moderate extent . . . Purer pleasures I never felt; nor have you, reader . . .

When I think of these times, and call back to my mind the grandeur and beauty of those almost uninhabited shores; when I picture to myself the dense and lofty summits of the forest, that every-where spread along the hills and overhung the margins of the stream . . . When I reflect that all this grand portion of our Union, instead of being in a state of nature, is now more or less covered with villages, farms, and towns . . . that steamboats are gliding to and fro over the whole length of the majestick river, forcing commerce to take root and to prosper at every spot—when I see the surplus population of Europe coming to assist . . . and transplanting civilization into its dark recesses —when I remember that these extraordinary changes have all taken place in the short period of twenty years, I pause, I wonder, and al-though I know all to be the fact, can scarcely believe in its reality.

At Louisville in 1808 there were no steamboats; commerce, though rooted, was blossoming but modestly. There, under the patronage of friends and relatives, Audubon set up a mercantile venture. It was the first of a series which would be hopefully be-gun, neglected in favor of ornithology, and allowed to fail, only to be refinanced by the long-suffering relatives—and to follow with-out variation the pattern of its predecessors.

Largest investor in these ventures seems to have been Ben-jamin Bakewell, wealthy and altruistic flint-glass manufacturer of Pittsburgh and uncle of Audubon's wife. Bakewell maintained many relatives in his day but even his ready purse could not keep the Audubons from the direst poverty at times during the next seventeen years upon the Ohio and in Louisiana. Bakewell's pa-tience and purse must both have been sorely tried before recogni-tion of Audubon's work finally brought European patronage, com-parative wealth, and eventually a reputation which placed him, as a patron of the birds of the air, only second to St. Francis of Assisi.

Seven years after the advent of Audubon—by then he had re-moved his mercantile operations downstream to Henderson, Kentucky—there burst upon the intellectual scene in the still exceedingly primitive West, the fabulous Constantine Samuel Rafinesque-Schmaltz.

A man so named could scarcely have been otherwise than distinguished: Rafinesque (he had abandoned the hyphened section of his surname before he arrived in the West) was that indeed. Born in a suburb of Constantinople on October 22, 1783, he was the son of a French merchant and a mother of German parentage (from whence the Schmaltz) who had herself been born in Greece. He, Constantine, was destined to become and to remain for a quarter of a century after his arrival the funniest and the most tragic, the most eccentric and the most amiable, the most productive but the least accurate and methodical figure in American science.

Between 1792 and 1796 Rafinesque had lived in Leghorn with his mother. His father, a merchant, died of yellow fever while on a business trip to Philadelphia in 1793, but there seems to have been no immediate financial pressure on the family; Rafinesque, writing of this period years later—and in typical Rafinesquian hyperbole—stated that "Before twelve years of age I had read the great Universal history and one thousand volumes of books on many pleasing and interesting subjects." He added that his interest in botany and zoology had already developed; that he wasted no time on dead languages (though in his maturity he spoke six or seven live ones in various degrees of colorful improficiency) but had spent his time "in learning alone and by mere reading ten times more than is taught in the Schools."

After 1796 he continued with his mother in Genoa, Pisa, and Marseilles, leaving her in 1802 to come to the United States as an employee of a mercantile firm—possibly a connection of his dead father's, since the firm's office was located in Philadelphia, where the elder Rafinesque is known to have transacted business and where, indeed, he had died and was buried.

One assumes, in view of his character in maturity, that young Constantine was no great business asset to the firm and that his attention to its affairs was at best divided. While employed in Philadelphia he found time to make botanical collecting excursions through the eastern states and in the course of his wanderings conceived an ambition to explore the Ohio. He became acquainted with members of the Tarascon family, who would set up as merchants and warehousemen at the Falls of the Ohio in opportune time to acquire a substantial fortune from wholesaling, storage, and transshipping of goods. In 1805 Rafinesque returned to Europe and settled in Sicily.

In Sicily he began to publish his scientific works. He formed a connection—sans clerical or civil sanction—with a Sicilian woman who seems to have wanted somewhat more in moral stature than would be indicated by even her flouting of custom in the matter of matrimonial form. The couple had two children. A son, Charles Linnaeus, the selection of whose name would seem to indicate that Rafinesque had now completely abandoned commerce for science, died in his first year, but his sister, Emily, survived to become a play-actress and singer in Palermo and to distinguish herself by bearing an illegitimate daughter to Sir Henry Winston.

Shortly after Emily's birth Rafinesque, then either on a trip to America or botanizing on the shores of the Mediterranean, was reported to his mistress as having been lost by shipwreck. That lady, after an infinitesimal period of mourning, married an Italian comedian—probably already an acquaintance—and together they managed to convert or dissipate what property Rafinesque held. Learning of these unhappy matters accomplished, Rafinesque turned his back on ill fortune and took up permanent residence in America in 1815.

After spending some time in Philadelphia, he decided to visit his former friends, the Tarascons, now well established at Shippingport, at the foot of the Falls below Louisville. He arrived there in 1818. The lower Ohio—still, then, with mile upon mile of uninhabited banks and teeming with a plant, insect, and aquatic life as yet untouched by any but the hastiest scientific surveys—inspired Rafinesque to a perfect frenzy of activity which continued as long as he stayed in the West.

He spent a few days in scrambling over the fossil-filled rocks of the Falls, peering into the still-flourishing bear grass in the river bottom about Louisville, and dashing in and out among the gigantic trees and the tangled underbrush of the Indiana shore, then started off downstream, every pocket filled with the spoils of his hunting, to call upon the ornithologist, John James Audubon, of whose work their mutual friends, the Tarascons, had told him.

Audubon would play many a shabby trick upon the gullible and overenthusiastic Rafinesque during their acquaintance. He would describe strange and impossible birds, beasts and fish for Rafinesque's voracious notebooks and numerous publications; upon one occasion, by report, he would sew together the remains of two decomposed specimens, piscatorial and mammalian, respectively, to produce a furbearing fish for Rafinesque's amazement and immediate acceptance—but Audubon also left what is probably an accurate picture of the man in his prime:

"His attire," says Audubon of Rafinesque, who had just disembarked at Henderson, Kentucky, on his first visit, "struck me as extremely remarkable. A long, loose coat of yellow nankeen, much the worse for the many rubs it had got in its time, hung about him loosely, like a sack. A waistcoat of the same, with enormous pockets and buttoned up to the chin, reached below, over a pair of tight pantaloons, the lower part of which was buttoned down over his ankles. His beard was long, and his long black hair hung loosely over his shoulders. His forehead was broad and prominent . . ."

His hat, his pockets, his knapsack bulged, as always, with specimens living and dead and in every stage of decay; he must, indeed, have been an impressive sight, even among the unconventionalities of garb and grooming common to the frontier.

It was Rafinesque's misfortune, perhaps because of the very plenitude of undescribed life on the lower Ohio, to develop a special mania for reporting new species. Because there was so much new, and because of the innate enthusiasm of his nature, *everything* became new; because of his utter lack of the methodical caution that must mark the true scientist he described each find as a discovery, renamed it—sometimes redescribed it, in its variations of season and development, as two, three, or even more fresh discoveries—and quarreled with men who differed with him. Soon fellow scientists discredited him completely because of his many errors, and a century passed before he, long in his grave, began to be appreciated for the really tremendous volume of sound scientific work he had produced.

Rafinesque's contribution to higher education in the West was as great—and as intermingled with faults exasperating to his fellow workers—as was his study of the natural history of the area. A former Philadelphia acquaintance secured for him an appointment to the faculty of Transylvania University during the brilliant but stormy administration of President Horace Holley over that pioneer Kentucky institution. He frequently missed classes—inspection of a strange fish reported gigged by a raftsman in the Kentucky River always took precedence over a scheduled lecture—and although it was partly the science he taught that aroused the ire of the more strict interpreters of the Scriptures in Lexingon and rendered Holley's administration stormy, Rafinesque's demonstration that there was learning beyond Greek, Latin, and mathematics available to those with a taste for it worked change in many frontier college curricula.

Among his other contributions, he introduced at Transylvania University the teaching of botany, geology, mineralogy, zoology, psychology, and modern languages at the college level for the first time in the new country west of the Alleghenies— some of these subjects, indeed, for the first time on the continent. (Modern languages were not among the continental innovations, since Thomas Jefferson had successfully introduced them at the College of William and Mary some forty years before when, as governor of Virginia, he sat upon the board of visitors of that institution.)

In 1820 Rafinesque published his *Ichthyologia Ohiensis or Natural History of the Fishes Inhabiting the River Ohio . . .*, the first separately published scientific work devoted exclusively to the river. It was a major contribution, but like his others it was well salted with obvious fallacies and inaccuracies and was, of course, the object of contemptuous criticism far in excess of what it deserved.

Rafinesque, in turn, was as impatient of his critics as he was of methodical research. As the clamor of criticism increased he seemed deliberately to permit more and more errors to mar his

work; to claim discoveries less and less believable—with results easy
to anticipate. Sweet-tempered as he had been, this resentment of
criticism, magnified and developed by the vast portion of that commodity which his scientific publications attracted, soon began to
prey seriously upon his mind. He developed a persecution complex
so strong that he left Kentucky in 1825. He even abandoned the
natural sciences in favor of the manufacture of patent medicine
and the promotion of a wonderfully complex co-operative investment scheme which promised ease and plenty for all. Rafinesque
first offered this plan to William Maclure, shortly to become associated with Robert Owen at New Harmony. But Maclure, besides
being already fairly well supplied with utopian visions of his own,
was an experienced man of business and refused to accept sponsorship. Rafinesque himself then plunged into its promotion but
the scheme could not rescue even its inventor from abject and
squalid poverty, in which death finally overtook him in Philadelphia during the year 1840.

Many naturalists, professional and amateur, soon began to investigate the river; some in search of commercially valuable plants
and minerals, some in pursuit of knowledge for its own sake. In
1825 came an influx, a mass immigration of scientists the like of
which had probably never trusted itself to one frail river craft
before.

Robert Owen had by then persuaded William Maclure,
wealthy Philadelphia patron of education and the sciences, to direct and to finance the educational features of the New Harmony
venture. The experiment had begun—was in fact well on its way
toward its noisy finale—when, in midwinter of 1825-26, a keelboat,
the *Philanthropist,* was laden at Pittsburgh with Maclure's educators, scientific pensioners, friends, and hangers-on and began an
arduous voyage down the Ohio to the town of Mount Vernon,
Indiana, where transfer was to be made to wagons for the short
overland haul to New Harmony.

Included in the passenger list of the *Philanthropist,* besides
several assistants and lesser figures, were Marie D. Fretageot and
Guillaume Phiquepal d'Arusmont, Pestalozzian teachers; she the
mistress, if not of William Maclure at least of his affairs, and he

the future husband of feminist Frances Wright. In the same list were also Charles Alexandre Lesueur, French artist and naturalist; Thomas Say, brilliant entomologist, conchologist, and zoologist; Dr. Gerard Troost, Dutch chemist and geologist; Dr. Samuel Chase; Captain Donald Macdonald, that cheerful assistant to Robert Owen whose diary, already often quoted, supplies the account of the voyage. Owen's son also was aboard—Robert Dale Owen, future literary figure, Indiana representative in Congress, and advocate of all the fads and reforms that his father and Frances Wright promulgated plus a horde besides of his own fantastic devising.

As might have been anticipated by a party less occupied with science, the arts, and the imminent regeneration of society, the boat became icebound after only a few miles' voyaging down the Ohio from Pittsburgh and remained so for some time. It was a slow journey but the passengers of the *Philanthropist,* chilled on one side by icy drafts through the crevices in the cabin walls and roasted on the other by the iron stove in the cabin, had plenty to discuss and the days and nights—they being mainly Philadelphians —may be assumed to have merged into one long "Wistar Evening."

Of the casually opportunist utopians drawn down the Ohio by Robert Owen's promises of a life of ease in a garden of inexhaustible plenty, some hundreds retreated promptly upstream in the spring that saw the bare bottom of Owen's purse; not so the passengers on the Boatload of Knowledge. They were not among these deserters of the West.

Partly because they were dependent upon William Maclure, whose purse was far from empty, but mainly because most of them were sincerely interested in the pursuit and dissemination of learning and found the West an excellent place to carry on those activities, the majority of the teachers, artists, and scientists who arrived on the *Philanthropist* stayed in the Ohio Valley after Owenism was dead.

Madame Fretageot continued to teach or direct the schools that Maclure maintained until, in 1833, she died while on a visit to William Maclure in Mexico. She was as competent a teacher of children as any found in Indiana for a century.

Lesueur studied the fish of the West—he was first to make a comprehensive report upon those of the Great Lakes—and taught drawing in New Harmony with an inspiration unknown to the West and little surpassed anywhere else in the country of that day.

Thomas Say, before his untimely death, published scientific studies in New Harmony and conducted surveys of the wild life of the West which won him lasting fame.

Dr. Gerard Troost, a professional geologist and mineralogist, had a most important part in opening the rich coal fields of Kentucky, Indiana, and Tennessee (as did David Dale Owen and his brother Richard).

Robert Dale Owen, until his delving into spiritualism during his later years rendered him mentally incompetent, contributed brilliantly to the cause of free education, women's rights, the establishment of the Smithsonian Institution, the rehabilitation of freed slaves after the Civil War, and to dozens of others equally worthy.

These passengers on the *Philanthropist* and their associates, who while not actually on the boat did arrive at the same period and under the same auspices, unquestionably brought culture and learning to the Ohio River below Louisville. They, and the impressive number of eastern and foreign scientists, scholars, educators, and writers they attracted as visitors (and sometimes residents) in New Harmony before the Civil War, firmly established a value and regard for liberal and scientific scholarship which left a mark yet to fade upon all but the most reluctant educational institutions in the valley.

But other men of learning were on the Ohio long before these pure scientists, even as soon, possibly, as the practicing engineers and cartographers. Surgeons came to the Ohio early, and by that same original vehicle—the military expedition.

There were probably French surgeons at Vincennes and Kaskaskia, both close enough to the Ohio to warrant inclusion, well before the end of the eighteenth century; they accompanied most major military forces—including, likely, such an elaborately organized one as Céloron's in 1749 and certainly Hamilton's in 1778-1779. Doctors of medicine were in the West almost as soon as the military surgeons—that dreaming of quick fortune and faraway places which still makes the average physician a ready prospect for gold-brick promotion caused some doctors to venture among the first land proprietors, even explorers, of the western country. Notable among these were Dr. John Ferdinand Dalziel Smyth, who described the lower Ohio in his book *A Tour in the United States,* published in London in 1784 (he was a charming and talented liar but he certainly had knowledge of the ground), Dr. Thomas Walker, explorer of part of Kentucky in 1750, Dr. Manasseh Cutler (he was also the *Rev.* Dr. Cutler and *Attorney* Cutler) a prime mover of the second Ohio Company, who led the party that founded the Ohio River town of Marietta in 1788, Dr. John Connolly, ambitious medical man who seized Fort Pitt for the colony of Virginia and its even more ambitious governor, Lord Dunmore, in 1774; even General James Wilkinson had studied medicine and had practiced in Maryland for a few months before he enlisted for the American Revolution.

There were many besides this sampling and soon some of them began at least to combine some medical practice with their exploring, land promotion, and military coups. The early practicing physicians on the Ohio were mainly eastern or British educated and were doubtless, by the far from rigid standards of their day, well enough prepared. On the other hand, in the vast and sparsely settled valley there was nothing to prevent any venturesome soul, with a smattering of Greek or Latin and a yearning to attain a lettered state, from awarding himself a degree in medicine and advertising for patients. There was no shortage of gentry of this ilk, but, since some of the best educated of physicians in the early day still found that most symptoms indicated only the need for bloodletting and a thorough purge, perhaps the self-appointed Aesculapius experienced a mortality rate among his patients little higher than that of his orthodox brother.

Too, there were new diseases—or old diseases aggravated by a new environment—with which to cope; and there was an emergency practice resulting from gunshot, knifing, bear clawing, snake poisoning, gouging, biting, scalping, axing, and tree falling such as had never presented itself before. Early nineteenth-century doctors, even given a sparse population among which to practice, could and did mature quickly in experience, and a surprising percentage of their adult patients survived—though another half century would pass before parents began to hope to rear much more than half the children born to them.

Two decades after 1800, the Indians being less a threat to Virginia and Kentucky and thus gaining something in romantic appeal, there arose in those parts a new school of medical culture, even more harrowing in its method and prescription than were the legitimate physicians and their imitators. The enlightened proponents of this new method called themselves "Indian Doctors" and the like of their wondrous prescriptions had not been seen before—for they added to the root and herb concoctions of the natural Indian certain refinements and elaborations of which his inferior imagination had not been capable.

Some of these Indian Doctors published the results of their study and experiments in books of medical prescription designed to provide the isolated settler with information sufficient to cope with the ills and accidents of his family: the prescriptions, while complex enough as to content, were basically simple as to preparation—they all consisted of combinations of roots, herbs, minerals, and animal matter steeped in something; generally whisky.

Dean, preceptor, and honored father of the Indian Doctors was Richard Carter, who came west from Maryland by way of South Carolina and settled eventually in Shelby County, Kentucky. His father (by Dr. Carter's account) was an educated English physician who, though he drank himself to death while Richard was an infant in arms, is reported by the latter to have instilled in him a burning thirst, not for liquor, but for knowledge. Carter's mother was an Indian woman, for whom drink also was the downfall, but with the double heritage, i.e., a turn for the healing science and Indian blood, Carter's future was destined. Orphaned, he underwent colorful experiences as a child, found his way to the Cherokees, studied their materia medica, did some serious drinking on his own account, was stricken down by the Lord and shown the error of his ways, and began the practice of medicine.

His first published work, enlightening as to title, came out in 1815. It was *Valuable Vegetable and Medical Prescriptions for the*

Cure of all Nervous and Putrid Disorders; ten years later the text was reissued in a new edition with an added volume of autobiography, verse, moral precepts, and attacks upon orthodox medical practise, as Dr. Carter's magnum opus: *A Short Sketch of The Author's Life And Adventures from His Youth Until 1818. In The First Part. In Part The Second, A Valuable Vegetable and Medical Prescription, With A Table of Detergent And Corroborant Medicines to Suit The Treatment of The Different Certificates.* The material—shall we say of a "technical" nature?—was largely taken from the previous publication but the biography was new. It contains far more than the dry data of a life well spent; it is a picture of the whole man; his philosophy, his creed, his opinions on the vagaries of a wayward world. Though he utilized prose in the main, Dr. Carter did not hesitate to break into a rash of poesy when sufficiently moved and any subject, from a sore throat to the problems of the lovelorn, was likely to move him.

The diseases most commonly demanding the attention of frontier doctors—Indian and orthodox—were the same but the prescriptions of the former class were so much the more colorful that they deserve preservation. A sampling, garnered from Indian medical guides published in the Ohio Valley by Carter, Father Peter Smith, and S. H. Selman between 1813 (Father Smith's book, printed at Cincinnati) and 1836 shows the common ailments and the most colorful cures to include:

AGUE

Commonly called "Ager" or "The Shakes."

This disease was peculiar to the inhabitants of shady, damp land and was characterized by fits of shaking chills and burning fever which might occur at intervals of one, two, or three days throughout the warm months. Whatever the interval at which the victim shook, it was always regular.

Ab Conkright shook at 11:15 A.M. on alternate days; his wife, Minervy, shook every day at 10:00 A.M.; his son, Eph, shook every third day at high noon, and so on, with exact punctuality, from early summer to early frost. There being a shortage of timepieces in the valley this was as much a convenience as not. Mealtime was regulated by the innkeeper's ague, stages were kept on schedule by the driver's attacks, and ma's spells indicated the hour to send the children to school.

Besides the rather obvious shaking, other symptoms were a yellowing of the whites of the eyes, a sallow skin, bluish fingernails, an air of overburdening lassitude, and—as a result of the round of knife-blade doses of calomel which the majority of the victims undertook sooner or later—a salivated mouth and complete loss of teeth added nothing to the pulchritude of the victims.

The "Philadelphia Doctors" (as regular practitioners were nominated by their Indian Doctor rivals) treated this malady with calomel, which, as already noted, might relieve the ague but was not without its drawbacks. Father Peter Smith had a better way; so simple that he could knock off a dozen ague cures before breakfast and never feel the strain:

Pepper, for Preventing the Fever and the Ague. If rightly taken, I suppose, pepper will commonly prevent all persons and families from taking the fever and ague. . . . At a season when you expect yourself exposed to this ague, keep fine black pepper by you; take about as much as will lie on the eighth of a dollar, and lick it up by itself if you please, or mix it in a teaspoonful of molasses or a little water, and swallow it. Do this every morning for about four mornings, before you go out, and then you may miss a week, and then take it again, and so go on until the sickly season is over . . .

THE RISINGS

The risings is a disorder, still reported to be common in the southern mountain region, with a name which should be self-explanatory. What caused its prevalence cannot be surmised unless a clue lies in a sample menu for one or all of the three daily meals of the time, which might consist, in prosperous families, of:

Salt pork, lightly fried. Dandelion and mustard greens cooked well down in the hog fat. Pone, made of corn meal and bear grease. Fried apples. Sassafras tea with sorghum molasses. With the introduction of baking soda and green garden sass this complaint became less common in the more enlightened areas.

SNAKEBITE

was a frequent occurrence and, if the victim had been prevented by moral scruples, or by the extreme distance to the nearest distillery, from loading up on the proverbial snakebite preventive and remedy before and after the attack, Dr. Selman offered a fancy cure:

> *The Bite of a Snake and Other Poisonous Animals: How Cured.* If a person be bit by a snake, beat black ash leaves and bind to the wound as soon as possible, and make a tea of the bark; this will cure any snake bite . . . or bind the liver and guts of the snake to the bite; or apply bruised garlic . . . or if the patient be far spent, put a poultice of garlic to the bottom of the feet, or bind salt and tobacco to the bite or take cucold bur leaves and bruise them, put in sweet milk, strain and drink the same.

Although Dr. Selman does not say so, the garlic poultice would also probably have the advantage of discouraging snakebites on the feet for a time in the future.

Do not get the idea, from these simple remedies, that Drs. Carter, Smith, and Selman limited their practices to ague and snakebites. Such an assumption would be irreverent to their memories and inaccurate in the extreme. They were ready to tackle anything, even

MORTIFICATION

which meant, in this case, "death of animal tissues" rather than "chagrin." It seems to have been a common complaint, for all the valley authorities treat of it, though Father Smith's remedy and preventive is the most interesting:

> *To prevent or counteract a mortification.* Take brimstone and allum, equal parts, (say the size of a thimble) put them with gun powder; then pulverize them separately and mix them well together. Take of this mixture, as much as can lie on the quarter of a dollar, make it into a potion, with vinegar and water (say half a gill) for a grown person. . . . Let the patient drink it without delay. For external application . . . a decoction of dogwood and sassafrass roots, to wash and poultice with, is very good . . .

If this did not effect a cure, and if, by 1836, the patient still harbored his mortification, he could try Dr. Selman's remedy:

Apply a poultice of flour, honey and water with a little yeast . . . or take some tar, feathers, brimstone and hickory coals and put in a vessel and hold the mortified part over the steam—this is wonderful.

It may have been wonderful to Dr. Selman—but it sounds as if he were either getting ready to bake up a batch of cookies or to give one of the Philadelphia Doctors the coating of tar and feathers that Dr. Selman probably thought he richly deserved.

The same authority had a further remedy which is of interest. It was, he stated, for the relief of "Incubus or Night Mare," and, while he classifies it separately, he might, in the light of modern knowledge, have discussed it under the head of "The Risings and Related Miseries."

Incubus, or Night-Mare. This is evidently a nervous affection, and comes on during sleep, with a sense of considerable weight and oppression at the chest, the person making many efforts to speak and move without effect, until, after many deep groans and much mourning, he at length awakes greatly frightened. . . . The causes which give rise to this complaint are chiefly anxiety, grief, despondency and intense thought; but it is sometimes occasioned by making use of food of a hard indigestible nature for supper . . . Those who lead an inactive, sedentary life and are lax of fibre, are most predisposed to its attacks. The remedies. The first thing necessary is to cleanse the blood; that is to get a handful of the bark of the yellow poplar, the same of dogwood bark, the north side; the same of wild cherry tree bark, the same of yellow sarsaparilla root, and the same of the roots of running brier, put these in a copper kettle and put a quart of water to every handful, and boil it slowly away to 2 quarts, then add a pint of whiskey, take a tablespoon full two or three times a day; let your diet be chicken, squirrels, beef, mutton and broths, not too highly seasoned.

With this presentation of incubus—the one ailment in the practice of the Indian Doctors in which the cure really sounds more pleasant than the complaint—we take our leave of Dr. Selman, not because we lack faith in his skill, but because he ignores *Teeth*— a subject in which his early predecessor, Father Smith, took an eager, sympathetic, and scholarly interest. The next prescription is, therefore, from the pen of that paragon of Indian medicine, Father Peter Smith, of the Miami Country:

The Tooth-ache, to cure by Sympathy. The process—The patient is enjoined not to narrate what is done to him, or the Tooth-ache will return, (but a repetition will restore the cure).

All the finger and toe nails are to be trimmed, the pieces off of each are to be laid on a rag or paper; to which, also is to be laid a lock of hair taken from the head; then the gum of the tooth is to be gouged, or pierced, to add some blood to the nails & hair; then the whole is to be wrapped together in some creek or gulley, at a place where no creature crosses. The operator may keep the putting away to himself, if he pleases.

Indians they say have queer notions. Ha! but I have tried this for perhaps fifteen years, on myself and many others, and seldom without immediate success . . . If you will try this you may rest assured it will do you no hurt.

Strangely enough, it might do some good if, as would probably be the case in more than half the toothaches, the tooth happened to be abscessed and the sufferer pierced his gum deep enough when he went after the necessary blood to complete the charm.

Transylvania University, at Lexington, Kentucky, instituted its Medical Department in 1799 and was offering a rather complete course within two years. Under the presidency of the dynamic Horace Holley, who added brilliant men to its faculty, it flourished gloriously for the decade after 1818 before pressure of religious groups in the town brought about the strife that ended in Holley's resignation and the consequent decline of the entire institution.

This was the first medical school of the West and no competitor entered the field until the organization of the Ohio Medical College at Cincinnati, which graduated its first class in 1821.

It must be recalled that the idea of medical education in formal college classes was a comparatively new one. For years to come the majority of hopeful students, even on the eastern seaboard, were accustomed to acquire what general education they could, to "read medicine" with an established physician (meanwhile rolling the pills, sweeping the office and currying the horse) until such a time as their mentor either wearied of their presence or judged that they had absorbed sufficient of the contents of the office library and the precepts of the master. After this preparation the student could repair to the nearest city supporting a charity hospital, buy tickets to the amphitheater wherein the presiding doctor lectured while operating upon those paupers who presented themselves or dissected those who had passed beyond the need for operation. Were the young man particularly ambitious and sufficiently affluent he could purchase an interest in a cadaver from those who dealt in such merchandise and join his fellow students in dissecting it. The student could, and frequently did, leave off his studies at any point after the first sketchy reading, go to another state, procure testimonials of one sort or another, and hang out his shingle. Usually there was no examining body, as there was in the case of law, to say him nay.

The Ohio Valley, then, with two well-organized, well-staffed medical schools of substantial enrollment before 1820, was in at least as good prospect of a rising crop of competent medical men as was any other section. Thereafter the history of medicine in the valley is one of great interest and considerable importance to the country at large: but it is already the subject of a substantial literature and need not be treated more fully here.

The law—even more popular profession than medicine in the first century of settlement in the Ohio Valley—did not have its representatives on the ground until later. Most of the earliest military expeditions, though they were accompanied by military surgeons, had little need for masters of jurisprudence.

But aspiring representatives of the legal profession could not have been far behind their brothers in medicine, for when settlement followed military conquest there must be someone upon the ground to witness documents, to hold in escrow, to be sworn before, to advise and mediate. The preparation of the representative of the law on the frontier in the days of the French and English trading post communities was not necessarily academic but he was most assuredly there and under monarchical government he had to be familiar with a great deal of form and ritual. After the Revolution the pioneer American judge or justice of the peace could, and often did, dispense justice largely upon the basis of his own convictions as to what was and what was not just, backed by the courage to support them.

Lawyers, too, are not immune to that weakness for speculation and venture in far places which has been mentioned as marking such a preponderance of physicians—though lawyers' speculations are, perforce, inclined to be somewhat less bullish and more carefully thought through in advance. Some of the largest of the land speculators upon the river were of the bench or bar—Judge Richard Henderson, for instance, and Judge John Todd, Harry Innes, and Robert McAfee in Kentucky; George Mason, of the original Ohio Company, though not educated for the law, drew up much of the most important of Virginia colonial legislation and of course Manasseh Cutler, of the second Ohio Company, was a lawyer—among other things.

The Miami Purchase—300,000 acres of the land between the Great and Little Miami rivers upon which now stands Cincinnati, all its suburbs, and many of its neighbors—had as its prime movers New Jersey Judge John Cleves Symmes, lawyer Elias Boudinot, and Jonathan Dayton. Dayton alone was not a lawyer by training but even he received an honorary Doctor of Laws degree from the College of New Jersey before, in 1807, he was charged with participation in Aaron Burr's conspiracy and retired from public life outside the confines of his own state.

The smaller fry of the bench and bar who participated in land speculation in later days upon the Ohio were legion indeed.

Fort Pitt, under its changing titles, had long held military court by royal authority for the correction of the erring soldiers upon the upper Ohio, while in the earliest years civilian offenders were tried in Bedford and later Westmoreland county courts of Pennsylvania. Pittsburgh was laid out in 1764 and a Pennsylvania county jurisdiction set up around it. Briefly, during Dunmore's sway, Dr. Connolly violently restrained the Pennsylvania justices and gave jurisdiction to Augusta County, Virginia. But Lord North ordered the ambitious Dunmore out of the colony and cases on the upper Ohio were afterwards disposed of there in the usual manner. First courts were held in Kentucky and the North-

The Ohio

west Territory, respectively, at Harrodsburg in September, 1777, and at Marietta in September, 1788.

With courts instituted which had jurisdiction over the whites of the entire Ohio Valley there was an opening for lawyers functioning, not as land speculators, but in their professional capacity. The call was heard and they soon began to respond.

Except for those upon the north shore above the Great Miami, the majority of early Ohio Valley lawyers seem to have been Virginians. This was to be expected in so far as the Virginia shore itself and Virginia's former county of Kentucky were concerned but it seems also to have been true of those parts which would come to be Indiana and Illinois: the Old Dominion apparently spawned lawyers in a crop as plenteous as that of ministers of the gospel in Massachusetts and Connecticut.

Training for the law in the late eighteenth and early nineteenth century followed the pattern of that for medicine: the student secured what he could in the way of general education, found a practicing attorney willing to let him "read law" in his office, either in return for a cash fee or for copying, fire building, and errand running, and studied until he could pass the local bar examinations and set up for himself. Passing his examinations guaranteed that his knowledge of the law was believed by his peers to be up to their beginning standard (and that standard certainly could and did vary) but only the student himself need judge as to whether his supply of general education and cultural experience was sufficient to ornament his calling. No one examined him in those matters.

This system sometimes produced lawyers, perhaps fairly competent ones in their knowledge of the law and its workings, whose preliminary schooling had included scarcely any of the arts of rhetoric and orthography—as witness the Indiana attorney who wrote to United States Senator John Tipton (a man of very similar literary attainments) in the thirties:

Our Enemis use our povrty as a means against us it is true that men who are some what under the wether looses in some measure there influences, although it shall nevr stager my independance, for if th

Earth was to clove assunder, and the clouds rain fier it should nevr change my coarse . . .

As in the case of medical education, however, the Ohio Valley was soon well up in competition with the eastern states in facilities for the formal study of law. Before 1800 Transylvania University opened a law department which flourished for three decades under a series of brilliant instructors—among whom was Henry Clay between 1805 and 1807.

Young men educated at Transylvania before narrow sectarianism brought about its degeneration in the late twenties were later cast for leading parts on the national stage: one of them, Jefferson Davis, taking a role which well-nigh rang down the curtain.

Early facilities for education in the law throughout the Ohio Valley, plus the immigration of eastern lawyers especially after the War of 1812, gave the West an ample representation of the legal profession which was thoroughly competent although not uniformly polished. By the time general settlement had brought about the inevitable tangle of litigation over land titles these gentlemen were ready to reap the harvest.

But not all learned men among the early land promoters of the Ohio Valley were doctors or lawyers: one, singularly dogged by misfortune, was a schoolteacher by profession.

He was John Filson and through the publication in 1784 of his famous book, *The Discovery, Settlement and Present State of Kentucke* . . . and of the even more famous map designed to accompany it, he was first to give widespread printed publicity to the virtues of the valley—and especially of "Kentucke"—as a rich and romantic site for homebuilding and fortune making. Thus, although he profited not at all in the matter, he had a great part in selling the lands of others long after he was murdered in the fall of 1788.

John Filson was born at the farm home of his grandfather on the banks of the Brandywine in southeastern Pennsylvania, probably in 1747. He is supposed to have had the advantage of whatever local schooling was offered and to have attended an academy presided over by the Rev. Samuel Finley, who later became president of the College of New Jersey. Thereafter Filson himself kept school in the neighborhood, probably farming meanwhile, since he inherited some land upon the death of his father in 1776.

At the end of the American Revolution Filson caught the western fever then epidemic through the medium of returning veterans of the western campaigns. Sometime in 1782 he made the trip over the mountains to Pittsburgh, down the Ohio to Limestone, and overland from there, through the forest, to Lexington. He is reported to have taught his first term of school at that place in the same year.

He was apparently well supplied with Virginia land warrants (which that state had issued to redeem her wartime paper currency, by then depreciated to near the vanishing point), for in 1783 he entered over 12,000 acres of Kentucky land and an undetermined amount in the country north of the Ohio, exchanging warrants evidently purchased from Clement Moore and John Boyd at a great discount.

Filson had spent his spare time, after schoolteaching and land locating, in gathering material from Daniel Boone, Levi Todd, James Harrod, and other pioneers for the book that he projected, according to its preface, "solely to inform the world of the happy climate and plentiful soil of this favored region." His map, first of Kentucky and, to that date, most detailed of the Ohio River where it forms the long northern boundary of that state, was wonderfully accurate even though somewhat out of drawing. It was a remarkable achievement based almost entirely upon rough sketches and verbal reports of the topographical observations those same pioneers had made while upon hunting and exploring trips in the district.

There being then no printing facilities west of the Alleghenies, John Filson took his manuscript back east, where he had the book printed in Wilmington, Delaware, and the map engraved and printed in Philadelphia—to be sold at two shillings six for the book and five shillings for the map—or less than $2.00 for the entire work. (One notes that Filson's land, purchased at probably 1 cent per acre, has increased in value an average of perhaps 10,000 to

15,000 times: a well-preserved copy of the first edition of his book and map, lagging far behind, is likely to bring only 700 or 800 times its original cost.)

Filson returned to Kentucky in the spring of 1785, making the trip from Pittsburgh to Beargrass Creek (at Louisville) by flatboat.

Back in Kentucky his troubles began. He had sold his Pennsylvania farm, possibly to pay for the printing of his book and the expenses of the trip east. He entered into some sort of trade at the Falls and made several trips to the Illinois country and to Vincennes in the course of which he had several lively adventures and some disastrous losses. Meanwhile, landpoor as he was, he became desperate for cash. He was sued upon several occasions by creditors and finally, a judgment having been awarded against him, he was found to have reached a state of almost complete destitution. When his property was ordered attached, the officers entrusted with the levy were able to find among his possessions only one battered sickle worth impounding.

In August, 1788, Filson contracted to join Matthias Denman and Robert Patterson in laying out and promoting a town in Judge Symmes' Miami Purchase: it was to be located on the north bank of the Ohio opposite the mouth of the Licking River. The site eventually proved to be quite eligible, the town now being known as Cincinnati.

There were 640 acres in the tract and Filson was to receive a one-third interest in return for a payment of £10 Virginia currency, the work of surveying, and that of writing publicity for the venture. He made the survey and he wrote the prospectus that was published in Lexington's *Kentucke Gazette*—but he did not pay his £10. In the fall of 1788, before profits could accrue to him from lot sales, he was killed, presumably by Indians.

It is not as an unsuccessful promoter, merchant, or even surveyor that John Filson claims our present attention—there were many others in all those categories—it is because, apparently, he was the first historian of the western country and the first professional teacher to follow his calling in the Ohio Valley. Probably he first taught in Lexington in the winter of 1782-83: it is certain that he operated an academy in Lexington and possibly had taught a few terms in Louisville before his death. In January, 1788, he had felt himself sufficiently conversant with the educational problems of the locality to publish in the *Kentucke Gazette* an article proposing the establishment in Lexington of a seminary upon a new plan which would not only give a liberal education and introduce the teaching of French but would also look to the morals and manners of the student. Had his plan been adopted he would have anticipated the introduction of the teaching of a Romance language in the West (finally brought about by Rafinesque) by thirty or more years: as it was, he permitted himself again to be led astray into the commerce at which he had already failed so spectacularly. This time his misfortune was complete and final.

The second professional teacher (he was also a minister of the gospel) arrived on the Ohio within six months of Filson's passing. He was Daniel Story, who came to the Ohio Company's land to guide the youth of that community in the way it should go, spiritually and academically.

Pittsburgh certainly had at least "Three R" schools before this time but the qualifications of their teachers do not appear to be of record. Within the second month of publication, the editor of the Pittsburgh *Gazette* was dreaming dreams of even higher learning—of an academy for the town, in fact. On September 2, 1786, he proposed that

The situation of the town of Pittsburgh is greatly to be chosen for a seat of learning; the fine air, the plenty and cheapness of provisions, render it highly favorable.

The editor was to be disappointed, for Pittsburgh would not take kindly to academies for some years to come. As late as 1814 Zadok Cramer, the city's most fervent booster, found it necessary to admit in print that "There is a public academy, but not in a flourishing state," although there were "a number of English schools where children are taught to read, write, arithmetic, grammar, &c.," and also "a seminary for young ladies, which is said to be well conducted . . ."

Seminaries offering some degree of advanced education were established early in the two locations in the Ohio Valley most favored, intellectually, in the qualifications of their early settlers. That of Kentucky's Bluegrass was opened near Danville, at the grammar school level, in the winter of 1784-85. It combined with another institution, moved to Lexington and opened as Transylvania Seminary in 1789. Even before its removal the school apparently had a competitor in the area, according to an advertisement in the *Kentucke Gazette:*

Notice is hereby given that on Monday the 28th of January next, a school will be opened by Messrs. Jones & Worley, at the royal spring in Lebanon town, Fayette county, where a commodious house, sufficient to contain fifty or sixty scholars, will be prepared. They will teach the Latin and Greek languages, together with such branches of the sciences as are generally taught in public seminaries, at twenty-five shillings a quarter for each scholar, one half to be paid in cash, the other in produce at cash price. There will be a vacation of a month in the spring and another in the fall, at the close of each of which, it is expected that such payments as are due in cash will be made. For diet, washing and house-room, for a year, each scholar pays three pounds in cash or five hundred weight of pork on entrance, and three pounds cash on the beginning of the third quarter. It is desired that as many as can would furnish themselves with beds; such as cannot may be provided for here to the number of eight or ten boys, at thirty-five shillings a year for each bed.

ELIJAH CRAIG

N. B. It would be proper for each boy to have his sheets, shirts, stockings &c. marked, to prevent mistakes.

Lebanon, December 27, 1787.

Muskingum Academy was organized at Marietta, Ohio, in 1797 and when it opened it presumably undertook the higher education of the youth whose early steps had been guided in that community by the Rev. Mr. Story.

Both the Virginia Assembly and the United States government had made brave gestures toward ensuring facilities for higher education in the West at the time of the granting or sale of lands to be opened for settlement. Though the plans did not mature as promptly as their proponents probably anticipated, they were sooner or later effective.

Virginia acted first when in 1783 the Assembly passed a motion making public lands available for the establishment of an institution of learning in Kentucky. This institution, aided by the Presbyterian Church, first appeared as the grammar school at Danville mentioned heretofore, became Transylvania Seminary, functioning in rented quarters in Lexington, and in 1793 moved to its own campus. After January, 1799, it became Transylvania University. A few years later the Presbyterian influence, so useful in the beginning, ceased to be an asset; other sects complained that Presbyterianism flourished at the expense of public funds and the reactionary element of the sect interfered with instructors and instruction. It was the culmination of this quarrel that finally robbed the school of the great President Holley and eventually dissension over the church and state connection caused Transylvania to cease to receive public aid at all.

As will be recalled, the Ohio Company purchase, made under the Ordinance of 1787, had carried the stipulation that one section in each township was reserved to be sold for the benefit of schools and two entire townships were to be reserved for the establishment of a university. Judge Symmes' Miami Purchase had been authorized on the same basis, except that only one township was there reserved for higher education.

Contrary to the common impression, the wise provisions of the Ordinance of 1787 did not provide the new West with a ready-made, immediately workable, system of free education. They only paved the way for the organization; well over half a century elapsed before action of the various states began to render the plan effective at the grade school level, though public-supported higher education came earlier.

Off to a later start than Kentucky, the settlements on the Muskingum and between the Miamis, Great and Little, were slower still in commencing their institutions of higher learning. Not until 1804 did the government of Ohio charter the institution (present Ohio University) in Athens, Ohio, which did not actually reach college status until 1822. In 1809 Miami University, in Symmes Purchase, was chartered; it opened in 1824. Both these institutions experienced difficulties only slightly less disastrous than those of Transylvania. Miami University even closed between 1873 and 1885 but it reopened under state support and is a great educational institution today. Even old Transylvania, after decades of precarious existence, enjoyed a renaissance after World War II and gives every promise of a useful future.

Of course there had been primary schools of sorts in the valley since the first few literate settlers with children arrived in the town of Pittsburgh, which John Campbell had laid out in 1764. There was teaching of some sort at Vincennes, a comparatively short pole-up from the Ohio on the Wabash, when priests began to visit the community well before the middle of the eighteenth century. Children certainly learned their letters in the blockhouses of Kentucky and Virginia (Mrs. William Coomes, John May, and "Wildcat" John McKinney took care of the education of the young sprouts in Kentucky's earliest forts) and amateur teaching was carried on with more or less enthusiasm depending upon the background and the attainments of the settlers.

The American settlers' first attempts at doing their duty by their young must have been uniform in that they were necessarily made in someone's cabin and were conducted by some interested party, male or female, with little or no previous teaching experience but with the ability to read, write, and cipher, at least to a limited extent. That, except in such highfalutin communities as Marietta and Lexington or in a town fortunate enough to include an educated preacher or lawyer among its very first arrivals, must have been the extent of the curriculum in the earliest day and so to have continued, in some localities, well into the nineteenth century.

In spite of the limited facilities in those cabin-fireside schools, there is every evidence that some of our pioneer ancestors, even when possessed of only a few weeks' tuition, spoke good, forceful, and often eloquent English, no matter how unorthodox their spelling.

In fact when one overlooks the spelling in the letters they

wrote, and when one reads-in the necessary punctuation, one often sees displayed a far broader vocabulary and a more telling choice of words and arrangement of phrase than is to be found in this day of compulsory education. Perhaps there was more careful listening to the speech of educated men because they were few; perhaps there was more conscious effort at self-improvement because it was difficult; perhaps (perish the thought!) a few weeks of schooling under pioneer adversity may have aroused a greater thirst for learning than eight to twelve years of enforced attendance in the modern schools.

The few letters in Daniel Boone's hand that survive are weak enough in spelling but, read aloud, they appear to be the product of a man who thought and spoke in good contemporary English. Whoever translated the speech of the Indian, Logan—whether he was John Gibson, Simon Girty, or some other—produced a masterpiece of the English language. Even the writing of the Indiana lawyer previously quoted, "our Enemis use our povrty as a means against us it is true that men who are some what under th wether looses in some measure there influences although it shall nevr stager my independance . . ." becomes, with only a bit of technical correction, "Our enemies use our poverty as a means against us. It is true that men who are somewhat 'under the weather' lose, in some measure, their influence—although that shall never stagger *my* independence . . ." and is thus a good and bravely spoken statement.

The opportunity for receiving formal education broadened rapidly. Soon after Filson and Story, professional schoolmasters began to move down the Ohio from New England, the middle states, and Virginia to open the subscription schools that continued until the middle of the nineteenth century. A school was advertised in Cincinnati's first newspaper in December, 1794, which planned to teach reading, writing, arithmetic, and surveying and there were certainly others, by then, which filled their enrollments at 75 cents to $1.50 per pupil per term without the necessity of advertising.

Some of the teachers were probably competent but their average proof of attainment must have been much diluted by the numbers who were too poorly prepared to gain patronage even in the uncritical back settlements of the seaboard states, who were too eccentric to exist in settled communities, or who had valid, but not necessarily admittable, cause for seeking a change of scene.

For some reason many of the early teachers were Irish, even as had been the majority of traders in British employ, and if the bulk of pioneer reminiscence is to be trusted, were vastly addicted to flute or fiddle playing and enamored of the jug and its contents. They were not necessarily skilled in the practice of the subjects they taught, although, from the published complaints of recently former patrons, it appears that their pedagogical disabilities were at least discovered after a term or so.

One such teacher (he happened to be spreading the light of learning in Wisconsin, though he probably had his counterpart on the Ohio) replied to a complaint with a letter which admits to a weakness for strong drink and bears its own witness as to its writer's qualifications in English composition:

> Your note in answer to mind . . . pleased me mush, as it maid me cum to my right sencess; in one part of my letter to you I returned thanks for your favour towards me, and in another that you abused me . . . I ment to say that you repremanded me several times in regard of the Blotter [a ledger or record book, one supposes] I kept last summer for you and my saying you was Jealous about my school I intended to mention that you gave me no answer to my note to you where I mentioned I would be quite happy if you would send you Children to school and I should charge onely one Dollar per Child instead of two—and about minding receaiveing person with spirits and Whiskey, I was half drunk and I maid Ceremonies to get quite so.
>
> In regard of making out my account against you, please to place to my Credit, your's and Mr. Frankes Children 7 months schooling $84— as to writing for you cupping invoices &c . . . I did not inteand to ask you aney thing—but as I am a poor reatch and Mr. Porlier has no Blanket . . . and my girl as not aney I would which you to Let me have one & ½ the Coloured Thread and one dressed deer Skin . . .

There was also, in those communities with predominantly New England traditions, the "dame school" type of establishment for beginning scholars. As the wilderness was penetrated and be-

ginners might be youths of fifteen or sixteen and of a bodacious
cast besides, these female-operated institutions grew few until later
years had brought refinement. For how could the most retarded
(and thus oversized) boy in school make up for his intellectual
disability by "whipping the master" when the master was a mistress? The roughest river town had its peculiar code of chivalry.

With the coming of the eighteen hundreds "academies," equal
more or less to advanced grammar school and high school level,
began to be organized in towns smaller and with population less
cultured in background than were Pittsburgh, Marietta, Lexington,
Cincinnati, Madison, and Louisville. Progress in this matter was
generally from east to west. These institutions increased in number
and quality of instruction with the growing western migration
of people of property concurrently with the retirement of the Indian and the bobcat, or as the circulation of real currency permitted leisure and a longing for eastern refinement.

The courses offered were progressively more varied as time
went on, reaching their ultimate elegance in the forties and fifties
—especially in those institutions dedicated to the purpose of ren-
dering the young female eligible for matrimony. While at the
beginning of the nineteenth century Mrs. Williams, presumably a
cultured and worthy matron, begged "to inform the inhabitants
of Cincinnati; that she intends opening a school in the house of
Mr. Newman, saddler, for young ladies on the following terms:—
Reading, 250 cents; Reading and Sewing, $3.00; Reading, Sewing
and Writing, 350 cents per quarter," vast progress had obviously
been made by 1842-1843 when, at the Methodist Female Collegiate
Institute of Cincinnati, the young lady of good family and sound
Wesleyan principles who happened to be completing her course
under the guidance of Principal, the Rev. Perlee B. Wilber, M.A.,
could occupy her attention with, in the First Term:

> Mensuration. *Day's.*
> Moral Science. *Wayland's.*
> Geography of the Heavens. *Burritt's.*
> Intellectual Powers. *Abercrombie's.*
> Moral Feelings. *Abercrombie's.*
> Logic. *Hedge's.*

(One assumes that after the course in Logic some reasonable explanation could be discerned as to why the courses in Moral *Science* and in Moral *Feelings* could not be integrated and combined.) The young lady took her departure, educated, after a Second Term, which offered:

> Botany. *Mrs. Lincoln's.*
> Evidences of Christianity. *Alexander's.*
> Rhetoric. *Newman's.*
> Elements of Criticism. *Kames.*

Throughout the course she had had "particular attention paid to Orthography and Penmanship" and—was father's purse capable of sustaining substantial extra charges—she could as well have been exposed to "Vocal and Instrumental Music, Drawing and Painting" and "Ornamental Needle Work."

Earlier, between the arrival of "the Boatload of Knowledge" at New Harmony and the death of Madame Fretageot, the West had seen some experiments in education which offered a curriculum somewhat more practical and interesting than that of the Methodist Female Collegiate Institute.

William Maclure had tried his varied ideas upon the available young of the Ohio Valley—with little immediate effect upon the general pattern of juvenile education in the region, but with a considerable one upon college curricula, vocational training, and adult education. Maclure, benevolent dictator-philanthropist, had some characteristics in common with Father George Rapp, but upon two points they differed widely: Maclure was an atheist, while Rapp proclaimed himself to be in direct contact with God; Maclure had a passion for educating his people; Rapp for keeping his in ignorance. In their aloofness, their strong wills, their business competence, the air of mystery with which they surrounded their doings, they were much alike.

Upon the final departure of Robert Owen, William Maclure began to promote his educational plans without interference. The institutions he projected included Maclure's Seminary, the Orphans' Manual Training-School, and an adult education scheme, the Society for Mutual Instruction. Frances Trollope (whose son, Anthony, was enrolled briefly in the seminary) characteristically viewed Maclure askance: "a venerable personage, of gentlemanlike appearance . . . a man of good fortune . . . who, after living a tolerably gay life, had 'conceived high thoughts, such as Lycurgus loved, who bade flog the little Spartans,' and determined to benefit the species, and immortalize himself, by founding a philosophical school at New Harmony." Much as one may hate to agree with the lady, it is to be feared that there was truth in part of her analysis.

The various schools, once established, operated under the direction of Madame Fretageot, as Maclure had departed in search of health in Mexico. There must have been good teaching in some departments, for Thomas Say, Charles A. Lesueur, Richard Owen, and Robert Dale Owen all took a hand, but Maclure's foibles included a study-and-work plan which seems to have erred as much on the side of overwork for the youngsters as Owen's had in the other direction for their elders. By 1831, when Madame Fretageot left to close her business affairs in France and to go on to Mexico to visit Maclure, the schools seem, all, to have undergone a process of reorganization which ended in the abandonment of the less practical of their experimental phases. Only the adult Society for Mutual Instruction, later changed to the Workingman's Institute, survived her death and Maclure's. It, as a model for the Maclurian Library Associations that were endowed by Maclure's last will, had considerable importance in introducing the first free lending libraries to the Midwest.

Bona fide colleges appeared quite early upon the immediate banks of the Ohio (they will be described elsewhere) but for some reason higher education, like the culture of the prehistoric mound builders, seemed to prosper first upstream upon the river's tributaries.

By the last quarter of the nineteenth century, education in the Ohio River states was free to all, though Kentucky and West Virginia had clung long to private education at the high school level in the tradition of the Old Dominion. The *quality* of this education varied greatly—varies greatly today—with the prosperity, culture, and enterprise of the neighborhoods both north and south of the river. Today, as throughout the past, education has not gone forward with uniformity, even by states: western Pennsylvania, Ohio, Indiana, Illinois, West Virginia, and Kentucky all have examples now of the most progressive of educational institutions —and all can also show, though the exhibition is not likely to be a willing one, schools more primitive than was either John Filson's in Lexington or the Muskingum Academy at Marietta. Certainly the Hoosier Schoolmaster, returning to this world, could step to the front of many a one-room school and take up his spelling book where he had laid it down, perfectly at home after a century's passing.

Chapter Fifteen

THE PROGRESS OF THE ARTS

The fine arts began to prosper surprisingly early upon the Ohio, although in the years immediately after settlement began their production should be judged with due regard for the uncultivated soil in which it grew as well as upon its virtue.

In the first quarter of the nineteenth century many an artist in this world was writing, painting, composing, or designing to a far better purpose than were the ambitious residents of the Ohio Valley: the great wonder is that the western artists made the effort at all. Their patrons should be respected because, few as they were and pressing as were the material needs of most of them, they still constituted an appreciative if not uniformly discerning audience.

The importance of that audience must not be overlooked.

There has always been a tendency to minimize the intellectual capacity of the pioneer; actually, when such cultural fare was available, even some of the commonest of his kind enjoyed good literature, good pictures, and good music. Remember those hunters reading Dean Swift during that hard winter in the wilderness on the headwaters of the Kentucky?

Expansive expression in fiction and poetry (particularly the latter) came earliest. Perhaps it was a manifestation of the feeling of loneliness that educated settlers must frequently have experienced among the worthy but not notably cultured backwoodsmen who—fortunately for the physical and economic improvement of the new country—predominated among the citizenry.

Whatever the cause, the earliest issues of pioneer newspapers, as journalism spread from its Ohio Valley beginnings at Pittsburgh and Lexington, gave space to contributed literary offerings. The volume of this contribution is proof enough that many made a try at writing, some with fair success and others with a glorious ineptitude which rendered their production more distinguished than was some of the passable but not brilliant work of their betters. These first emigrant littérateurs were mainly Virginians and New Englanders, the same stocks which pressed hardest for qualified education in the valley. That poem already quoted from the Pittsburgh *Gazette,* which concerned itself with the progressive dissolution of the unfortunate Eliza, was typical in style and quality —if somewhat unusual in subject matter. Generally contributions to newspapers published on the south bank and therefore affected by Virginia influence were likely to tend more to pure romance; those on the north to moralizing.

On either north or south bank simile, hyperbole, and actual or pseudo-classical allusion bloomed luxuriantly; what the writer fancied to be a well-turned phrase took precedence, any time, over sense.

As more newspapers began publication—the Pittsburgh *Gazette* (1786), the *Kentucke Gazette* (1787) at Lexington, the Cincinnati *Centinel of the North-Western Territory* (1793), the *Indiana Gazette* at Vincennes (1804), the *Illinois Herald,* Kaskaskia (1814)—and local contributors of creative writing more often achieved the thrill of seeing themselves in print, the next step toward a home-grown literary production was obvious and easily taken.

One of the profitable side lines of the early newspaper publisher was that of printing the acts and collected statutes of the local legislative bodies. These appeared as pamphlets or bound volumes, either of which formats lent itself readily to the reproduction of poems, literary sketches, or essays. Any author who could write one "piece" for the newspaper obviously could write more and, when he had laid by a sufficient stock of them, the printer who could put together a book of statutes could as easily turn out a volume of poems, opinions, or reminiscences. Production was entirely at the author's expense, of course, and if the printer was in his right mind work began only after the receipt of a substantial down payment.

Hugh Henry Brackenridge is usually credited with being the Ohio's first novelist. Born in Scotland in 1748, he was brought to America in his fifth year, worked his way through preparation for college and entered Princeton. He received his degree in 1771 but while still an undergraduate he had collaborated with Philip Freneau in *The Rising Glory of America,* which was published in 1772. He did editorial work, studied theology, and served as a chaplain in the American Revolution, writing a play, *Bunker Hill,* while still in the service.

After the war Brackenridge studied law, was admitted to the bar and opened a law office in Pittsburgh in 1781. He became a leading figure upon the Ohio, a contributor to the press, active in all the controversies in the new country, and finally, after 1799, became a judge of the supreme court of Pennsylvania. He had some literary reputation before he moved to the Ohio and he took advantage of this fact to introduce the West as the setting and contemporary events as subject matter for some of his later writing.

In 1795 he published his *Incidents of the Insurrection in Western Pennsylvania,* an account of the Whisky Rebellion. His novel *Modern Chivalry, or the Adventures of Captain Farrago and Teague O'Reagan, his Servant,* published in Philadelphia in 1796, is the first book-length fiction known to have been written upon the Ohio. That, unfortunately, is its chief distinction; the judge might better have occupied his very great talents as a publicist and shaper of opinion in the public weal.

But he was far from a home-grown product—a Scot by birth, educated upon the Atlantic Coast in the improving society of such bright young men as Freneau—he was after all only an educated and talented man who had happened to come west. Shortly the Ohio Valley was to have a novelist actually bred on its own soil, his first novel home-produced while the powder smoke of Indian warfare still hung over the land.

The author was Jesse Lynch Holman, who was born October 22, 1783, in a cabin in Mercer County, Kentucky. There was no question as to the authenticity of his western background; his father had been killed while helping to defend a blockhouse in which the family was besieged by Indians, his uncle underwent one of the more famous of the Indian captivities, and Jesse Lynch Holman himself picked up education as that commodity offered in the back country and read law under Henry Clay. His book, a historical novel, *The Prisoners of Niagara, or Errors of Education. A New Novel, Founded on Facts,* appeared in 1810, having been so carelessly produced by a Frankfort, Kentucky, printer that the author's name is even misspelled on the title page. It is a treasure, nevertheless—as rare as a book can be and still be known at all. Probably the rarity results from the fact that the only edition must not have exceeded a hundred or so copies and the author, grown a little self-important as a judge of the Indiana supreme court in his maturity, decided that "the morals of the book were not suitable" and bought up and destroyed such copies as he could locate.

Holman probably wrote the book while practicing law at Port William (now Carrollton), Kentucky, before removing across the river to Indiana.

Most early Ohio Valley writing—whether it appeared as contributions to newspapers or periodicals or in book form—was of a deadly serious cast, or was at least aimed by its authors to be so. Writing in the lighter vein had begun in America long since and had gained an audience, but perhaps the people of the new West were still a bit too jealous of their dignity to risk such a departure. In any case the light touch had to await 1806, even in the sophisticated Bluegrass of Kentucky, to appear in book form. It was not until then that William Littell, intriguing figure on the early Kentucky scene, essayed satire in his *Epistles of William, Surnamed Littell* and another eight years passed before even that independent spirit was ready to venture humor on the printed page —in his *Festoons of Fancy,* published in Louisville in 1814.

Upriver residents continued to write meanwhile and shortly commercial publishing along currently practiced lines began to flourish. Bookstores existed in Pittsburgh and Cincinnati before the twenties and presently booksellers began to purchase some manuscripts outright or to publish books by western authors on shares. (It was not necessary to make recompense in the case of English or even Atlantic seaboard authors; pirating their books was far more profitable.) The increased settlement of the thirties, the greater leisure that resulted, and the advent of such enterprising booksellers as J. A. and U. P. James and Robert Clarke permitted Cincinnati to proclaim itself, and to be in fact, "The Literary Emporium of the West."

Beginning in the twenties there developed a school of writing between Pittsburgh and Cincinnati which was not only rather peculiar to the Ohio Valley in its volume but was, at the same time, most valuable to it historically. It began to occur to other writers (as it had to H. H. Brackenridge) that a straight narrative of events in the *recent* past, a description of the *local* scene, a report of *current* activities might find an audience, and the success of the new historical writing proved their judgment to be correct. A considerable body of narrative popular history, accurate in detail and palatable to the lay reader, resulted.

Zadok Cramer, already so frequently quoted, managed to include a great deal of this sort of local history within the technical framework of his guides to navigation of the river, as did, later, Samuel Cumings and other of the authors of pilots' guides. But most distinguished of the amateur historical works were offered to the public when Alexander S. Withers published his *Chronicles of Border Warfare, or A History of the Settlement by the Whites of North-western Virginia* at Clarksburg, Virginia (now West Virginia), in 1831. Joseph Pritts, of Chambersburg, Pennsylvania, compiled his *Mirror of Olden Time Border Life* and published it at Abingdon, Virginia, in 1849. Dr. Joseph Doddridge, a prominent Episcopal clergyman, not only brought out in 1824 his *Notes on the Settlement and Indian Wars of the Western Country in 1763-83* but also did a couple of charming period pieces laid upon the upper Ohio—*Logan . . . A Dramatic Piece* (first published in Virginia in 1823) and *The Dialogue of the Backwoodsman and the Dandy,* which he composed in 1821 and in which, you many be sure, the "Dandy" comes out second best.

No section of the country has been more fortunate in the preservation of its history by contemporary scholars than the Ohio Valley. For instance, James H. Perkins began in 1846 the work entitled *Annals of the West,* a history of midwestern settlement as scholarly as could be expected or desired. James R. Albach and John Mason Peck continued and enlarged it through its many later editions. Charles Cist, who arrived at Cincinnati about 1827, having come from Philadelphia by way of Pittsburgh and the Rappite town of Harmonie, Pennsylvania, developed a passion for Ohio River history. At Cincinnati in 1843 he began publication of a newspaper the *Western Weekly Advertiser* which was devoted almost entirely to Ohio Valley history: the name was later changed to *Cist's Weekly Advertiser* and the paper continued with interesting content but scant financial success until 1853. Cist also published several books describing Cincinnati and reprinted a selection of articles from his newspaper in a work entitled *The Cincinnati Miscellany* in 1846.

Cincinnati proved to be especially fertile soil for the culture of both amateur and professional historians. Benjamin Drake, a native of Mason County, Kentucky, born in 1794, came to Cincinnati with his famous brother, Dr. Daniel Drake, about 1815. There he studied law and was admitted to the local bar ten years later. His successful books began with his joint authorship of the descriptive *Cincinnati in 1826* and continued to include lives of Tecumseh, William Henry Harrison, and Black Hawk before his death in 1841.

The Rev. Mr. Timothy Flint, Massachusetts-born Harvard graduate, spent ten years on the Ohio and Mississippi, 1815-1824,

which resulted in a history of the area, a geographical description
in two volumes, *The Indian Wars of the West,* published in 1833, a
life of Daniel Boone, and two or three novels with scene set in the
new country. Flint also edited a periodical, the *Western Review,*
which was published in Cincinnati, 1825-1828.

Most successful and most talented of the midwestern writers
of the period was unquestionably James Hall, an officer in the War
of 1812, lawyer, editor, and judge in Illinois, one-time treasurer
of that state, and later cashier of the Commercial Bank of Cincinnati. As the West's leading editor of periodicals, he continued his
ephemeral *Illinois Monthly Magazine* as the important *Western
Monthly Magazine.*

Between 1829 and 1847 Judge Hall authored a dozen books on
the midwest, descriptive and historical, including two novels. He
joined with Thomas L. McKenny in publishing the monumental
History and Biography of the Indians of North America with the
colored portraits of Indians which are now so sought after by
collectors of Americana.

Moses Dawson, editor of the *Cincinnati Advertiser,* made a
significant contribution to the history of the Valley in 1824 with his
*A Historical Narrative of the Civil and Military Services of Major-
General William H. Harrison* and Dawson's subject, General
Harrison himself, wrote *A Discourse on the Aborigines of the
Valley of the Ohio* in 1838. Wellsburg, West Virginia, and Indiana
share in one of the most scholarly of the amateur historians, John
Brown Dillon, whose great *Historical Notes on the Discovery and
Settlement of the Territory . . . Northwest of the River Ohio* was
published in Indianapolis in 1843. After 1822 Illinois had a clear
title to John Mason Peck, Baptist minister, whose *A Guide for
Emigrants,* 1831, and *Gazetteer of Illinois,* 1834, were important
descriptive works.

Even with due consideration for the quantity of history that
was enacted upon its shore, and the early cultural activities in the
Bluegrass and Beargrass regions, Kentucky supplied at least its full
share of on-the-scene historians of the early day. Some of the most
significant of the early Kentucky works were William Littell's
Political Transactions in and Concerning Kentucky, Frankfort,

1806, Humphrey Marshall's *History of Kentucky,* Mann Butler's *History of the Commonwealth of Kentucky,* Samuel L. Metcalf's *Indian Wars of the West,* Robert B. McAfee's *History of the Late War in the Western Country*—which came out in Lexington in 1816, before the buttons on the uniforms of Kentucky volunteers had had time to tarnish—and Dr. H. McMurtrie's *Sketches of Louisville,* published there in 1819.

And then—whatever nefarious motive may have prompted its writing—there is always General James Wilkinson's *Memoirs of My Own Times,* which finally came out in complete form in 1816 and is a valuable contribution to the history of the American Revolution and a rather devious apology for certain less reputable activities in the West afterwards.

Sound writing by amateur historians did not stop with the passing of the pioneers. It continued through the nineteenth century from Pittsburgh to the Mississippi and some of it was of great importance. Peculiarly so was that of a group of professional and business men of Louisville. There in May, 1884, ten prominent citizens met and formed a society for the purpose of collecting and preserving the history of Kentucky. Appropriately, they named the organization the Filson Club, after John Filson, the state's first historian. The organization has grown through the years, publishes a quarterly journal, and maintains a fine library. Especially interesting are the early publications—handsome quarto volumes to which the members of the club contributed scholarly studies of great significance. Similar work has been carried on by the Historical and Philosophical Society of Ohio at Cincinnati and, under state direction and support in recent years, the historical societies of Pennsylvania, Ohio, West Virginia, Indiana, Illinois, and Kentucky.

There were great journalists on the Ohio—from Charles Hammond, who began with the *Scioto Gazette* in 1811, defending the tactics of General Arthur St. Clair, and continued in the profession to bring the Cincinnati *Gazette* to power in the valley during his editorship between 1825 and 1840.

The alternately fiery and mellow George Dennison Prentice served a term on the staid *Connecticut Mirror* before coming west, eventually to edit the Louisville *Journal* from 1830 to 1860 and to continue as its mischievous, entertaining, and at the same time fearless contributor until 1868, when its editorial chair was filled by Henry Watterson—the legendary "Marse Henry" of modern newspaper lore—who consolidated it with the *Courier* to produce the influential and important Louisville *Courier-Journal,* which flourishes today.

Though born in Washington, D.C., and not arrived in Kentucky until his twenty-eighth year, Henry Watterson in later life came to be *the* Kentucky colonel to some hundreds of thousands who knew him, or knew of him, or had only seen his picture. He was, indeed, the colonel complete—with flowing mustache, goatee, a vast appreciation of and capacity for Kentucky bourbon, and a never-say-die fire-eating editorial spirit reminiscent of what one supposes to be that of the heyday of the duello and the julep in the Bluegrass.

The writing of fiction on the Ohio, off to a start with the early efforts of Brackenridge and Holman at opposite ends of the river for all practical purposes, continued to prosper moderately with its most enthusiastic audience pretty well confined to the valley. One of Timothy Flint's historical novels, *Francis Berrian,* elicited, by some kind of blood-from-a-turnip miracle, kind words from Mrs. Frances Trollope. The truth is that Flint's fiction, like James Hall's, is stilted and tedious stuff which compares in no way with the quality of the work both did when their subject was history, travel, or description. As a matter of fact the Ohio Valley boasted no widely popular novelists until Emerson Bennett's western tales, some of which were written during his residence in Cincinnati, achieved success near the middle of the nineteenth century.

Fiction other than that of Bennett's blood-and-thunder school was likely to be thin and mawkish stuff through the seventies on both sides of the river. Sample titles of what passed for literature through this general period in Ohio, Indiana, and Kentucky (the Illinois and Virginia shores were mercifully uninspired at the time) tell the story: *Married, Not Mated; Grace Truman, or, Love and Principle; First Quarrels and First Discords in Married Life; Mt. Echo, or, The Mother's Mystery; Zoe, or The Quadroon's Triumph; The Drunkard's Child; Willie Elton, the Little Boy Who Loved*

Jesus and *Black Steve, or, The Strange Warning* (both the latter by Martha Finley, who later committed the Elsie Dinsmore books); *Early Engagements; Hannah, the Odd-Fellow's Orphan—*

In the late seventies some of the output began to improve, at least by the standard of the current national average. Lew Wallace published *The Fair God*, Edward Eggleston gained international fame with *The Hoosier Schoolmaster, Roxy, The End of the World,* and *The Circuit Rider*: William Dean Howells became editor of the *Atlantic Monthly*—although his novels were not written until the next decade—Lafcadio Hearn was working on Cincinnati newspapers, and Ambrose Bierce was reaching maturity.

The writing of literature—good, bad, and indifferent—became a full-flowered passion in the region during the eighties and nineties, recognized and encouraged by William Henry Venable's *Beginnings of Literary Culture in the Ohio Valley,* which appeared in 1891. From then on, apparently, half the residents of the valley, especially in Indiana, Ohio, and Kentucky, undertook to write for publication. West Virginia still refused to be stampeded into pursuit of literary glory and the writers of Illinois, while many, appeared then and thereafter to concentrate not upon the Ohio but upon those sections of the state whose waters drained into the Mississippi or Lake Michigan. No matter, however; there were plenty of littérateurs in the other three states.

First, from the valley there were the Eggleston brothers, Wallace, Howells, Mary Hartwell Catherwood, Maurice Thompson, Charles Major, James Lane Allen, Annie Fellows Johnston—not all great writers but all greatly read—and if you preferred homely verse which sold in quantities there were, in the nineties, well-established James Whitcomb Riley and rising Paul Laurence Dunbar.

Then, as the 1900 mark was passed, came the deluge of novelists, biographers, and historians whom any reader of fiction or biography of more than thirty years' standing must recognize after only a glance at a few of the names. Indiana held allegiance of the greatest number of the best selling—Booth Tarkington, Theodore Dreiser, George Ade, George Barr McCutcheon, David Graham Phillips, William Vaughn Moody, Joaquin Miller, Kin Hubbard, Gene

Stratton Porter, Meredith Nicholson, and a little later, Albert J. Beveridge, Claude Bowers, Frederick Austin Ogg, Charles and Mary Beard, Albert Edward Wiggam, George Jean Nathan, Elmer Davis, Lloyd Douglas, Leroy MacLeod. After 1900 Kentucky offered John Fox, Jr., Irvin S. Cobb, and Alice Hegan Rice. Illinois's great were, as has been mentioned, concentrated on Lake Michigan and Mary Roberts Rinehart was best seller enough to uphold the honor of the river in its short run through western Pennsylvania, but Ohio was second only to Indiana in fecundity of successful twentieth-century writers. She had Brand Whitlock, Zane Grey, Sherwood Anderson, Louis Bromfield, and Earl Derr Biggers—even William Sidney Porter, though not an Ohioan, owes something to the state, for he began to write while an inmate of the Ohio State Penitentiary under (I'm sorry; I can't resist the impulse) the pen name O. Henry.

After 1900 writing—like steel and agriculture—truly became Big Business and the production of best sellers by natives or residents of the valley states at times approached the status of a national monopoly. Alice Payne Hackett's book, *Fifty Years of Best Sellers,* lists the ten titles with greatest sales for each year since 1895. Identifying the American authors of these successful novels from 1900 to 1940 by the states in which they were born and giving ten points for the first book on the list, nine for the second, etc., the total allowable to each of the first ten states is

    1. New York 218
    2. Indiana 213
    3. Pennsylvania 125
    4. Virginia 102
    5. Kentucky 94
    6. Missouri 80
    7. Ohio 73
    8. Michigan 70
    9. Minnesota 67
   10. California 64

Even with West Virginia and Illinois failing to place, the other four Ohio River states, Indiana, Ohio, Pennsylvania, and Kentucky,

muster a total of 505 points against the remaining six states represented in the first ten with a total of 601.

There's no if or and about it—yarnin' held its own on the
Ohio long after the tall-tale fabricators of the keelboat era had gone
to rest.

Residents of the Ohio Valley have been decidedly laggard in
composing music in the classical tradition, even though the playing
and singing of tunes of one sort or another has been a principal
expression of art since the time of discovery. The cities in the
valley, the educational centers, can point to many who have
atttempted composition in the higher strata; each will name a few
local musicians or teachers who have essayed the flight—but the
fame of few of those proclaimed in one city has progressed as far
as its next neighbor. Classical music, superlatively well performed,
is to be heard in many of the cities of the valley and is appreciated
by a far larger than normal percentage of American population:
but little classical music of importance has originated there.

During the past century, however, the valley has produced
something more than its fair share of music popular with all the
people: another indication, perhaps, of the solid foundation of its
claim to recognition as America's heartland.

The composition of widely popular music began when Thomas
D. Rice composed his "Jim Crow" song at Louisville, Pittsburgh, or
someplace between. Dan Emmett, author and composer of "Dixie,"
was an Ohioan by birth. Of course Stephen Foster, America's
greatest popular composer, came from Pittsburgh and developed
his remarkable talent entirely upon the Ohio, and Paul Dresser,
of "On the Banks of the Wabash" and "My Gal Sal," spent his
youth and young manhood on the Wabash and Ohio before moving
on to New York. Whatever may be the artistic merits of "My
Darling Nellie Gray" and "Down By the Old Mill Stream," no
quartet, drunk or sober, can harmonize ten minutes without render-
ing one or both of them and their respective composers, Benjamin
R. Hanley and Tell Taylor, were Ohioans both. Ethelbert Nevin
was a resident of Sewickley, Pennsylvania, and some of the finest
popular music produced in recent years is that of two native
Hoosiers, Cole Porter and Hoagy Carmichael.

Music, homemade or professional, vocal or instrumental, was popular upon the river from the beginning of history—although for several decades after the Great Kentucky Revival the more hard-shell branches of the Baptist, Methodist, and Campbellite faiths frowned upon any but religious tunes—and music still continues a favorite form of artistic expression in the valley.

Jenny Lind and Ole Bull were received in Ohio cities with as much frenzy as their talented press agents stirred up for them elsewhere and the minor artists of the nineteenth century were patronized in proportion to their talents and those of their advance agents. But wide popularity of orchestral and operatic music in the high classical tradition had to await the arrival of the cultured German refugees who established themselves in Pittsburgh, Cincinnati, and Louisville. There is no question as to the responsibility for this development—a glance at the names of those who sponsored public performances of classical music in these cities, and in others in the valley, establishes the credit beyond the shadow of a doubt. Particularly in Cincinnati, which received the greatest number of these immigrants, does the tradition thrive.

An even wider appreciation of music has resulted in the past two decades from the enormous local enthusiasm for the teaching of music in the public schools. High school band and orchestra concerts and contests attract crowds as great as came to the county fairs a generation back and communities scarcely large enough to support a post office enthusiastically uniform and equip their school bands and boast of the virtuosity of their saxophone players and drum majorettes with all the unreasoning loyalty Brooklyn is reputed to give to its baseball club. Even though this enthusiastic support might seem, to the casual observer, to be more appropriately shared between the teaching of music and, perhaps, that of English grammar and mathematics, there can be no question but that it is having a profound cultural effect upon the valley—as support of professional music and changing taste in phonograph records indicate quite plainly.

Like that of the writers of the Ohio Valley, a study of the painters who were born within its limits—though many were forced by necessity to move to older-settled regions in order to flourish—

is a work in itself. Painters, after the beginning of the nineteenth century, were a crop as plentiful as were writers a century later. Only a glance at those who were a product of the valley and who visited it in the early days is possible here.

George Jacob Beck was probably the first white artist to paint upon the Ohio. Beck came to Cincinnati in 1790 as a member of the army General Anthony Wayne brought west to avenge the defeats suffered by Harmar and St. Clair. Whether Beck was allowed time off from his military commitments or was discharged from the service is not known, but in either case he was soon painting landscapes in the neighborhood and is reported to have decorated an elaborate houseboat for General James Wilkinson. Beck is said to have found small market for his canvases.

Doubtless the neighborhood of Lexington, Kentucky, was first in the West in which a painter—particularly a painter of portraits —could earn a dependable living. Many of those who arrived within the first thirty years of Kentucky settlement were former residents of the Virginia and Maryland seaboard, where portraits had long been a necessary adjunct to family pride. As the Kentucky branches of such families prospered they were naturally ready to follow the tradition. Perhaps a crude—and one trusts unflattering—miniature of John Filson drawn upon the flyleaf of one of his books was among the first portraits made in the state. If that was the case, the quality was to be remarkably improved within a decade.

Soon after the turn of the nineteenth century Kentucky began to be served by a native portrait painter. Matthew Harris Jouett was born in Mercer County, Kentucky, in 1787. His talent early recognized, he was sent to Boston, where he studied under Gilbert Stuart for some months before returning to Lexington to set up his studio. There and at Louisville—for short periods also at Natchez and New Orleans—he painted hundreds of portraits (350 are recorded) before his death in 1827. William Edward West, born in Lexington a year after Jouett, enjoyed his greatest success in the East and in Europe.

Audubon himself painted portraits upon the Ohio and Mississippi during the second and third decades of the nineteenth century —though one assumes that his heart was not in the work and that he did it only for cash to buy bread and butter and paints for his bird pictures and their incidental landscape backgrounds. After 1827, Charles Alexandre Lesueur was sketching and painting landscapes, wild life, and the goings-on around him in New Harmony, and young John Banvard, in the same town, was painting his first panorama; Thomas Cole, later to become one of the most meticulous of the Hudson River school of landscapists, was teaching himself to paint with hog-bristle brushes while wandering through Ohio.

Two other movements on the Ohio in the twenties and thirties made a contribution both to art and to history: the traveling artists who "reported" the valley in pictures and the three or four discerning souls who visited the valley states for the purpose of recording the appearance and life of the Indians.

Of the former, there were Captain Basil Hall, whose work has already been mentioned, August M. Köllner, William Momberger, later, and others of widely varying talent. Of the latter—those who specialized in painting Indians—first, apparently, was the lithographer and actor, J. O. Lewis, whose *Aboriginal Portfolio* was published in parts in Philadelphia, 1835-1836. George Winter, who came to Indiana to study the few Indians who still remained in the state in the thirties, made his living as a portrait and landscape painter—and his beautiful water colors of Indian life were finally published in 1948! Greatest of the artists who painted Indian life was Karl Bodmer, a Swiss, who accompanied Prince Maximilian of Wied upon the tour in 1832 that included the Ohio, a visit to New Harmony, and to the Mandan and other of the Plains Indians beyond the Mississippi. Besides the handsome illustrations that accompany Prince Maximilian's resulting book, *Journey Through North America,* published 1838-1842, Bodmer did several American historical scenes—including the stirring "Simon Butler's [Kenton's] Ride," which is of the very essence of river history.

The infants arriving upon the Ohio scene 1800-1820 were as liberally endowed with artistic talent as any, but most of them had eventually to seek fame and fortune in the East or in Europe, for a slump had come in the market for art upon the river by the time they reached maturity. Even the austerity that resulted in the neglect of art might be attributed in part to lingering effects of the Kentucky Revival, though a contributing factor was probably an increased preoccupation with making money, which developed after 1830, and the fact that the Old Virginia tradition, even in Kentucky, was being diluted by the arrival of numbers of hardheaded Yankees and hard-handed foreign immigrants who held no truck with such foolishness as having their portraits painted.

When these materialistic new arrivals did let their sentiment overcome them, they usually patronized the traveling artists who carried ready-painted figures, ready framed, in which a reasonable facsimile of any patron's face could be inserted on short notice for a small fee. Many of the "Portrait artists" who advertised in the city directories of Pittsburgh, Cincinnati, and Louisville were doubtless temporarily static specimens of this class. By that time not even the painting of tavern signs offered much of a market, for styles had changed and tavernkeepers grown less choosy: tavern signs had been staple assignments for artists from Hogarth down and only a few years before even such a painter as Chester Harding, who did striking portraits of Daniel Boone, Robert Owen, and a good many far more prominent Americans and Englishmen, had not hesitated to paint a head of Washington upon the signboard of the Columbian Inn at Shawneetown.

There were plenty of artists born in the valley during this period who later made distinguished names for themselves: Aaron H. Corwine in Mason County, Kentucky, in 1802; John Insco Williams, Dayton, Ohio, 1813; Miner K. Kellog, born in Cincinnati about 1816; and Joel T. Hart, Winchester, Kentucky, 1810. These had the misfortune to reach the height of their powers a bit too late to have the advantage of the first thriving market for art in the West and they were slightly early for the second. Only a little later, Samuel Woodson Price, born near Nicholasville, Kentucky, in 1828, was able to make a successful career as a painter in the West before it was interrupted by service in the Civil War—as a brigadier general. There was still more opportunity for the talented born in the forties: T. C. Steele, born in 1847 in Owen County, Indiana, succeeded in the West, and Frank Duveneck, who first saw the light of day in Covington, Kentucky, in 1848, studied at Pittsburgh and in Munich but returned to Cincinnati, just across the river from his birthplace, to enjoy a distinguished career as painter, sculptor, and etcher.

With the eighteen-fifties came a new and awe-inspiring development in what was invariably referred to as Art—with the capital A—it began to be widely pursued by the young ladies in process of being finished in the myriad of seminaries throughout the valley. Probably the teachers were not too well qualified but the production, judging from the quantity of examples of the students' work that has survived, must have been prodigious.

That was a sentimental age, preoccupied with hopeless love, lingering illness, death, and the moldering grave: it fairly reveled in sorrow manifested in the fabrication of wreaths worked from the hair of the living or dead, or in crewelwork mottoes of somber theme and color. These were comparatively innocuous, but when the art of this school descended to crayon illustration or portraiture it could be depressing in the extreme. Emmeline, deceased daughter of the feudin' Grangerfords of Arkansas upon whom Huck Finn reported, was not alone in her devotion to funeral tributes and what Huck said

they called crayons . . . They was different from any pictures I ever see before—blacker, mostly, than is common. One was a woman in a slim black dress, belted small under the armpits, with bulges like a cabbage in the middle of the sleeves and a large black scoop-shovel bonnet, with a black veil, and white slim ankles crossed about with black tape, and very wee black slippers, like a chisel, and she was leaning pensive on a tombstone on her right elbow, under a weeping willow . . . and underneath the picture it said "Shall I Never See Thee More Alas."

She had many kindred spirits upon the Ohio, had Emmeline, in her unwitting revival of the death cult of the Middle Mississippi mound builders.

Where before young artists of more serious pretensions than were the maidens of the seminaries had had to depend upon apprenticeship or individual instruction, art schools of more or less permanence began to operate in the seventies. And shortly, when the steel and coal boom brought great individual wealth, private collections, sometimes open to the public, began to be assembled, especially in Pittsburgh, Cincinnati, and Louisville.

These collections were not invariably of the most fortunate selection; one prospering gentleman of a small town in the valley took his wife to Europe, in the sunset of their lives, and returned to saddle his fellow citizens with a collection of what is referred to by the present janitor-curator as "genuine hand-painted Old Masters pictures." The old gentleman's intention was of the best, and had there been more connection between the names signed to these works and the artists who bore those names in life, the gallery's present contents would rival in value that of the vault at Fort Knox.

Mostly, however, the Ohio Valley collectors fared very well, and as they added to their wealth and finally met the end to which all collectors must eventually accede, they employed their collections as a nucleus and their money as an endowment to establish—particularly in the three great cities—public galleries and public schools of art.

The cause of painting, after a promising beginning followed by a long artistic drought, is again in a flourishing state upon the river.

Chapter Sixteen

OLD TIPPECANOE TO JOHN BROWN OF OSAWATOMIE

THE years of America's forties and fifties saw strange and porten-
tous developments in politics, commerce, nationalistic spirit, and
the social and domestic life of Americans. Seventy-four years later
Meade Minnigerode named the first half of that period the "Fabulous Forties," and the second decade might well have been included:
fabulous the forties were and the fifties were no less so—especially
upon the banks of the Ohio.

The forties opened auspiciously enough for the valley as
William Henry Harrison, favorite son, was elected to the Presidency
in a campaign to which may be traced the genesis of modern

publicity methods, political and commercial. No one seems thus far to have troubled to learn the name of the genius who planned the details of the Log Cabin and Hard Cider assault on the electorate but he should be enthroned as the patron saint of modern press-agentry to receive whatever recognition that position merits.

Harrison had proved himself a competent administrator of Indiana Territory, a capable officer during the final phases of the War of 1812 in the West, and had given evidence, through formulating an improved method for selling public lands, that he was an able representative in Congress. But his publicized qualifications for office were based upon his survival of the minor Battle of Tippecanoe, upon the fictitious allegation that he had been born in a log cabin, and upon the dubious claim that he preferred hard cider over all other available beverages.

Actually General Harrison's strategy at Tippecanoe had been wide open to criticism and there was considerable doubt as to whether his troops had triumphed or the Prophet's Indians had simply quit. Harrison had been born at Berkeley, the comparatively elegant plantation house in Virginia that was owned by his father, Benjamin, a signer of the Declaration of Independence. If he preferred hard cider to port, brandy, rum, or Monongahela whisky, his taste was considerably inferior to that of the average born and bred Virginia gentleman.

The fabrications concerning the candidate suited the purposes of his backers much better than did the facts, however, and they played them to a finish. Most of the media employed were new. Of course pamphlets were issued—there was nothing original in the bare fact of that—but many of *these* pamphlets were profusely illustrated by woodcuts depicting episodes in Harrison's life. Naturally the illustrations showed him in the thick of battle, shaking hands with a veteran, and patting an adoring dog simultaneously (that one was still good for Messrs. Truman, Dewey, and Wallace), sitting in front of a log cabin which bore no resemblance to his own fine brick mansion at North Bend on the Ohio, and reproving Tecumseh at Vincennes, apparently to more effective purpose than witnesses had reported of the actual occasion.

These same woodcuts, or variations of them, were mailed to

newspapers throughout the country (and *that* was an innovation), while, in districts which did not already possess papers whose editors could be depended upon for wholehearted support, new ones were started and continued through the campaign. Literally dozens of these papers were published, especially in the Ohio Valley, and were supplied with both illustrations and text. The names of all the campaign papers were similar and mail pouches were jammed with the *Log Cabin Advocate, Tippecanoe Banner and Spirit of Democracy, Tippecanoe Calumet and War-club,* and *Harrison Banner, Harrison Republican, Harrison Democrat, Harrison Flag,* and others, all containing identical news and illustrations but with various date lines; there, at least in effect, was the beginning of newspaper syndicated feature service!

There were almanacs too—*Log Cabin Almanacs, Hard Cider Almanacs,* and many more, again with those woodcuts—and there was the sheet music for dozens of campaign songs, printed bandanna handkerchiefs, tablecloths, coverlets, and table china, jugs, and flip cups. Best of all there was that catchy slogan— "Tippecanoe and Tyler too!"

Naturally William Henry Harrison was elected and the West had its first president.

The West did not accept the honor calmly or with modesty befitting. Washington groaned under a great accession of minor politicians bearing Ohio Valley mud on their boots who took the president's folksy campaign propaganda seriously and moved in for a bait of hard cider or whatever other reward might offer. Harrison, actually a man of rather formal bearing, was not amused to any noticeable extent. It was not only the small-time politicians who took advantage of Harrison's success, however; the West as a whole was feeling its oats to a rather alarming extent and was giving hearty voice to its exuberance.

As it happened, President Harrison died in the month following his inauguration and a series of misfortunes visited the Ohio which should, but did not, have a sobering effect:

The serious financial depression that had beclouded the country since 1837, and which had helped, with thinking people, to bring about Harrison's election, continued into 1841; Charles Dickens toured the valley in 1842 and found only Cincinnati unobjectionable; the winter of 1842-43 was one of the long, cold ones that held honored place in the recollections of the Oldest Inhabitants during the half century following; banking scandals continued and serious anti-Mormon conflict came to a climax in Illinois when Joseph Smith was killed by an armed mob at the town of Carthage in June, 1844. In 1845 the immoral influence of young men's singing clubs had become so evident in Pittsburgh that the parents of one youth were preparing to separate him from danger of being lured into minstrelsy by shipping him off to Cincinnati to work for his brother, the junior partner in the firm of Irwin & Foster. The same year nearly a quarter of Pittsburgh went up in flames—although not, according to evidence, through any overt act of the singing clubs.

But obviously these matters could only be classed as petty annoyances; the high destiny of the country had already made itself manifest to all forward-looking citizens, especially to those of the valley. The territory in the Great West, which remained in foreign hands—present California, Arizona, New Mexico, and the Republic of Texas in the south, and present Washington and Oregon in the north—was all the logical heritage of the Union, said they. The Oregon Country (between the 30th and 50th parallels to the Pacific, according to the more optimistic) should be added at once—by the citizens north of the Ohio River unaided, if necessary!

In the autumn of 1843 President Tyler began negotiations designed to accomplish part of the ambitious program, the annexation of Texas. The Texans were willing: many, perhaps most, of their leaders were originally Tennesseans and Kentuckians from Ohio River tributary country—and they had resolved in favor of annexation for their republic in 1836. In April, 1844, Tyler submitted a treaty to this end and at the Democratic National Convention that year candidate James Knox Polk was pledged to carry on the campaign when elected. Polk *was* elected and Texas was admitted to the Union on December 29, 1845.

These Texan developments were enormously interesting and encouraging to the slave states, especially the western ones where large-scale plantation operation was reaching its zenith, to Tennessee and Kentucky, which had furnished so much of Texas's population —and to Illinois, Indiana, and Ohio, because Texas was a vast undeveloped market which could use their farm and manufactured products and employ their shipping facilities and because they, more brash and more innocent of knowledge of the true extent of European power than their seaboard neighbors, already felt a premature call to international leadership. As a writer of the period put it, the annexation was a measure regarded as "a link in a chain of events that will girdle the North American continent with a wide belt of illumination: which has given an impulse to the extension of Anglo American institutions, that cannot be arrested until the circuit of empire shall have been completed on the shores of the Pacific."

Sad as it is to say, the beautiful Ohio played her full part in the disreputable Mexican War. Those who sullied her waters were not more than evenly divided between Washington politicians and opportunists who lived on her banks, notwithstanding the pronouncement of freshman Representative in Congress A. Lincoln, of Illinois, who stated that in his opinion it was solely a politician's war.

Apologetic historians may blame bungling Mexican diplomats, the failure of the Mexican government to restore American property seized in the preceding Mexican Revolution, and Mexican resentment of the annexation of the "free Republic of Texas" (in the freeing of which, they imply, the United States could have had no possible part). But the facts remain that the United States had long coveted California and New Mexico, that strong southern interests were anxious to acquire slave territory to balance the expanding free North, that the northern Ohio River states were not averse to extending their markets, and that many military men, amateur and professional, wished for a good, cheap, easily won war in which to demonstrate their attainments.

When, after his election, President Polk secretly ordered General Zachary Taylor to approach the Texas-Mexican border with a view to provoking a convenient "incident," Taylor was willing and eager. A large but ill-equipped and poorly trained Mexican force under General Mariano Arista crossed the Rio Grande (the reaction

Polk and Taylor anticipated) and American territory was invaded! National honor was assailed; conquest threatened! Desolation! Annihilation!

General Taylor and his 2,228 men were remarkably alert: an artillery duel began at Palo Alto from which the Mexicans retreated, having little artillery and being quite unskillful in the use of what they had. They made a stand, next day, at Resaca de la Palma, but Taylor attacked and the infantry battle that ensued accounted for Mexican losses of 547 against United States casualties of 122 killed and wounded. The Mexicans fled back across the Rio Grande.

On May 11—correctly assuming that if a battle had not already taken place it shortly would—President Polk delivered a message to Congress announcing the invasion and informing the nation that a state of war existed. He was promptly authorized to act and a remarkably convenient war was shortly declared.

The Ohio River's indisputably prominent part in this conflict began almost at once, as troops began to move. Railroads in the West and South being the independent and unconnected stretches of track they then were, canal traffic being interminably slow, and horse-drawn wagons being beneath the consideration of a people now grown steamboat and railroad conscious, Ohio River transportation was the quickest and most direct means for reaching the seat of war from the home of something like two-thirds of the population.

The war began, for the majority, as a sort of light-opera affair colored by the sentimentality that had recently generated from the romantic novels, the saccharine poetry and songs popular at the time, and—to no small extent—from the sad, sweet lithographed prints produced by Nathaniel Currier (later of Mr. Currier and Mr. Ives), which were already decorating even the humblest sitting rooms.

Young men who cherished ambitions, who sought repute for political purposes, or who merely thought their appeal to the opposite sex would be enhanced by the elegant uniforms provided, enlisted to defend their country's honor, to "plant the Stars and Stripes in the Halls of the Montezumas," and to get away from plowing. They came from all sections, since the slavery and anti-slavery issue was pretty well forgotten as soon as war was declared.

Companies were raised in Pennsylvania, Ohio, Virginia, Indiana, Kentucky, Illinois, Wisconsin, and Michigan. New York and the New England states are said to have lagged somewhat and even no farther east than Pittsburgh the *Daily Commercial Journal* sniffed: "The Recruiting Sergeant was out yesterday with drums beating and colors waving their invitation to glory and broken bones." But all sections eventually produced volunteers and shortly the troops that were not taken around to the Gulf by sea were being shipped down the Ohio, inspired by the promise of 160 acres of land as a bonus, $7 per month pay in cash, and a clothing allowance of $47 per year.

The fact that most of them got no farther southwest than the Gulf towns or the rallying points between the lower Mississippi and Texas, steaming in the summer heat, made no difference to the throngs who cheered their flag-draped steamboats and accompanying barges down the river.

The military was waved on its way by the fluttering handkerchiefs of sentimental damsels who leaned from the upper windows of young ladies' seminaries and private residences along the river's bank. Attired in tight-bodiced, voluminous-skirted gowns, shoulders exposed to the limit of the extent their mothers' compunctions and the last issue of *Godey's Lady's Book* dictated, shawls draped mantillalike over their slicked-down front hair and corkscrew back curls, they watched the troopboats out of sight before retiring to their boudoirs (Godey had already substituted *boudoirs* for bedrooms) to ward off fainting spells with sips of elderberry wine and to commit horrible occasional verse to tear-stained paper. (Not all bad music and verse was produced by amateurs, however; Stephen Foster, clerking at the Cincinnati steamboat booking office, contributed "Santa Anna's Retreat From Buena Vista," a quick-step far below the standard he had already established for himself.)

Jingoistic patriotism ran high, particularly in the Ohio Valley, and exhibited itself in new manifestations and revivals of old ones. There appeared columns upon columns of chip-on-shoulder editorials, such as had not been seen since the War of 1812, daring any and all European powers to make something of the aggression against Mexico; there was jockeying for places of command between officers from northern and southern states; a new eye to the exploitable possibilities of the Far West and, as victorious skirmishes with the underfed Mexican forces began to be reported, there developed a gauche and overweening national cockiness which is hard to forgive after even a century.

Came also something rather amazing in the way of commercial activity: iron and coal were required of the river from Pittsburgh to Hanging Rock; hardtack, guns, harness, and various fabricated supplies came from Cincinnati and Louisville, drinking liquor from the Monongahela and the Kentucky limestone region, and flour, corn, and hog meat from everywhere on the Ohio's lower half. There is no preserved indictment against a lack of quality in the coal, ironware, and leather but of the refreshments from the Ohio, liquid and solid, there still linger memories of the fabulously unfavorable current descriptions.

For many years before the forties these Ohio River comestibles —from Monongahela whisky and Rappite beer to Kentucky hams and Ohio and Indiana beef and corn—had borne high repute in New Orleans and the South; the first commerce of the river had been founded upon them, plus tobacco, and they had held their place by their pure excellence. That wide reputation for quality made the sudden change in them all the more evident as they appeared when they reached the encampments of the American soldiers in the South.

It was apparent to the Ohio Valley soldiers who drew the rations that something was amiss: these were definitely not the vittles they had known in mother's kitchen. When letters began to come from home reporting the surprising rise of So-and-so, who had slopped a half dozen hogs before, to wealth as a pork packer, and the affluence of Such-and-such, whose wife formerly baked pies for the stevedore trade, in his new role as a manufacturer of army bread, peculation began to be suspected by the volunteers.

It had to be a combination, of course, made up of So-and-so and Such-and-such in association with the army service of supply and the licensed sutlers but that did not mitigate the terrible havoc wrought upon the digestive systems and even the lives of the volunteers.

Hardtack moved about, when placed on a smooth surface, and its quickening could not have been caused by Mexican jumping-bean flour, since commerce with Mexico had ceased—could it have been moved by the locomotive power of resident weevils or maggots? Bacon, singularly and uniformly white in color, was likely to melt and run in reasonably cool air, making up in the strength of its aroma for its almost total lack of fiber. Whisky froze on an average winter night as far south as Arkansas. In short, the part Ohio Valley commerce played in the Mexican War was at least no more honorable than was any other feature of that conflict.

In 1848 the war was formally ended and California and New Mexico were taken over at about half the moderate price President Polk had been willing to pay for them in the beginning. Some hundreds of soldiers had been killed in action—some thousands were dead or dying from the effects of vermiculose food, barracks life in a strange climate, rioting, and disease contracted from disreputable places of entertainment and equally disreputable camp followers.

But the national honor was vindicated, there was a vast new territory, acquired at a bargain cost, to be explored and settled by restless young men and capitalized and developed by astute old ones.

Nine days before the formal transfer of the new land, on January 24, 1848, James W. Marshall discovered that California had what turned out to be considerably more than enough gold to pay the purchase price of the whole new land and the costs of the war as well. The Ohio ran toward the west; gold and lands were in that direction and travel on the river picked up immediately.

In providing means for the first stage of the route to the Pacific the Ohio made a great contribution to the settlement of the Far West. It helped immediately, thus, to make the United States a country which faced the Old World upon both east and west, with vast resulting power—and responsibilities which did not begin to make their true gravity manifest until ninety-five years had passed.

From the beginning of westward movement the Ohio had been the chief highway of emigration but during the first six or seven decades it had served chiefly to populate its own tributaries and neighboring lands upon the Mississippi. Then, in the thirties, travel to Texas began. That being largely an emigration of single men, they had employed every sort of craft from skiff to steamboat and the California gold rush was of the same character except that, in it, speed was a vital consideration; it was a rush indeed, for every participant was impelled by a fear that all the gold might be mined before he reached the workings.

The principal route (except for that by ocean from eastern ports either to the Central American isthmus or around the Horn) made its first leg down the Ohio to the Mississippi before its variants branched off on that river to Independence, Missouri, or to the Arkansas River and thence overland, or to New Orleans and transfer by oceangoing ship to the Isthmus. Thus the Ohio had the lion's share of the traffic whichever route was finally selected. Irwin & Foster, the Cincinnati steamboat agents, chartered the *South American* and ran her on special trips to Independence for the particular convenience of gold seekers, and their young clerk, Stephen C. Foster, booked passengers and listened to them singing parody lyrics of "Oh! Susanna," anthem of the gold rush, which had been first sung in public at Anderson's Eagle Ice Cream Saloon in Pittsburgh during September, 1847. Few knew the name of the song's composer and he himself was not sufficiently proud of it to care.

Considerable travel to the Oregon Country (the Pacific coast between 42 degrees and 54 degrees 40 minutes north latitude) began early in the forties, but much of it was overland, since these emigrants were farmers and under the necessity of transporting their households to live permanently. The Oregon Trail was used most commonly but as emigration increased, river travel on the Ohio, Mississippi, and Missouri began to be used by the single or prosperous for the first stage. As the forties progressed, so did emigration. A new generation of land-hungry young men had been born and had grown up in the Midwest and they now wished for the same sort of chance their fathers had improved on virgin land. The Oregon fever had caught and there was no stopping its spread. Only a few of its more conservative victims were deterred by the fact that settlement was forbidden and no legal title could be

secured; soon a start was made at overcoming even this difficulty.
An entirely unauthorized Oregon Convention was held at Cincinnati in July, 1843; oratory flowed and a resolution was adopted
which called for a cessation of British claims to all land below north
latitude 54 degrees 40 minutes. Northern members of Congress
took up the cause and pushed it concurrently with their southern
colleagues' moves in favor of expansion in the Southwest; soon
there appeared to be danger of provoking a third war with Great
Britain—a somewhat more dangerous foe than that which was
contemplated by the plan of the Southerners.

Senator Edward Allen Hannegan mounted the platform in the
courthouse yard at Covington, Indiana, dived into his whisky-lubri-
cated intellect, and came up with one of those felicitous phrases for
which he was already famous: the British-American line in the
Northwest, said the Hon. Mr. Hannegan, should be "Fifty-four,
forty—or fight!" The eloquent William Allen of Ohio used the
phrase in a speech in the Senate in 1844 and it shortly became a
popular slogan—especially north of the Ohio, where Stephen A.
Douglas and Lewis Cass lent their able talents to its popularization.

Negotiations with the British were carried on more successfully
than the country had any right to expect and in 1846 the rather
thumb-handed but astute President Polk accepted a compromise
which set the 49th parallel as the border. The titles to Oregon lands
were cleared, the antislavery element pacified in advance for what
has already been described as happening on the Mexican-Texan
border and the appetite of the slave states was whetted still more
sharply for southwestern expansion.

Typical of those who encouraged the movement to the North-
west, first and last, were three Ohio Valley residents, Overton
Johnson, William H. Winter, and Ezra Meeker: Johnson and
Winter as explorers and trail blazers and Meeker as leader of mass
emigrations and promoter of settlement.

Johnson and Winter were Wabash College students, Winter
ready for his sophomore year when the western fever struck him
in 1840 and Johnson, a sophomore of full standing. The college
sophomore is recognized by those who have spent time in and
about institutions of higher education as being possessed (possession

is temporary, happily) of certain traits of character, philosophy, and temperament which make him a being apart. It is so today; it was so in centuries gone. He is likely to be in revolt against the status quo, a yearner for the distant scene, an enemy of order; he is a ready apostle of any movement which appears ill-conceived; he is tremendously restless and wonderfully important. He is, in fact, a sophomore, and the term "sophomoric—inflated in style or manner" sprang from the observations of his vagaries by educators in ancient days.

Johnson and Winter were restless, as sophomores of any day would be with thousands upon thousands of uncharted miles stretching westward from their hypothetical doorsteps: in 1840 they looked at Missouri, cast their eyes farther, and in 1843 went to Oregon and California. They kept journals of their experiences and when Johnson returned two years later to resume his college work he combined their observations in book form. The extreme rarity of copies of *Route Across the Rocky Mountains, With a Description of Oregon and California* . . . published in Lafayette, Indiana, in 1846, vouches for the scores of emigrants its comprehensive directions guided to the Far West.

Ezra Meeker was distinguished in another way: his specialty was leading family emigration to the Oregon. An Ohioan by birth, he was reared in Indiana, married before he reached voting age, packed up his bride and drove his ox team to Missouri, wintered there and waited while his wife had her first baby, packed both back in the wagon and went on his way, becoming leader of an emigrant train en route and later going back east to lead many more such trains to the new country. There were several young Ohio Valley men who followed the same program; the only difference was that Ezra Meeker made his first trip in the fifties and was still busy encouraging emigration to the Oregon in the second decade of the twentieth century, making many missionary trips in an ox-drawn covered wagon and his last tours to the East by auto and finally, before he and his long white hair and flowing beard were laid away, one trip by plane!

But all the gold that flowed in the Ohio Valley at the end of the eighteen-forties was not the product of California, of Mexican

War supply swindles, nor of trading in Oregon lands: 400,000 hogs were slaughtered in Cincinnati alone in 1849, and though that city—"Porkopolis" then—led in the industry, plenty of smaller ports in Ohio, Indiana, and Illinois made sizable contributions. The coal and iron business from Pittsburgh to Wheeling and farther down-river in the Hanging Rock region, the fabricating of finished products at Cincinnati, now making full use of the arriving thousands of German artisans (Cincinnati had a German-speaking population of 20,000 by 1848), and wholesaling and distributing activities at Louisville had lost none of the impetus gained during the Mexican War.

Along the lower Mississippi and middle stretches of the Ohio's tributaries, the Tennessee and Cumberland, the South was prospering as never before. The growing demands of the cotton processors in both old and New England increased the market and kept up the price of the raw product while still farther south sugar and rice raising increased and became more profitable. While most of these southern products were shipped to manufacturers and consumers from New Orleans by sea, the planters received their food-stuffs and most of their other purchases except imported luxury goods by steamboat down the Ohio and Mississippi. Provisions alone made up an important article to northern producers and to shipping; the production of cash crops was far too profitable to encourage the planter in the Middle South to permit his slaves to waste time growing their own rations: an economy thus developed which has its unfortunate victims yet today.

With the augmented income on the rivers came a great increase in passenger traffic on steamboats, especially of the more luxurious type, on the lower Mississippi and the Ohio. Prosperous planters began to bring their families up the two rivers to escape the summer heat (which in less prosperous times, before and after, they simply endured) and foregathered with wealthy manufacturers, merchants, and promoters and their connections at the resorts—"Springs" of some kind, almost all of them were—in western Virginia, North Carolina, and eastern Kentucky. The steamboat trips and the stay at the resorts were considered equally enjoyable, equally opportune occasions for demonstrating social polish and displaying new-found wealth.

It was probably the early fifties that marked the absolute peak of luxury and elegance in steamboat travel. For a couple of decades before there had existed some rivalry between the captains of boats in the matter of the variety and quantity of the menu offered to cabin passengers, now new money in the purses of the travelers permitted higher fares to be charged and competition in speed, fitting, furnishing, and serving grew fiercer in proportion.

Records for speed were being lowered: the *Telegraph* ran the 141 miles from Louisville to Cincinnati in 9 hours and 52 minutes; the *Pittsburgh* made the trip from its home port to Cincinnati, 490 miles, in 24 hours and 15 minutes; the *Eclipse* ran the 1,440 miles from New Orleans to Louisville in 105½ hours—all during the years 1852 and 1853. But elegance of appearance and accommodation was the chief object of the pride of captains and owners.

Steamboat architecture achieved its gaudiest; became in fact *the* "steamboat architecture" that has now become a recognized school. Utilitarian and mechanical features were concealed behind a wooden shell trimmed in tortured "wooden lace" and fretted beams which combined the front porch, still beloved by Ohio Valley residents, with the second-floor gallery of the South. The craze for the Swiss cottage that was sweeping the country contributed the beams; the many windows derived from it and also the neo-Gothic school, but other standard items of decoration seem to have been pure invention of steamboat builders—or at least new combinations and permutations from varied sources, including the Moorish. Most steamboats with pretense to elegance were painted a dazzling white trimmed in gilt with as many colorful scenes and landscapes enclosed in panels and medallions as could be crowded on. Income derived from painting these probably replaced that from the execution of tavern signboards, which had once bolstered the economy of the western artist.

Inside, the staterooms and common rooms were finished with elaborate woodwork—usually inspired by the Greek revival of the fifties—in white, ivory, cerulean, maroon, and yellow paint with more, far, far more, gilt. Furniture (steamboat chairs are valuable antiques today) was walnut or mahogany or gilded finish upholstered in what was apparently mandatory Turkey red. Everywhere there were mirrors, and in the main saloon hung crystal chandeliers dripping prisms which tinkled with the forward surge of the boat.

Of course there was a bar, its brass and mahogany and crystal combining to lure the majority of the male passengers from care and responsibility, and there was food such as perhaps will never again be set in such profusion before common man in this now crowded world. It was all included in the cost of the cabin passengers' ticket; it was on the table, or on one of the sideboards that stretched interminably along the dining saloon walls, and if you couldn't reach it there were two waiters ready to answer every call. It was no longer country fare; fashion already demanded at least a few French titles—even a British dish or two—but it still had a frontier plenty and the diner in good health and spirits failed of a proper attitude toward his host, the captain, if he did not sample at least two or three dishes in each course before staggering from the saloon. As an example there was offered, one day in the fifties on an Ohio River boat of moderate pretensions:

SOUPS

Bean Soup   Sago Soup

FISH
Baked Pickerel

BOILED	ROAST
Corned Beef, and Cabbage	Beef
Leg of Mutton, Caper Sauce	Ham, Champagne Sauce
Chickens and Pork	Stuffed Chickens
Tongues	Pork, Apple Sauce
Jole and Cabbage	Leg of Mutton
Ham	Veal

Escalopes of Pork, Brown Sauce	Fillets of Mutton, British Fashion
Potted Veal with Vegetables	Boiled Beef, Piquante Sauce

COLD DISHES

Roast Beef   Veal   Chicken Salads   Roast Chickens

SIDE DISHES

Fricandeau of Veal with Onions
Croquettes of Potatoes, Parisian Style
Ragout of Mutton, Country Fashion
Bass, a la Bordeaulaise
Beef Kidneys, Provincial Style
Queen Fritters, Rum Sauce
Beefsteak Pies, American Fashion
Calfs Head, Fried in Batter

RELISHES

Tomato Catsup   Horseradish   Cole Slaw   Worcestershire Sauce
Pickled Cucumbers   Apple Sauce   Cranberry Sauce   Pickled Beets

VEGETABLES

Plain Boiled Macaroni   Boiled Potatoes
   Hominy   Fried Oyster Plant   Cabbage   Boiled Rice
   Beets   Mashed Potatoes   Baked Potatoes   Onions

GAME

Venison Steaks

PASTRY

Custard Puddings   Lady Fingers
Peach Pies   Apple Pies   Farina Pies   Frosted Eggs   Jelly Tarts

DESSERT

Brazil Nuts   Raisins   Vanilla Ice Cream   Apples   Filberts   Almonds

In the bar were offered sixty-one imported and nine domestic wines, five brandies seven to twenty-seven years old, six liqueurs, sixteen champagnes, and ten ales and porters. Needless to say, though unmentioned in the interests of elegance, there was a proportionate selection of whisky, gin, rum, and maybe, stuck away under the bar where its display would not mar the elegance of the general effect, a plebeian jug or two of hard cider, cherry bounce and natural, unadulterated fresh corn likker.

Always there were servants, usually free Negroes on the Ohio, since a slave could scarcely be expected to resist the invitation of the free territory ports on the north bank, and an army of firemen and deck hands, with a hierarchy of officers arising, for show, to the captain, for real authority, to the pilot. Though the captain appeared to the passengers to command, shippers of freight and those directly connected with the steamboat business knew it was the pilot who in the last instance ruled from his glassed-in pilot-house above as the Almighty regulates the sparrow's fall.

There is certainly much of interest in the details of meeting the exacting problems that face the pilot of a river boat, but most of the essentials of the calling were carefully and beautifully explained by Mark Twain in his great *Life on the Mississippi* more than sixty years ago and writers ever since have been quoting, plagiarizing or paraphrasing his work unsuccessfully. His report is, in truth, of piloting on the Mississippi but problems upon the Ohio were not dissimilar: on the lower Ohio the banks caved and changed and washed away and made cutoffs on only a smaller scale than did the lower Mississippi, and the upper Ohio presented some peculiar problems of its own which resulted from a narrow, rockbound channel and which did not pester rivermen on the main stretch of the Mississippi.

Even the yarns of river pilots of today seem to hark back to Mark Twain—which appears to indicate either that all possible hair-raising experiences which pilots might encounter were foreseen by him or that his experiences at the wheel and among the bells were so typical and so comprehensive that there remained no new crises to face. To know the pains and problems of piloting, then, read *Life on the Mississippi* and let it go at that; there can certainly be no more pleasant way in which to gain the knowledge.

Increased travel in the late forties and fifties invited greater heights of elegance and that, in turn, invited more travel—which undoubtedly encouraged business and caused money to circulate more freely in the West. Every river city felt similar effects to that which the Cincinnati *Gazette* reported in 1846: "The River Packets, Stages and Railroads are bringing in about one thousand per day . . . many come here from the interior towns below us, where they can select their mode of travel, and depart for the East either by River, Railroad, Stage, or Canal. All the Hotels are constantly full."

One of several of the watering places already mentioned as having drawn seasonal travel from the South was reached by a short run up the Kentucky River. It was the establishment set up at Drennon's Lick and the fifties were great days for it, typical as it was of the summer watering places that drew the prosperous and gregarious, the penniless but ambitious, from the length of the valleys of the Ohio, Mississippi, and their more sophisticated tributaries.

The lick had been early discovered, having a salt and sulphur content and thus being a converging point for game traces. It had belonged at one time to George Rogers Clark and had fallen into the hands of one A. O. Smith by the late forties.

Inspired by the example of the various Virginia springs developments, now for some time satisfying the needs of prosperous mamas as a place to exhibit their marriageable daughters and have the results of overeating and hard drinking boiled out of papas at the same time, Smith put up a hotel, the North and South House, in 1849. He surrounded it with bathhouses, cottages for families, and cabins for the hired help and it took the fancy of the hopefully elegant almost at once, claiming to have entertained the governors of thirteen states at one function without apparent strain on its facilities.

As a necessary adjunct to the social activity at Drennon's came, each season for a time in the fifties, a hairdresser who finally incorporated her behind-the-scenes memories of that place as well as Cincinnati, Louisville, Madison, New Orleans, Paris, and London in an anonymous volume published in Cincinnati in 1859.

The book, *A Hairdresser's Experience in High Life,* ghost written as it probably was, is full of lively matter concerning the period. With a few brief paragraphs on Drennon's the keen-eyed and sharp-tongued coiffeuse points some significant trends of the day: the new taste for what was considered to be Society in the European Manner; the increasingly lavish scale upon which lived the newly rich of river commerce and the plantation system; the growing appetite of pampered belles and young blades for romantic adventure, and the perfectly defenseless state in which the whole world, above stairs and below, still found itself when stricken by epidemic disease.

Early in the season, hearing that great preparations were making at Drennon's, I concluded to again spend the season there. On arriving I found that a large number of visitors were expected. Gambling rooms, billiard saloons and ten-pin alleys were fitted up, and every arrangement made that could add to the amusement and excitement of the pleasure seekers at a watering place.

The principal building was very large, and adjoining on either side was a row of smaller buildings, with family rooms; then around were numerous little cottages where families from Cincinnati, Madison, Louisville, Lexington, and a great many Southerners were accommodated. At the foot of the hill were cottages without number, called Texas, where the servants and commoner classes of people resided . . .

There was also, back of all this elegance, although the circumstance was not notable to the hairdresser's contemporary eye, a complete absence of sanitary toilets, an alley behind hotel and cottages into which the servants of the Cincinnati and Madison families and the slaves of the others sloshed the contents of their masters' chamber pots, no thought of inspection in the kitchen, a system of food preservation in which spring water served as the only cooling agent in the steaming Kentucky summer, and no barrier anywhere against fly, mosquito, flea, louse, or bedbug which might desire to fly in upon its own wings or ride in upon the person of servant, slave, or casual transient!

The hairdresser claims that Drennon's was "attended by all the fashionables" and she describes some of her customers—who in the case of two or three neurotics might better be described as her patients.

I now had an opportunity to display my talent in hair-dressing; combing a young bride who, in a freak of passion, cut off her hair to vex her husband . . . She was at this time alone with her husband and from some cause they seemed to be in a continual excitement . . . There was . . . a great fancy ball; she went, dressed very simply as a flower-girl, and looking very sweet and pretty; her husband came into the ball-room and wanted to take her out; she would not go, and they had quite a fuss. Some gentlemen there asked him out and threatened him with what they would do, if he did not behave himself. She soon after left the ball-room and going into the cottage of an acquaintance, she begged protection of the lady and gentleman till her father could be written to . . .

Father returned from the East and set to work to get a divorce for his daughter. (Divorces, phenomena in that day, were in most states granted by special act of the legislature.) But the capricious little lady wasn't really angry, after all, for—

A few days before the divorce was to be granted, the elder sister gave a polka party, and while the guests were dancing and enjoying themselves in the front of the house, the sister slipped out of the back door and joined her husband again. . . . This was not the last of their separations. Every once in a while they separated and lived apart a short time. Nevertheless the lady has raised a pretty little family, and is herself one of the handsomest women in Kentucky.

Married ladies seemed destined to experience extremely tough times at Drennon's, one way and other:

Another of the notables . . . was a famous married belle, whose delight it was to fascinate all men, married and single. There was . . . a gentleman from the South, with his wife, a mild, delicate, lady-like person. This belle took a great fancy to the Southerner and flirted with him on all occasions, greatly to the displeasure of his wife. They used to walk about in the evenings, and sit till late at night, even under the very window where his wife was waiting. One night, about one o'clock . . . she heard this belle say to him: "Let us sit till three, and see if she will wait up for you." Was this not trying to a delicate wife's feelings? The next season the belle was back again, but the gentleman and his family did not make their appearance.

—the gentleman's wife being, one gathers, less mild than she appeared.

Not all these affairs ended with such felicity. Sometimes they were fatal, as in the case of the "married belle from Georgia":

This lady's husband was a man of high standing, and very wealthy, but in very poor health . . . She would leave her husband coughing and bleeding at the lungs, her children fretting and crying, dress herself, and go down to the ball room, where she would stay, it may be, till one or two o'clock at night . . . before the end of the season death came and rid her of him. The next I saw of her was in a hotel in New York, reclining on a sofa, elegantly attired, covered with diamonds, and every-thing about her exquisite. Ostensibly she was under the care of a physician, but I learned the physician was but a cloak to cover her long stay from her second husband; and occasionally a gentleman from New Orleans came to see her, making business in New York the excuse to his wife and family for his absence. The last time he visited her, on his return home he found his wife had taken laudanum and destroyed herself. Full five hundred such scenes have come under my notice since I've been a hair-dresser.

The young men at Drennon's were wayward too, as many pampered sons of new-rich fathers were wont to be in that distant day. Especially so were the southern youths, constantly reminded of the ancient chivalrous splendor of the Virginia Tidewater and the baronies of the Carolina coasts and currently inspired by Sir Walter Scott and his feeble local imitators. (Those "local" novelists in the South and on the Ohio in the fifties formed a school of letters the product of which was wonderful to contemplate and worthy of more scholarly study than it has yet received because of its horrors of style and plot and of the social trend it indicated.) There may have been two explanations for the form that young men's amusements took at Drennon's—because certainly occasional penniless adventurers did worm their way into such establishments in search of alliances with young ladies of fortune—but it is to be feared that adventure and not financial gain was the chief purpose.

During that season there were many exciting scenes . . . A number of young men took to robbing, and got taken up. The evidence was clear, and they tried to get one of the young men to tell where the money and valuables were concealed. To make him confess, they put his hand in a vise. His screams were dreadful but nobody minded him. This young fellow had made his haul, buried his share of the plunder, and was coming back after more when taken up. They were far from any city or officer, and as he would not confess, they had to let him go free. He was . . . seen to go after his plunder, get it and leave.

The season closed at Drennon with a grand fancy ball . . .

Even the opening sentence of the hairdresser's account of the next summer hints at a difference in the custom of the fifties from that of the less prosperous decades before, and from the twentieth

century. Fashionable gentlemen, in that gaudy and rather dissolute era, sought holiday entertainment away from their families
and in cities which offered indoor entertainment less innocuous
than fireworks:

It was on the 3rd of July, and all the gentlemen had gone to
Cincinnati, or other neighboring places to celebrate the Fourth, leaving
their families behind. Those that remained had fire-works and various
other out-door amusements, to amuse themselves and gratify the villagers
around. In the evening was a grand ball. . . . I retired to my room, as I
did not feel well, having drank considerable sulphur-water during the
day. A little after twelve the ball broke up. Just then the carpenter was
taken very ill. They said it was from eating cherry pie and drinking
milk, but during the night the proprietor and his family all had a slight
attack, but, for fear of frightening the boarders it was kept quiet . . . At
five o'clock . . . a man named Allen, from Cincinnati, was dead . . . of
cholera in its severest form. . . . There were five deaths in all during the
first twelve hours . . .

Servants and guests alike were stricken, the hairdresser was
called to attend one of the chambermaids with whom she had
talked only five minutes before:

On entering the room I found her all cramped and black around
the eyes and mouth . . . I gave her a large dose of laudanum and brandy,
put a large mustard plaster to her chest, feet and hands, and staid with
her till I got her into a perspiration; then I left her, went into my own
room, and got everything I had ready to put into my trunks. During the
short time I was out of the room, an old doctor, who was good for
nothing but attending on babies, went in and gave her . . . an emetic.
Knowing there was no physician near, and finding the book-keeper and
this old man were trying the most desperate experiments, I became so
alarmed . . . I crammed in my trunks what I could easily, and gave the
rest away, and was ready by the time the stage came along.
Some of the ladies that I had worked for came to the gallery and
said: "Good bye Iangy; go in peace and sin no more." My reply was:
"I wish you would go with me, as I fear when you get ready to go you
will not be able."
There were fourteen in the stage that left Drennon that morning;
out of this number but five lived. . . .
Being afraid to go to Cincinnati, as the cholera was very bad there,

I went to Madison. One of the wealthiest families in that place took me to their house and gave me one of their best spare rooms, where I was very sick for several days. After I recovered I learned some of the horrors of Drennon . . . From twelve on Thursday to twelve on Friday there were twenty-three deaths . . . In their sore sickness they had no help, no aid, no physician . . . In their agony and helplessness many of those who had laudanum and morphine took it, and slept themselves away. . . . They had few lights . . . The sick had to take care of the sick, and dying bury the dead.

The proprietor and his family fled, taking with them all the medicines. . . . There were full forty or fifty deaths occurred in that little place, and only the notice of about a dozen was given. There was neither physician nor coffin within fifteen miles . . . All this occurred within a short distance of four of our most populous cities. After this it was found impossible to make a watering place of Drennon's and they turned it into a military school.

—thus a view of the life of fashion and opulence of the fifties not usually displayed in nostalgic modern dramas set in the period, any more than descriptions of those Cincinnati and Louisville establishments in which the gentlemen from Drennon's spent the Fourth of July found their way into the home-grown romantic novels that the period produced in such lush profusion.

At the end of the fifties the Ohio Valley was riding high, wide, and handsome, despite plagues and the growing pains arising from too quickly found affluence. It was not quite over; there would be a few more months as riotous, as fabulous, and then the crash would come.

Already *Ivanhoe* had become the bible of the South and Dan Emmett, the minstrel man, wrote "Dixie' and first coon-shouted it with Bryant's Minstrels in 1859. A touch of something that resembled the old fanaticism of the Puritans was developing among the militant abolitionists above the river and much of the North was tending to identify itself with New England's self-appointed guardians of the nation's morals and manner of being.

A hot-eyed fanatic named John Brown was assembling a collection of guns and ammunition on a rented farm near Harpers Ferry, Virginia, that summer too.

Chapter Seventeen

A LINE OF DEMARCATION

THE Ohio had been crossed in stealth and for purposes unlawful and militant since the days of historic Indians—probably since the advent of predatory man of any race in its valley.

Indian raiders had carried their captives back across it to their towns and to the headquarters of their British allies during and after the period of the American Revolution. Up the Ohio's northern tributaries—the Muskingum, the Scioto, the two Miamis, the Wabash—the captives had met with treatment which varied with the current whim of the Indians, the tension of the Indian-white relations at the moment, and the personal traits or appearance of the unfortunate individual. He or she might be adopted as an honored son or daughter of an Indian family, claimed as a husband or wife, or as readily and as unpredictably, enslaved, tomahawked, or roasted at the stake with whatever tortures the Indians could devise to accompany, extend, and elaborate that interesting ceremony.

A great body of narrative accounts of these captures and subsequent adventures were published and, according to their testimony, most of these captives were removed north across two favorite stretches of the river. Earliest, because of earlier settlement, the Indians usually carried their human and commodity loot through the wild Virginia (now West Virginia) and Kentucky countryside and across the Ohio between the mouths of the Great Kanawha and the Little Miami river, where old traces led to the heart of the Indian country through land which was satisfactorily rough for concealment. A little later, after there came to be a sufficient number of Kentuckians farther down to make a raid upon them a paying venture, the sparsely settled section of the Kentucky and Indiana shores between the mouths of the Great Miami and the Kentucky River became a popular crossing place.

Even today there are still thinly settled areas, especially above the mouth of the Kentucky, where within a few miles of modern cities and industrial centers the traveler along the river at dusk gets the impression that a war party might still make a safe crossing and glide off toward the north with its captives were it not for the facts that the Miami trail nowadays crosses so many four-lane highways and the scalps are no longer a commodity salable at Detroit.

After the end of the eighteenth century a significant number of trans-Ohio passengers of a different color began to use these old favorite localities. Actually this was not then a new movement—only increased—for slaves had fled across the Ohio to join the Indians since the earliest slaveholders had moved within striking distance of its banks. More enterprising or more desperate slaves had run north in greater numbers after the Territory North-west of the River Ohio was declared "free" in 1787, for though the slaveowner was permitted to follow and reclaim his runaway chattel the chances against their meeting in the wilds north of the river were great indeed.

The value of the north shore of the Ohio as a sanctuary for runaways was somewhat decreased by the penalties written into the federal Fugitive Slave Act of 1793 but the opportunity for escape was still worth trying. That act provided that a runaway could be arrested by his master or an agent, that he could be taken before a magistrate and, upon due proof of ownership, could be removed back across the river; that a person who concealed or aided a fugitive slave could be fined $500 and in addition could be sued for damages by the slaveowner, should loss to him result. But the act was not completely effective, for though New England's slaves had been emancipated only a couple of decades before—partly because they were of little use in the local economy—New England people immediately developed an exceedingly self-righteous antislavery attitude and those who came to the Ohio brought their sentiments with them. Magistrates sometimes demanded impossibly comprehensive proof of slaveownership (during the gathering of which an intentionally careless jailer might let an imprisoned slave escape) and in any case, magistrates were few and far between in the territory. Of course an owner could easily carry his slave back across the Ohio without benefit of process of law if he had a sufficient force with him but, on the other hand, a runaway, reinforced by a few of his fellows or by friendly Indians, could dispose of his pursuers in the wilderness without leaving a trace of their passing. A slaveowner sank as readily in the Ohio River as anyone else, once he had been struck down, had his belly slit open and his viscera replaced with rocks, and the gigantic catfish and crawfish found white flesh as palatable as red or black.

As settlement of the Territory North-west of the River Ohio increased and as it was divided, along the river, into first territories, then the states of Ohio, Indiana, and Illinois, the passage of fugitive slaves increased year by year. Many of the people who had come to settle north of the river from slave states had done so either because they had not prospered in their old homes sufficiently to own slaves or because they were not in sympathy with slavery as a system. Thus many emigrants from the South became abolitionist in sentiment, especially after a generation or two of residence north of the river in association with the militant antislavery workers—the New Englanders around Marietta, Ohio, the Presbyterian "home missionaries" in Indiana, the Quakers upon the Whitewater and Wabash, the New York and Connecticut settlers in northern Ohio, or the English emigrants at New Harmony and Albion. The general complexion of the north side of the valley grew to be strongly antislavery.

Canada was now the ultimate aim of the runaways. The first slaves escaped there immediately after the War of 1812, but when the states between the border and the Ohio became fully settled it was the only permanent refuge. Sometime, probably in the forties, the best routes began to be well defined and persons friendly to the slaves began to form loose organizations by which Negroes could be moved along from one safe station to another.

A southern owner, whose slaves had disappeared completely as he pursued them across Ohio, is said to have originated the "Underground Railroad" title and it was quickly adopted by the operators. Places of hiding became "depots" or "stations" and those who guided the runaways from one to another were "conductors." These routes covered the North from New York and Pennsylvania west to the Mississippi—but no more extensive passenger business was carried on anywhere than was that across the Ohio within the two old favorite passes of the Indians.

The comparative number of escaping slaves was not great—less than 1 per cent of the total per year—but most of them came from the three northern of the slave states where the consequent loss bulked large. As much as two or three million dollars' worth of property was being lost by southern slaveowners every year—and in truth by means as unlawful as if they had been taken at the point of a gun. Something obviously had to be done and slave-state pressure brought about the passing of a new law, much more strongly implemented, the Fugitive Slave Act of 1850.

This measure provided that the claimant of a fugitive slave in free territory could establish ownership by his own affidavit alone; that citizens must assist him in securing his property, that harboring a fugitive or aiding one was an offense for which the culprit could be fined $1,000, imprisoned for six months, and declared liable for $1,000 personal damages payable to the slave's owner. The act included one dangerous and unjust provision: United States commissioners delegated to enforce the law were to be paid $10 for every warrant issued for a runaway but only $5 for discharging a Negro proved to be free—the abuses to which such terms could lead are readily discernible; and all those abuses were soon put into practice.

The kidnaping of free Negroes was thereafter a simple matter for the northern racketeer or the southern slave dealer; thieving commissioners began to reap a harvest of fees—and in retaliation the business of the Underground Railroad grew by leaps and bounds.

Every conceivable ruse for moving fugitive passengers was employed, some of them most ingenious; all of them desperately dangerous for the conductors. Simplest system was that of flouring a black woman's face and hands, concealing her hair and features under a sunbonnet, surrounding her with white children borrowed for the purpose, and hauling her in a wagon; a hearty Negro man could always be led through swamps and thickets by night—but either method was easy of detection. New methods had to be originated constantly by the underground but ingenuity supplied them.

Negroes were hauled from place to place in bales, boxes, and even—in the case of a central Indiana tombstone cutter who also dealt in other funeral goods—in coffins. Houses of underground workers were constructed with secret passages and cupboards, moving panels, and all the equipment of the modern mystery thriller: these places were not merely occasional phenomena; they were common. Most sizable towns in Ohio and Indiana had at least one. Communities on more frequented routes had dozens. They are still to be seen along the Ohio and north of it in considerable numbers. The Underground Railroad was a major industry, though one which offered no profit and great peril to its operators —75,000 slaves are supposed to have escaped through its service by 1861; perhaps to a total value of forty or fifty million dollars!

The movement offered a peculiar anomaly in ethics. It was backed by people of personal integrity and of high principles who had a burning desire to serve humanity and believed generally in the behests of Mosaic law and the unquestionable pains of hell-fire that awaited its violators. Of course a lunatic fringe bordered the abolition movement and there was a handful of individuals who made a paying career of agitation. These few, however, were the scattered exceptions. The direction of the movement rested in normally honest and exceptionally God-fearing and charitable people.

Yet abolitionists north of the Ohio, through the Underground Railroad movement, did not hesitate to break the law to lure away the most valuable chattels of their southern neighbors; to connive, to plot, to subvert justice, to lie, to steal, to perjure themselves, or to shed blood in order to free slaves. The fact that the holding of human beings in involuntary servitude violated the principles of both Christianity and democracy—by the fifties many a slaveholder agreed to that—only complicated the picture. For slaves were held under the laws of the land and the unlawful act of freeing them also posed a question of Christian ethics, was equally contrary to the principles of democracy.

The Civil War and the defeat of the Confederacy eventually brought an end to the problem both North and South had inherited but it was an end that was not immediately a solution and it left scars which are still unhealed today. The thinking man, northern or southern, white or Negro, is likely to wonder why, once providence had permitted the establishment of the slave system on New World soil, a way could not have been found to end it in some less violent and bloody manner.

Back in the fifties, however, the Underground Railroad movement flourished in the West. Besides the old favorite stretches of the Ohio, new ports of entry were put into use. The main lines led north from the river at Marietta, which had been New England settled, and up the Scioto River passed a road which led to Oberlin, Ohio, where rabidly abolitionist professors of little Oberlin College were always ready to welcome runaway slaves and to defend their liberty with impassioned oratory, barrages of pamphlets, or pitchforks and musket fire if the occasion demanded. Yellow Springs, Ohio, on the Little Miami, was a safe haven—home of an Owenite community in the twenties and seat of Antioch College after 1853, it was and is always ready to sponsor advanced movements. A favorite route, perhaps most used of all in the Ohio Valley, was that which ran from the neighborhood of the mouth of the Great Miami along the Indiana-Ohio state line following the north star toward Canada. Alternate routes branched from this up the Whitewater River to the Indiana Quaker settlements around Richmond and Centreville (Quakers were militant figures in this movement in spite of the pacific tenets of the faith); northwest across Indiana through the college towns of Franklin, Bloomington, Greencastle, and Crawfordsville, where New England-reared professors were usually abolitionists in sentiment although—fortunately for the peace of those communities—less violently inclined than their brothers at Oberlin.

It was the Kentucky country above Newport that Harriet Beecher Stowe, sister of the handsome and magnetic liberal Presbyterian minister Henry Ward Beecher, and wife of bumbling Calvin Stowe, made the scene of *Uncle Tom's Cabin*. The thrilling literary and theatrical spectacle of Eliza crossing the river—traditionally at Ripley, Ohio—upon the floating ice soon made this *the* locality, in the minds of American and English readers, at which any fugitive slave of orthodox background and pretension to respectability among his kind *must* have crossed. Of course those whose familiarity with *Uncle Tom's Cabin* stems from the theater rather than the book carry a mental picture of a beautiful octoroon flying over blue calico waves on ice cakes bearing a close resemblance to white-painted soapboxes located a convenient step apart, a pair of lethargic and moth-eaten great Danes pursuing. Perhaps it is not exactly photographic representation of either a typical runaway slave, an ice breakup on the Ohio, or pursuing bloodhounds —but that was no fault of Mrs. Stowe's.

Cincinnati itself was never an entirely peaceful sanctuary for abolitionists: it "faced south," commercially, and Mrs. Stowe and her connections led no easy life there. Her father, the Rev. Lyman Beecher, had been made president of Cincinnati's Lane Theological Seminary in 1832 under a stipulation attached to a gift of the annual interest on $20,000 donated by the New York abolitionist Arthur Tappan. That circumstance itself put Beecher in immediate disfavor with some of the population of the city. His daughter Harriet resigned her position as a teacher in her sister Catherine Beecher's Boston school for girls and came to Cincinnati. There she presently married the Rev. Mr. Calvin E. Stowe, her father's professor of Biblical literature.

Within a few years Lane Theological Seminary was in an uproar. Some of its students, led by Theodore Dwight Weld, had become active workers for abolition and were involved, with a part of the faculty, in a battle with the trustees, many of whom were either cautiously nonpartisan or proslavery in sentiment.

Slovenly, eccentric, but powerful President Beecher rode the fence as best he could in the early days of the trouble but when local sentiment against the abolitionst element in the seminary took the form of insults to the faculty members and their wives in the city streets and the pelting of professors with what was currently described as "offal" he sided with the trustees. The radical elements of both faculty and student body were driven out—most of them took to the road in the cause of antislavery or sought sanctuary at other colleges which were unanimously abolitionist— and President Beecher held his post, actively or in an advisory capacity, until his death in 1863.

Harriet Beecher Stowe and her husband left Cincinnati in 1850, when he accepted a place on the faculty of Bowdoin College. There, drawing upon observations made during her occasional visits to the Kentucky side of the Ohio in the past, tales told by fugitive slaves whom Lane professors had harbored in their homes —and possibly, to some extent, her imagination—she wrote *Uncle Tom's Cabin*. It appeared serially during 1851 in the *National Era* and shortly, with phenomenal success, in book form. That work, more than any other single factor, brought antislavery sentiment to jell in the nation at large. Whatever one may think of the book, its tremendous influence must be admitted. Even President Lincoln, who was harried almost as much by abolitionists as were the slaveowners, admitted the power of Harriet Beecher Stowe: introduced to her during the war he exclaimed, "So this is the little lady that caused all the trouble!"

Of course Mr. Lincoln was indulging in a bit of harmless social exaggeration but Mrs. Stowe's novel did have a good deal to do with bringing the trouble to a head, as did several other contributing elements which were scarcely noted by even the keenest observers at the time: there was the fact that the poor competition available in the Mexican War had failed to exhaust the yearning of some individuals for military glory; the provincial arrogance of slaveholders in the Deep South—and of northern clergymen and educators (especially those of New England origin); the befuddled romanticism which Sir Walter Scott's novels had engendered in the South; the sneaking suspicion of many a hard-working northern farmer that his southern counterpart led a rather more easy life under this deplorable slavery system than he himself could hope to achieve; the unsettled state engendered in the minds of the slaveholders by fears of Negro insurrections at home, and by an inward conviction that there *was* a basic wrong in slavery; the lawless oral and physical violence of the lunatic fringe of the abolitionists—and the equally lawless depravity of some slave traders and slave stealers. All these things contributed as much to the beginning of the war as did the apparently major matters of slaveholding, states' rights, and the interpretation of the Constitution.

With the coming of war the Ohio became a positive line of demarcation in theory, though much, much less so in practice. Though all states north of the Ohio River were committed as units to the Union cause, in practice the river was by no means the sharp divider it appeared on the maps to be. There were loyalists and even abolitionists aplenty to the south of it and there was no shortage of nigger-haters and southern sympathizers to the north —a fact which made the rebellion in the Midwest the singularly complicated, bloody, and historically and sociologically interesting episode that it was.

Chances are that the majority of average citizens on the north shore of the Ohio could not convince themselves that the election of Abraham Lincoln would inevitably lead to war, even to the general secession the South threatened. Lincoln's campaign did not compare in organization or in excitement aroused with that comparatively unimportant one which the partisans of William Henry Harrison had promoted twenty years before. The chances are that many of those who voted for the Railsplitter had little hope of his

election—and one certainly gets the feeling that Lincoln himself
was mightily surprised by it.

But Abraham Lincoln was elected president of the United
States and thus a typical product of the Ohio Valley was chosen
to lead the nation through its difficulties. Lincoln's life had followed a pattern common to the early nineteenth-century Midwest:
born in Kentucky near the headwaters of the Green River, reared
to manhood a few miles north of the Ohio in Indiana, and grown
to maturity in central Illinois, his personality was colored by the
nature of all three sections and his contacts had extended among
most of the principal stocks that had settled the valley. His life,
before his election, had been a rather unhappy experience but one
peculiarly well designed to develop intellectual strength and balance and a sympathetic understanding of his fellow man. These
acquired traits combined, in his maturity, in a character ideally
composed to fulfill its destiny. Lincoln was little known at the
time of his election and his competence was generally and justi-
fiably doubted by many, even among those who supported him.
In retrospect, however, it is plain that no man of his day could
have taken his place during the next four years.

News of his election touched off the southern powder keg.
South Carolina called a state convention and seceded from the
Union on December 20, 1860, and within four weeks she was
joined by Mississippi, Florida, Alabama, and Georgia. Delegates
met and drew up a constitution for the Confederate States of
America and Jefferson Davis of Mississippi was elected president
on February 4. Before the inauguration of Abraham Lincoln as
president of the United States on March 4, 1861, Texas and Louisiana had joined the Confederacy.

The war, as had secession, began in South Carolina. Upon the
secession of South Carolina, Major Robert Anderson, a regular
army man from Kentucky, realized that his command at Fort
Moultrie in Charleston Harbor must soon be in danger. He spiked
his guns, burned the fort, and retired to nearby Fort Sumter, a
newer and stronger fortification, on December 27. The fort was
occupied peacefully for some time but finally the hotter heads of
South Carolina recommended a move against it. President Davis

forbade such action but his order was disregarded and Fort Sumter was fired upon by secessionist troops on April 12, 1861. Anderson, isolated as he was, held on until the 14th, when he surrendered. Next day President Lincoln called for 75,000 volunteers to suppress the rebellion, rallies were held in northern cities and the South was no less stirred. Virginia, Arkansas, North Carolina, and Tennessee joined the Confederacy within the next few weeks and the pattern of hostilities was set.

Set, that is, except for the people of western Virginia and eastern Tennessee—where the mountainfolk held few slaves and had no great love for their more prosperous fellow citizens of the valley who did—and for the border states of Missouri and Kentucky.

Kentucky's dilemma was many-sided; typical in some ways of the whole length of the Ohio River, in others peculiarly its own. Internally, throughout the war, the state was torn by a conflict more harrowing than that of many of those in the Confederacy which were the scene of extensive fighting. Kentucky had furnished much of the stock that had settled Indiana and Illinois, and which in a generation had become antislavery in sentiment or at least against the extension of slavery to the West; thus abolitionist brother was pitted against proslavery brother; father against son. Kentucky's commerce, aside from the raising of tobacco, hemp, corn, and hogs and the processing of these commodities, consisted of the wholesaling and distribution of northern products to the South and southern products to the North: friendship with both localities was vital. Kentucky was not a great slaveowning state yet those who did hold considerable numbers of slaves—the people of the Bluegrass, of the Beargrass, and of the Barrens—were among the state's most substantial citizens. That class presented another anomaly, for many Kentucky slaveholders were violently pro-Union in sentiment and action. Kentucky had been (except for the aberration of a few of her people at the time of the Spanish Conspiracy) intensely patriotic; she had mothered many military men who became prominent in both Union and Confederate armies; both President Abraham Lincoln of the

United States and President Jefferson Davis of the Confederate States had been born within her borders!

Union and Confederate forces were equally anxious for her allegiance and both set up fortifications on her borders—one north, one south—to take advantage of her decision. But there were plenty of men in Kentucky who had heard from their fathers what it meant to be a buffer between forces; who remembered that throughout the early settlement of midwestern America Kentucky had been a "Dark and Bloody Ground": it is no wonder that Kentuckians tried to maintain neutrality.

Neutrality might have worked to a limited extent had it not been that the men of western Kentucky—from Owensboro to the Mississippi—were strongly secessionist and rather violently inclined individualists anyway, and that the configuration of their part of the state offered ready access from both north and south. New Orleans was blockaded by the Union in 1861 and commerce on the Mississippi virtually ceased. By fall the Confederacy was gathering its strength on the Cumberland and Tennessee rivers south of the Kentucky border, a hitherto unsuccessful professional soldier and merchant named Grant was pacing the floor of his quarters in the St. Charles Hotel at Cairo, busy with plans for moving through western Kentucky, while Flag Officer Foote's headquarters down the street planned the Western Gunboat Flotilla. Federal troops were poised on the north bank of the Ohio at Cincinnati and Jeffersonville, Indiana, and the Confederate General Felix K. Zollicoffer was moving into Kentucky from the east, shortly to be defeated by Union General George H. Thomas. In the fall of 1861 General Albert Sidney Johnston led Confederate troops from Tennessee into his native Kentucky and General Simon Bolivar Buckner led that portion of the Kentucky State Guard which wished to fight for the Confederacy down to meet and join him.

Since the state was professedly neutral, all partisanship had to be carried out in some degree of stealth:

Nearly 10,000 "Lincoln Guns" were judiciously distributed to "Union Men" of the state by way of the rivers . . . Soldiers traveled in small bands and with much caution through the gaps of eastern Kentucky mountains to join the Confederate Army because of their fear of death at the hand of grim mountain patriots whose idea of defeating the Confederacy was to shoot volunteers before they reached the army. These "bushwhackers" were independent citizens, who, in some cases, acted under commands of Union officers . . .

In Tennessee, immediately south of the Kentucky border, the Confederate forces rallied at Fort Henry on the Tennessee River and Fort Donelson on the Cumberland, at the towns of Hickman and Columbus on the Mississippi, and at Bowling Green on the Barren River. All three towns were in Kentucky and they, with the two Tennessee forts, formed an east and west line about 150 miles in length, which controlled the Cumberland, Tennessee, and lower Mississippi rivers.

Some 60,000 troops manned this line when General U. S. Grant began to move south from the Ohio in the winter of 1861-62. With his staff on one of the hastily converted gunboats, the general had already gone up the Ohio from Cairo to Paducah, where he arrived on the morning of September 6, 1861. His troops—some 5,000 men—had been brought up the Illinois shore and a bridge, laid upon barges constructed or commandeered for the purpose, was thrown across the river slightly below the town. By midafternoon, Union troops were marching into Paducah to occupy it.

Grant issued a shrewdly phrased proclamation to the citizens —the majority of whom he recognized as being of Confederate sympathy—in which he said: "I have come among you not as an enemy, but as your fellow citizen . . . An enemy, in rebellion against our common government, has taken possession of and planted his guns on the soil of Kentucky . . . He is moving against your city. I am here to defend you . . . I have nothing to do with opinions and shall deal only with armed rebellion and its aiders and abetters . . . Whenever it is manifest that you can defend yourself . . . I shall withdraw the forces under my command."

That same evening the general steamed back to Cairo to carry on the larger business of moving enough troops to southwestern Kentucky to clear the Confederate forces from their Tennessee and Cumberland River strongholds.

That campaign is not too closely related to the Ohio River. It is another story, in which, on February 16, 1862, General Buckner surrendered Fort Donelson, the last of the five points, to his West Point classmate, General Grant, and the bloodiest phase of the war moved farther south to the horrors of Shiloh, Chickamauga, and Kennesaw Mountain.

Things were happening on the upper as well as the lower Ohio in 1862. The people of the mountainous part of western Virginia did not take much interest in the possible abolition of slavery (President Lincoln had clearly stated that he aimed to save the Union first, regardless of whether or not slavery continued) nor, for that matter, did most of them waste much thought on the abstract question of the rights of man. But definitely they resented the real or fancied snootiness of their slaveholding fellow Virginians in the Tidewater and a majority of them had retained a sort of unreasoning patriotism from earlier days. Those factors now led them to follow the Stars and Stripes and to separate their rough but wildly beautiful and minerally rich district from the Old Dominion. The campaign in northeastern Virginia and southern Pennsylvania being in the precarious and uncertain state it was, parent Virginia could not say them nay, and a new state, West Virginia, was born. The Ohio River was more securely Union by the margin of the long and intricately directioned shore line of this new state, and the safe shipment to the southwest of vital munitions from the great arsenal of which Pittsburgh was the center was further assured.

Strangely enough, some of the cousins of these same independent Virginia and Kentucky back-hill people, located in southern Ohio, Indiana, and Illinois, were the first who, likewise resentful of the prosperity of *their* neighbors on more fertile lands, followed the Ohio spellbinder, Clement L. Vallandigham, into the Copperhead movement.

These divergent elements in the states of Virginia and Kentucky complicated the scene enormously—for both sides—but the fact that sabotage by southern sympathizers was held to a minimum north of the Ohio plus the quick and successful action of Grant in the West soon decided the fate of the river. The Ohio

River was established as the chief and safest artery of supply and transport for the Union forces west of the mountains; its towns became the shipping points and depots for munitions and provisions and its boatyards built gunboats from the keel up—including James B. Eads's first federal ironclads—and reconditioned and armed the purchased or captured river packets by the score.

There was to be no significant combat upon the Ohio—although, unfortunately for their peace of mind, its inhabitants had no assurance of that happy circumstance at the time—but in the dry statistics of the quantities of troops, arms, corn, hogs, and equipment and munitions that were transported upon it and the amount of food and munitions produced, processed, and shipped from its valley lies one of the principal reasons for the success of the Union in the East as well as the West.

Safe as they were to be, dwellers on the middle and lower Ohio were by no means left undisturbed by threats; a colorful Rebel by the name of Morgan saw to that.

In the fall of 1862 central Kentucky—and, by readily believed inference, Louisville, Cincinnati, and southern Indiana and Ohio—were threatened by the advance of the Confederate Generals Kirby Smith and Braxton Bragg—the former through the Cumberland Gap and the latter from a point on the Kentucky-Tennessee line south of Bowling Green. General Smith captured Lexington but Bragg (whose mania for collecting provisions from the countryside prompted him to overload and slow his own supply train) veered away from weakly garrisoned Louisville to join Smith at Lexington, where he wasted precious time in setting up a Confederate government for the state.

His government had but little time to function, for General Don Carlos Buell, who had pursued Bragg up from Tennessee, reached Louisville, received steamboatloads of Union reinforcements which had been rushed down the Tennessee and up the Ohio River and marched toward Lexington. The Confederates moved south.

There followed a comedy of errors. Neither Buell nor Bragg was overly competent and each was criminally negligent in the matter of securing information as to the movements of the other.

When they finally made contact at Perryville, Kentucky, they found to their mutual surprise that they were prepared to camp within two miles of each other.

Seven thousand men were killed or wounded in the four-hour battle that followed but it was far from a victory for either force. After the engagement, Bragg withdrew through the eastern mountains while Buell retreated southwest to counter an attack which, still befuddled, he expected Bragg to make on Nashville in west Tennessee.

Large-scale warfare actually in the lower Ohio Valley was now at an end, but the sole military invasion of the states of Indiana and Ohio and the only significant crossing of the river by enemy troops was still to come: General John Hunt Morgan, with his unorthodox methods and his loyal men, would return from Tennessee.

Morgan, though born in Alabama, had been brought to Kentucky in his fourth year and had lived in Lexington until 1861. (Another Kentuckian! In spite of their official neutrality, one soon gains the impression that the Civil War could not have been waged without Kentuckians.) Morgan had served in the Mexican War and after it had engaged, prosaically enough, in the manufacture of bagging in Lexington. His first wife was dying when the Civil War began, but upon her death Morgan raised a company of southern sympathizers (they were as thick in Lexington as anywhere in Kentucky) and joined Simon Bolivar Buckner's regiment. His was not a cavalry company; it was infantry then and mounted infantry later, when it gained its fame.

He served under General Braxton Bragg in the latter's ill-executed sortie against Louisville—which, had their positions of command been reversed, would probably have been successful. Their respective tactics could not have differed more; Bragg's passion for a full cupboard caused him to gather up all provisions in sight and add them to the baggage of his slow-moving train; Morgan's men traveled with nothing but arms and saddlebags, living on the country, dividing, reassembling, shadow boxing and feinting, stealing horses as they went, and indulging in tomfoolery which included the sending of telegraphic commands and fictitious reports of their position and objects to Union officers and assessing a levy of both pies and cash against captured towns. In addition they accumulated silver plate and jewelry for the benefit of the Confederate cause.

"Morgan's Raiders" supplied stuff for yarns in Kentucky and southern Indiana and Ohio for half a century following the war. Probably their exploits suffered no desiccation in the telling—but they must have been juicy enough even in their barest essentials. The first of the raids was made while Morgan's men were serving as an advance party for Bragg's meandering column in July of 1862. It extended only to northern Kentucky but through the genius of Morgan's aide, the British soldier of fortune St. Leger Grenfell, and the triumphs of wild imagination originating with Morgan's telegrapher, it produced great alarm as far north as Indianapolis, some two hundred miles from the actual northernmost point of penetration.

A year later—in July, 1863—John Hunt Morgan actually crossed the Ohio River. He had not more than two thousand men, at most, and no serious military purpose could possibly have been served by such a force, but he certainly jangled a merry tune upon the nerves of the good people of southern Indiana and Ohio—already taut and frayed as those vital filaments were by rumors of Copperhead plots and Confederate invasion. Possibly Morgan's chief aim was to encourage the northern Copperheads by example and to add some booty in the shape of cash and horses to the Confederate treasury rather than to dash to Indianapolis and release the Confederate prisoners of war held there, as was supposed by Governor Oliver P. Morton and his advisers to be the case. The news of his approach was no less alarming.

Called Copperheads or Butternuts by others (always with an unprintable prefix), the organized Confederate sympathizers called themselves "Knights of the Golden Circle" or the "Order of American Knights." They constituted an element potentially dangerous to the Union cause, but the actual extent of their threat was exaggerated in the public mind in about the same proportion as was that of Morgan's raid. Copperhead membership was mostly of the rabble—under the leadership of a few opportunists who

escaped scot-free when the day of reckoning finally came. The majority of southern partisans of principle had gone south, early in the war, to fight or work openly for the cause in which they believed: those who joined the Copperheads were of no such honorable stamp—they were not particularly anxious to fight on either side nor to be brought into intimate contact with work for any reason. They and their organization were very like the Ku Klux Klan of the nineteen-twenties and of today; recruits mainly enlisted from among the sneaking and the cowardly with a thin sprinkling of the woolheaded and fanatical.

The mystery surrounding the ritual of his order gave the Copperhead a strength he could not otherwise have mustered. Protected by secret meetings in lonely, well-guarded gullies in the woods, and guaranteed anonymity even among his fraternity by his mask and his assumed name, he might be anyone—though usually he was a nobody. In the language of a satirical poem published in 1863, "Ye Sneak Yclepid Copperhead"—

> He wired in, he wired out,
> Leaving the people still in doubt,
> Whether the snake upon the track
> Was going South or coming back.

Upon the ordinary citizen the effect of this hooded menace was as might have been expected; with no means of knowing certainly who among his neighbors might be a member of the fearsome clan, ready to burn his barn or blow up his dwelling, the average man suspected many. Probably there were never over a dozen actual members of the organization for every hundred to whom membership was imputed.

John Hunt Morgan's possible aim in rallying the strength of the Copperheads need not be taken as an inference that he was of their stripe: such was most decidedly not the case. Morgan was a brave soldier—even though a rather piratical and unorthodox one—and had probably seen few Copperheads in the flesh. If he intended to encourage them it was a matter strictly of business; even as the Union was distributing guns to presumably noncombatant mountain toughs for the purpose of "protecting their homes."

Morgan's five-day raid, while it terrified the north bank of the Ohio at the time, appears in retrospect to have partaken about equally of the characteristics of a Victor Herbert operetta and a Mack Sennett comedy, in spite of the fact that it is reported to have cost Indiana and Ohio each a cool half million dollars in burned buildings, purloined public and private funds, blasted railroad tracks, and stolen horses, silverware, and yard goods.

After a dash up through Kentucky from the Tennessee line, Morgan installed himself in his old friend General Simon Bolivar Buckner's house on the hill above Brandenburg, Kentucky, and from that excellent observation point commanded his men in crossing the Ohio River to Mauckport, Indiana, in a couple of commandeered steamboats.

Governor Morton of Indiana had alerted troops when there was little danger the year before; now, with Morgan actually over the river, all was surprise. There was no one to bar the Kentuckians except the "Home Guard," an organization of boys and inexperienced oldsters as ineffectual as could be imagined for any service more militant than a Fourth of July muster—who, in spite of their incapacity, fired some shots and did their duty.

From Mauckport the raiders drove straight north toward Corydon, the former state capital which Indiana had abandoned forty years before in favor of Indianapolis. They acquired good Indiana horseflesh and liberated farmers' eatables as they went. There was some sniping from the knobs and the Corydon Home Guard showed fight outside the town but outnumbered five to one they surrendered after killing or wounding some forty raiders. Soon Morgan was in full possession of the town and ready to talk business. While his men were looting the stores, he assessed the merchants and manufacturers up to $1,000 apiece for the privilege of having their buildings left unburned.

Governor Morton was rallying a militia force in Indianapolis and General F. H. Hobson's federal cavalry was on Morgan's trail from behind when Morgan, instead of going north to Indianapolis, turned to the northeast.

The ineptitude of the home guardsmen at Salem, Scottsburg, Vernon, and Versailles, Indiana, could not have been quite so

profound as their neighbors later reported. It is conceivable that
an excited guardsman might drop a hot coal in his shoe while trying
to apply it to the touchhole on a cannon; that an untrained company
might charge a herd of cattle in the dark and fall en masse down
a bluff—but even the least enlightened guard commander could
scarcely have mistaken Confederate gray uniforms for Union blue
and have asked military advice of Morgan's cavalrymen! Whatever
the caliber of the resistance, Morgan's men moved on toward Ohio,
stealing what they could lay their hands on both individually and
officially, levying tribute on business, confiscating public treasuries
and private strong boxes, and blasting railroads and public works.

But the end was in sight. The Indiana militia was entraining
at Indianapolis; Hobson's cavalry was moving even faster than
Morgan's; word of the Battle of Gettysburg had been received and
some of Morgan's men were discouraged while others, preoccupied
with this opportunity to recoup their personal fortunes, were out
of control. Morgan's strategy, whatever it had been, was hopelessly
befouled.

The Confederates crossed into Ohio at the town of Harrison
and fled east and south to by-pass Cincinnati and angle down to an
Ohio River crossing into West Virginia. They almost made it; they
had reached the Ohio shore at Buffington's Island (near Burlington,
Ohio) when they were intercepted by federal cavalry. The battle
lasted overnight and into the next day and ended as Morgan and
twelve hundred of his men broke through the Union lines. Next
day he was captured at Salineville and confined in the Ohio
Penitentiary; from which, presumably through Copperhead con-
nivance, he presently escaped to raid Kentucky yet again before he
was shot by Union officers in Tennessee in 1864.

In the very month that Morgan crossed the Ohio toward the
north, the Confederate doom was sealed. The blockade of the
Atlantic ports had already become almost totally effective; Admiral
Porter had taken the Mississippi Squadron (developed from that
Western Gunboat Flotilla which Foote had assembled and Eads
had armored at and above Cairo) south to assist Grant at Vicks-
burg. That city fell on July 4, the Union controlled the Mississippi
Valley and the Confederacy was split in two. Next month Farragut's

naval force entered Mobile Bay and the last considerable southern port was closed. The war still had almost two years to run but, as it touched the Ohio immediately, it was as good as ended; even transportation upon its waters was now of secondary importance.

On April 9, 1865, General Robert E. Lee surrendered the Confederate Army of Northern Virginia to General U. S. Grant under liberal terms and the Confederate hope was gone; by May 26 the last Confederate command had surrendered.

The war was over though troubled times, in most sections and especially in the South, were only well begun: but in the United States the section to suffer least of all in the next five years would be the valley of the Ohio.

The war had brought a terrific inflation to the river but, it now being the most nearly self-sufficient area in the land, inflation in its neighborhood was less disastrous. As a result of the production demands put upon its farms, manufactories, and mineral resources, there had been a change in its basic economy which was destined to have far-reaching importance.

Petroleum, for instance, was no longer only a nuisance to those who were drilling for salt water, or a product to be skimmed casually off creeks and bottled as a cure-all: the upper Ohio was enjoying America's first oil boom.

Coal—the utility of which had long been understood but which had not been economical for use in heating, manufacturing, or locomotive power as long as vast forests pushed near most town limits—had also come into its own. Five years after the close of the Civil War, more than seventeen million tons were mined, mostly on tributaries of the Ohio. Through the war years manufacturing and processing plants had turned to the use of coal almost unanimously, railroads had made the change even before the war, steamboats had followed suit, and coal had been largely adopted for heating, especially in the cities.

The loyal territory, cut off from its smoking and chewing tobacco supply by war, had increased its acreage; some of this planting ceased after the war, but Kentucky never again released its lead as a grower of burley and southern Ohio, Indiana, and Illinois continued as important factors in commercial tobacco raising, rather than as areas in which a man only planted tobacco for the family use, as before the war.

The heavy industries employing iron and coal on the Ohio from Pittsburgh to Cincinnati cast about to find outlets for the greater production capacity and know-how with which they found themselves possessed after their wartime expansion. Militant selling campaigns began to open markets for such items as new heating devices, ornamental "iron lace" for building, structural materials, lawn furniture and tools, plows and vehicle parts.

Farm machinery was not produced in quantity upon the Ohio, but parts and materials came from the river. Manufacturers of reapers and threshing machines had achieved good production figures well before the war and such equipment had been sold largely in the North—it has been remarked that these aids to food production were no small factor in the outcome of the conflict—but after the war their annual production and sales began to multiply with each decade. The cause was partly more efficient manufacture and distribution, partly the newly developed art of creating consumer demand, and partly—even in that comparatively hand-fought war—an increased experience with machinery by citizen soldiers during service and a resulting adaptability to its use once the men had returned to peacetime pursuits.

Another innovation with a more intimate and far-reaching effect upon American domestic economy was that in the preservation of foodstuffs. In its first years the Civil War was fought by soldiers supplied with the same old indigestible menu of Mexican War days—rancid salt beef and pork, weevil-infested flour, moldy hardtack, sanded sugar, and roasted navy-bean coffee—and through the machinations of profiteering contractors and scheming federal job holders, and because of the War Department's characteristic opposition to progress, many soldiers continued upon this diet throughout the contest. This time, however, there was a difference; in the Mexican War these had been the only staples available in the field and they varied only in quality. By a wonderfully rapid development, item upon item of edibles were now added to the available menu. Between 1861 and the end of the war some of them were being introduced if not to the mess halls of enlisted men, at least here and there in that of the officers (who usually "found" their own victuals), and in some of the better-operated Union hospitals.

Condensed milk came on the market in 1861, fruit was being canned commercially shortly afterwards, with oysters, vegetables, and corn soon following. Salmon packing began on the west coast in 1862. Even pickles, jellies and preserves were being put up by factory methods and in wholesale quantities before 1864 and at least two of the great modern food-processing companies were founded in the war period. While none of these new departures had originated in the Ohio Valley, the agriculture of the area was quick to participate once the war was over. Its contribution thereafter played one of the major parts in varying the American diet and emancipating the housewife. The new processing industry made another contribution in which the cereals, fruits, and vegetables of the valley played a major part: it not only varied the diet and furnished a leavening element to the fried menu of the farm hand and laboring man, which was then as uniform in the North as it continued to be for many years in the South, but it did something toward assuring uniform quality and palatability.

The fact is unpublicized by the food industry, which must sell the housewife, but it is certainly true that much so-called "home cooking" is still hopelessly bad even in this enlightened day: before the advent of canned and packaged food, difficult to spoil in preparation, it must have been deadly indeed.

Most experienced diners-out grant the wisdom of avoiding the public eating place that offers "home cooking." Dishes "like mother used to make" are not the gustatory delight of song and story if mother happens to be the uninspired or slovenly cook that she all too often is. There exists no evidence to indicate that she averaged better a century ago—the assumption being, rather, in view of the small variety of available edibles and the limited kitchen equipment, that she was probably much worse. The memory of the generic mother's cooking must have become a much happier one after the Civil War when progress in commercial canning enabled those of her kind who were less talented at the cookstove to offer at least an occasional appetizing dish: she should have been as grateful for this opportunity to gild her memory as she was for the saving in labor—whether or not the food-processing industry may wisely advertise the fact.

After the war President Lincoln, with his lenient plan for reconstruction, was gone, and whatever charitable motives President Johnson may have had were effectively squelched by vindictive leaders in Congress. The South was a shambles, physically and spiritually; New England shipping and maritime commerce was stagnant, the financial world of New York and Philadelphia was skittish, the trans-Mississippi West, long cut off from normal commerce, was in a financial slump—safest, most prosperious, part of the country was the Ohio Valley and the country north to the Great Lakes. The Ohio Valley had finally taken the place for which nature had destined it—as the evenly pulsing, stabilizing heart of the nation.

Chapter Eighteen

"THAR'S GOLD
IN THEM THAR HILLS!"

THE wildest dreams of the most imaginative travelers on the upper Ohio in the early eighteenth century could not have anticipated the wonderfully varied sources from which the valley's wealth would spring in the century ahead.

Those who first examined the valley with a view to settlement —Gist, Washington, Boone, Harrod, the others—saw game trails which led to springs showing traces of salt; "stone coal" outcroppings were noticed but not greatly valued, though Christopher Gist carried some samples home in his knapsack to show to the elder Washington brothers and Governor Dinwiddie; iron ore was recognized as was an occasional deposit of limestone especially suitable for building purposes. Clay, thought to be possibly useful to the potter, and rocks of suitable texture for mill stones and whetstones

485

were located. As soon as there began to appear to be a possibility of settlement in the near future the variety of timber and the eligible mill sites were carefully reported. All these natural assets of the new country were more or less appreciated and classified in the minds of those whose eyes were turned to the great West—*but only because of their value to the farmer-craftsman-small merchant economy that was necessarily the limit of early eighteenth-century vision.*

Prospectors for new homesites were glad to locate supplies of salt water which could be easily evaporated for table use and pickling meats; limestone would be needed for laying up the house walls, lime for mortar, and wood for buildings and boats and for use as fuel. The local blacksmith could even use a little of that stone coal in forging the iron for the settler's horseshoes, gun parts, and tools, though charcoal would serve as well as stone coal for the purpose. A millsite would always attract a thriving farm community and if the miller could cut his own stones on his own land he might be able to grind flour and meal more cheaply. Such matters as those, plus the goodness of the land itself, summed up the virtues of the Ohio as they were seen by all—from the least imaginative Pennsylvania farmer to young George Washington!

Even as these advantages to agrarian economy were being discussed, and as, through the years, political events were bringing about the moves and countermoves, the wars and treaties that would open the Ohio to permanent habitation by whites, new forces were shaping man's destiny to an end in which the riches of the Ohio would be put to new and hitherto undreamed-of uses.

The first important stirrings of the industrial revolution in Europe took place during the period of greatest activity in driving the Indians from the Ohio. The movement soon made itself felt in America. Before the Indians were all gone and settlement of the whole river was under way the steam engine was available, a new and practical source of power for both locomotion and producing goods. An asset fully as important to American development was the anxiety of the young and democratic nation to prove its virtue to the world. There resulted a stimulation of interest in invention and the improvement of industrial processes at all levels of society, from apprentice to employer. In a land of varied and plentiful resources—and especially upon the Ohio—this enthusiasm for progress worked wonders indeed.

In 1775, General George Washington and General Andrew Lewis, commander at the Battle of Point Pleasant, had located land around the "Burning Spring" ten miles or so above the great salt lick on the Kanawha. The tract interested them chiefly, as Washington said, "on account of a bituminous spring which it contains, of so inflammable a nature as to burn freely as spirits, and is nearly as difficult to extinguish." At that time George Washington believed that the colonial capital would shortly be moved west to the neighborhood of the Kanawha (an event which, largely through his own efforts, never came to pass) but no one would have been more surprised than he could he have foreseen that his heir, Dr. Lawrence Washington, would eventually sell the tract to the firm of Dickinson & Shrewsberry, who would drill in 1843 and strike the largest flow of natural gas discovered to that date in the region.

Chief sources of salt in the West until the Pomeroy, Ohio, and later the Saginaw, Michigan, regions were developed was the salt spring region on the Kanawha above the mouth of Campbell's Creek. This had been an Indian saltmaking place for centuries. Boone lived near here, on the banks of Crooked Creek, around 1788-1790 and gave a futile passing thought to setting up a saltworks before he removed to Missouri. Neither Indians, Boone, nor early commercial operators could have had the slightest presentiment that these springs would one day become the source of chemical by-products as valuable as their salt.

Of course the mining, smelting, and manufacturing industries did not spring full blown upon the Ohio, although those industries may have developed more rapidly there than they had in any locality before: the obscure genesis of many of these valley industries may be traced back far in time—sometimes before even the historic recording of time—but in most cases they have proved of importance along lines which their early operators could never have anticipated.

*   *   *   *   *

Salt was the first mineral product of the Ohio Valley to engage the serious attention of whites, even as it had attracted the giant sloth, the mammoth, the elk, deer, buffalo, and the earliest of the aboriginal human inhabitants. Salt had made those springs where it seeped from the ground in solution the favorite prehistoric meeting places of man and animals alike.

Salt-producing springs were common in the present confines of Virginia, Ohio, Illinois, Kentucky, and Indiana—in about that order of importance as far as potential salt production was concerned. Some of those earliest reported, and at first most widely known, eventually proved to be impractical of commercial utilization and were abandoned in favor of other springs of less historic interest but a greater salt content. The Big Bone Lick in Kentucky, for one, attracted the interest of the entire scientific world because in the sheltered valley that surrounds it were miraculously preserved the bones of early denizens of the continent which bore every mark, to the early eighteenth-century mind, of belonging to those anticipated Beasts of the Apocalypse—a surmise which was strengthened by the corpselike gray-green color that the sulphur content of the spring water imparted to the surrounding landscape. But famous as it was, the Big Bone Lick proved to be only a snare and a delusion to those pioneer salt evaporators who undertook to develop it.

The failure of the United States Saline Works near Shawnee-town, Illinois, came only after centuries—probably tens of centuries —of operation but for all that past performance it was none the less unable to compete commercially in the nineteenth century. The springs in the Shawneetown area were noticed by early travelers not only because, as usual, they were the converging point of animal trails but because on their sites existed a phenomenon reported only in a much lesser degree at other points: the fields around them were liberally sprinkled with pieces of ancient pottery. Such remains were occasionally found—usually buried—around other large salt-water springs but their abundance on the surface at this place was so remarkable that it encouraged the first workings of the springs by the whites. Incidentally, as casings were sunk in the muck, pots were found which grew progressively more crude with the passing of lower levels—indicating to even the unscientific minds of the early white workers on the job that here was an industry venerable indeed. Shortly after 1850 the production in this area reached five hundred bushels of salt per day and nearly a thousand people owed their living to the business, but new sources in Ohio and Michigan and new developments of the West Virginia fields produced more cheaply. By 1875 the United States Saline Works at Shawneetown were abandoned and today the site of this once-great activity is reached by a country lane, which after passing a couple of shanties winds through land apparently as wild as when the primitive Indian potsherds were first discovered by the whites.

The earliest large-scale commercial salt production west of the Alleghenies must have been that on the Kanawha River in present West Virginia, which has already been mentioned. Here Elisha Brooks leased some springs from Joseph Ruffnér and in 1797 built a furnace for boiling brine. Brooks was not particularly successful but shortly Ruffner and his brothers took over the business and under their management it prospered. Within twenty years the immediate area was producing six or seven hundred thousand bushels of salt a year—more than all the rest of the Ohio Valley saltworks combined.

Salt was hard to handle and store and expensive and risky to ship, but it was a necessity and in the early years of settlement, while labor and wood to fire the furnaces were still cheap, every spring which offered the slightest return was worked to meet the demand. Most successful of the early operations north of the river were those on the Muskingum and Scioto in present Ohio. They helped to bring prosperity to their neighborhoods but they were not sufficiently productive to hold their own with the competition of the Kanawha River works, which presently employed coal as a fuel and after 1835 made use of a patented steam furnace. Cheaper production resulted in a lower price and, although the salt interests made an unsuccessful effort to protect themselves, the price continued to fall. All Ohio Valley salt contained impurities which made it less than satisfactory for preserving meat and, that being its chief use except as a seasoning, the market for it slumped seriously as soon as improved transportation enabled merchants to bring in a better grade.

After the development of the Saginaw springs in Michigan, the salt industry in the Ohio Valley was dead, to all intents and purposes, except on the Kanawha. Even there production fell but the business survived through consolidation of plants and improvements in manufacturing technique until the twentieth-century increase in manufacturing of chemicals brought an enormous growth in demand for salt and its by-products and gave the locality a prosperity well worth the century of struggle.

Next mineral after salt to be produced commercially on the Ohio was iron—and it was probably the discovery of the extent of the iron ore deposits that gave the settlers of vision the first hints of the industrial possibilities of the valley.

The iron business was nothing new to Americans at the time of the Revolution; in fact the question of limiting American iron production had been one of the bones of contention between the British crown and the colonies.

The production of pig iron in North America began successfully at Lynn, Massachusetts, in 1644. By 1700 the seaboard was contributing one-seventh of the world production; seventy-five years later the North American colonies were producing more than any other single country—about one-seventh of the world's total tonnage.

At the time of the Revolutionary War iron had been mined and forged around the headwaters of the Monongahela by a plantation-type economy in which the owner and his family, employees, and slaves lived together on the farm which included the deposit of iron ore, the wood for charcoal, the furnaces, living quarters, gardens, fields, and pastures.

"Forges," these little communities were generally called, and "Forge" forms a part of the name of many a Virginia and West Virginia hamlet today—frequently with no evidence of industry within its limits except for the crumbling remains of an ancient stone chimney and a wasting pile of slag. The owner and his hands upon these farms burned the necessary charcoal and ran the smelter in winter and raised garden sass and corn for the stock in the summer; upon occasion, of course, they also repelled Indian raids and served enlistments in the colonial and later the Continental armies.

With the settlement of Pittsburgh and the towns below, iron deposits on and near the Ohio began to be developed, some retaining the plantation features of the old plan, especially on the Virginia side where slave labor was available. The industry prospered immediately—if modestly—because of the ease with which pig and bar iron could be floated downstream to the blacksmiths of the newly opening West or to New Orleans for transshipment by sea to eastern ports or foreign lands.

Shortly after 1800 two thousand flatboats and keelboats were carrying as much as five million dollars' worth of cargo to New Orleans in a year and the percentage of it that was iron grew steadily.

By the end of the first decade of the nineteenth century the stretch of the river from Pittsburgh to Wheeling was well set in the general industrial pattern it follows today. (In the June, 1810, issue of the *Portfolio,* Alexander Wilson, the American ornithologist, wrote that he was "Bidding adieu to the smoky confines of Pitt"!) In the new country with its fabulous resources, the enterprising iron men seized and developed foreign improvements as they appeared. The puddling furnace and rolling mill, inventions of the Englishman John Cort, were introduced in western Pennsylvania in 1817 and within five or six years this method of operation was common in the vicinity.

In the eighteen-thirties the center of the iron industry began to move downstream, temporarily, to the place on the Ohio called Hanging Rock. There, in new ore and coal fields some 1,800 square miles in extent on both sides of the river, was beginning a tremendous iron production under the inspiration of new methods.

The ore deposits in the Hanging Rock district had been discovered before 1826. In that year John Means set up a charcoal furnace in the vicinity of present Ironton and began to make pig iron. Means's furnace is incorrectly said to have been the first of its kind north of the river; at least there had been a furnace opposite the Licking River in present Cincinnati in 1812—Means's was, at best, the first of continued success. John Campbell, who would become one of the nation's greatest ironmasters, came to Hanging Rock presently and constructed a furnace along new lines; he placed boilers and hot blast over the tunnel head to utilize waste heat and—although there seems to have been the customary doubt on the part of the self-appointed authorities on the business—his idea worked. He tried it again, and there was soon a general change in construction of furnaces to follow his plan.

The boom at Hanging Rock continued through the Civil War, during which the high-grade ore from the neighborhood was preferred by the arsenal at Pittsburgh for casting Union ordnance, but deposits were becoming exhausted and thus more expensive to work, a counterboom had begun in the Pittsburgh-Youngstown district and—within almost exactly a century after its beginning— the Hanging Rock field was dead in that peculiarly final sort of death which only exhausted mining areas seem able to achieve.

In the meantime iron production elsewhere on the Ohio had been the subject of interesting experiments which were responsible for auxiliary developments.

From its first production in exportable quantities iron earned quick and generous profits for all concerned and opened new sources of revenue not only to those directly engaged in the business but to their neighbors as well. Naturally craftsmen had gathered in iron-producing regions to fabricate the metal close to its source; when iron was first produced in western Pennsylvania the rifle and the ax were items essential for maintaining life downstream, and first the Pennsylvania rifle, shortly the Kentucky, achieved fame as quality products. As early as 1807-1808 Christian Schultz observed, while touring the Ohio: "There are a number of rifle manufactories established in this country, but the best and handsomest I have seen are to be procured in Kentucky and Tennessee where they are made in every size . . . and the price from fifteen to a hundred dollars." Local craftsmen, from Pittsburgh on down the river, modified their production to fit the needs of the progressing settlement, and they, and the ironworks, profited.

Besides the profits that resulted from fabricating, the actual production of pig or bar iron itself was responsible for a varied prosperity. Under the earlier method of manufacture about four cords of wood were required to produce 137 bushels of charcoal. This fuel, plus something over 300 pounds of limestone and 5,000 pounds of average Ohio Valley ore, were needed to make a ton of pig iron. This process thus gave part-time work at woodcutting, quarrying, and hauling to many besides those regularly employed at mine and furnace.

The experiments carried on in connection with iron production in the valley were of world-wide importance. The first manufacture of crucible steel to any significant extent in the United States began at Cincinnati in 1832 under the direction of Dr. William Garrard. Garrard, an Englishman, found in what is now West Virginia a deposit of clay suitable for the necessary pots and discovered by experiment that charcoal iron from Missouri would serve as well as the Swedish product, until then considered essential in England and Europe for making cast steel. The doctor had been apprenticed to a bricklayer in his youth and was thus able to design and build a furnace exactly to his liking; he began to turn out saws, axes, files, and cutlery of the finest quality. Unfortunately his financial management was not equal to his inventive talent and he failed during the panic of 1837—after he had proved that such manufacturing was possible.

The greatest spurt in the iron production of the upper valley probably came with the discovery, in the forties, that anthracite coal from eastern Pennsylvania could be substituted for charcoal. Already wood had become scarce and relatively expensive in the neighborhood of ironworks but there was plenty of anthracite coal and even more bituminous, the latter deposited close to the ironworks. The true extent of the deposits of bituminous coal in the valley was beginning to be suspected as a result of geological surveys and the sinking of deep wells for salt water and mine shafts for iron ore. Naturally experiments were made with the object of employing bituminous coal and it was found to be satisfactory. Eventually coke, made from it, was found to be the ideal fuel.

Steel manufacturing developed meanwhile. In England Henry Bessemer received his patents for decarbonizing molten metals in 1856, but before that date an Ohio Valley resident, William Kelly, had conceived some ideas of his own upon the subject which evidently played a rather significant role in the development of Bessemer's invention.

Kelly was born in Pennsylvania and, after having served some sort of apprenticeship in the eastern iron industry, settled in Eddyville, Kentucky, a few miles up the Cumberland River from its junction with the Ohio. The story is that he, as proprietor of a small iron furnace, began to experiment about 1851 and discovered the principle that resulted in the rapid decarbonizing of iron by an air blast. Two Englishmen employed at his furnace assisted in his experiments but as these began to show promise, Kelly's assistants tried to discourage him. One night they departed without notice and Kelly's subsequent inquiry revealed that they had returned to England.

Within a few years Henry Bessemer's patents were announced in the United States and Kelly came forward with his claims to prior discovery of the principle. By formula the story should end as did that of the other Kentuckian, John Fitch—with Kelly not believed and haunted the rest of his life by sorrow and frustration. Happily it did not; Kelly was granted American patents and, by agreement, these were combined with Bessemer's improvements for American operation—indication enough in those day of uncordial relations between the two countries that William Kelly had prior claims sufficient to substantiate his contention.

The beginning of farming on the tributaries of the Ohio upon a scale incomprehensible to residents of the eastern seaboard had much to do with the increasing demand for iron and steel products after 1840. Agricultural implements, even in their early stage of development, required castings, blades, wheels, gears, rods, bolts, and nuts beyond any demand which could have been anticipated at the end of the War of 1812. The homestead market was by no means confined to farm machinery, for it carried with it the growing interest of the housewife in iron stoves, lamp frames, and grinding and dicing devices for the kitchen, and a thousand and one other items of utility indoors. Growing familiarity with machinery in and about the home brought ever-increasing use, and experimentation by amateur mechanics resulted in improvements and new departures.

Such a preoccupation with the uses of iron not only stimulated the production but also the fabrication of the metal. The pattern-making, foundry, casting, and milling industries followed the mining and smelting and progressed as rapidly in efficiency of operation. As would continue to be the case until well into the twentieth century, there was an unrecognized division in production in this field; even in the eighteen-fifties, -sixties, and -seventies the Pittsburgh-Hanging Rock area tended to specialize in the production of heavier goods, the Cincinnati neighborhood in lighter—the latter furnishing, as one interesting contribution, most of the printing presses and type that helped to enlighten the West.

One market alone was slow to welcome American iron: the railroads, whose development was causing such a fervor of excitement in the United States after 1840, were dubious of the local product.

By 1840 the railroad had achieved the general recognition of American capital as a practical mode of transportation and a sound investment. Lines began to be extended in short and sometimes apparently purposeless stretches throughout Pennsylvania, Ohio, Indiana, Illinois, Kentucky, and to a lesser extent, western Virginia. At first the western roads felt constrained to use iron from England or Wales, as had the eastern lines in the beginning. Railroad iron had been developed in the British Isles; they were the traditional source and American iron smelters found great difficulty in convincing the western promoters, often entirely inexperienced in railroad operation, that the metal could as well be fabricated closer home. After a bit, however, some daring—or perhaps impecunious—railroad entrepreneur gave the cheaper American rails and gear a chance and before the middle fifties American manufacturers were secure in the business.

The mineral-utilizing industries on the upper Ohio really came into their own during the Civil War, and aside from the capricious service of the still-primitive railroads, always subject to capture, demolition, and sabotage, the Ohio River was the *only* route by which arms and heavy supplies from the area could reach the Union forces in the West and Middle South. Because of its mineral assets, the skill of its workmen, the comparative efficiency of its manufactories, and the fortuitous direction of the Ohio's course, the neighborhood of Pittsburgh was in a position to seize the Civil War market, especially for heavy ordnance: and it did just that.

After a short breathing spell following 1865 all the volume of wartime production was regained and was multiplied, year by year, as southern and midwestern railroads were rehabilitated and bridges were built. Iron began to be introduced into the building industry; by the seventies railroads were pushing westward from the Mississippi and thousands upon thousands of miles of track were to be laid and supplied with rolling stock and motive power. Thanks to its enterprising spirit and its natural assets, the upper Ohio industry had the lion's share of this business also.

The years after the Civil War witnessed the rapid substitution of steel for iron in the nation's economy. The consolidation of the Bessemer-Kelly patents was completed in 1866 and processing improvements were rapidly introduced which adapted American ores to steel manufacturing. The open-hearth process was introduced in 1868 and it, with the Bessemer, continues in use.

The development most important to the steel industry—and to American manufacturing as a whole—was the growing tendency after the seventies to increase production and lower costs through greater industrial efficiency and the control of raw materials and marketing. Naturally the movement toward consolidation was viewed askance by alarmists but, as always when better use is made of natural resources in quantity production, prices went down while profits and wages began a steady increase.

The nineteenth-century "captains of industry"—whose careers were targets for the venom of political radicals and the worship of ambitious youths—were definitely in the saddle. They sometimes came a cropper at the hurdles they chose to try; some of them did not hesitate to ride down the crops of the peasantry and the more daring of their kind caused momentary panics in which they and the nation suffered. Nevertheless, they built an economy and introduced a way of life in the United States which has already enabled this country to see the world through two wars.

Pittsburgh and the upper half of the Ohio were the very heart of the new development during the last quarter of the nineteenth century. None can object to the selection of Andrew Carnegie as typical of those men who led the country to industrial greatness. In the public mind for a good many years the name of Carnegie was symbolic of Pittsburgh and steel—that it eventually came to symbolize free libraries and advancements in teaching and education is a satisfactory testimonial to the virtue of the system that first made it great.

That old plantation type of economy under which iron was first produced in the Ohio Valley had some advantages. Ore mined on the owner's land, a furnace fired with charcoal burned on his ground from his timber, employees in houses belonging to the owner of the works and purchasing rations from his storerooms, all these features were rediscovered and put back into practice later at Hanging Rock, where coal, iron ore, and lime used by a furnace were likely to be mined or quarried on land under the same ownership. The operations at Hanging Rock were efficient and profitable for their day—until the ore ran short.

Obvious next step for the progressive industrialists was to salvage the best features of the older plans—the assurance of a ready supply of materials, and labor maintained at hand—but for the future upon a scale large enough to give insurance against failure because of a temporary market slump or the exhaustion of resources in any one particular locality.

A general movement to merge holdings began in the eighties and it culminated around the turn of the century with the consolidation of some dozens of companies controlled by Andrew Carnegie, Henry C. Frick and E. H. Gary in the vast United States Steel Corporation, which continues a giant among industries. From the beginning the corporation controlled its own ore, coal and oil fields, fabricating plants, and considerable rail and river transportation; in addition it had, through the inspired financing of J. P. Morgan, ample capital to protect those facilities and to provide for expansion.

Not all Ohio Valley coal comes from the headwaters. Today there is also great coal-mining activity upon the lower river. The spreading Illinois Coal Measures (so called by geologists) reach the Ohio along the Illinois shore and extend east well across the Indiana line and into a large pocket of western Kentucky. Thus, found as it is in good quality and workable quantity almost from one end of the river to the other, coal is the most valuable geological asset of the valley—second only to rich farm lands.

By the time the Ohio began to be fully settled coal had long been established as a fuel in cities and towns in the British Isles and on the European continent. In America, however, as has been noted, there was some resistance to its use because of the dirt that resulted from its burning and because it did not readily demonstrate its superior qualities in supplying heat and holding fire when burned in a fireplace not especially designed for its efficient combustion. Too, the timber on newly cleared land was a nuisance; it had to be removed before cultivation could begin, so there was little sense in mining and hauling coal when wood could be easily cut to the proper length for burning and had to be destroyed some way in any case.

Notwithstanding its lack of general acceptance, coal was early used upon the river; almost as early, in fact, as there existed a settled population with manpower to bring it in: "A coal mine," said the ubiquitous Zadok Cramer, "was opened in the year 1760, opposite Fort Pitt on the Monongahela, for use of that garrison." By "mine" Cramer meant simply what would now be called a "working"—there was still plenty of surface coal in 1814, when Cramer wrote, to supply the demand without the necessity of shaft mining. It seems safe to assume that coal was used elsewhere on the Ohio's Pennsylvania and Virginia shores before the end of the American Revolution and on the north bank as soon as the Ordinance of 1787 had opened it to settlement. Even the earliest travelers usually reported its outcroppings—calling it "stone coal" to distinguish it from the far more familiar charcoal.

The first steamboat on the river, the *New Orleans,* was fired for at least part of its maiden voyage with coal picked up at a free deposit located below Louisville on the Ohio; possibly coal from the Pittsburgh area had also been utilized on the first stage of the trip downstream. Coal was used in the steam engines developed in England and its lower cost and superior efficiency seem to have been recognized in the United States, even though it would not be generally used by steamboats, except on the upper reaches of the river, until shortly before the Civil War. In 1827, for instance, $1.50 worth of Pittsburgh coal (12½ bushels at 12 cents per bushel) produced power equal to that from a cord of hickory wood sold currently along the river at $2.87, but coal was not then offered for sale on the lower river and the high freight rates made it poor economy to load a packet with coal enough for a long trip when small quantities of wood could be taken on at short intervals and a greater payload carried—such were those carefree days of superabundance of natural resources.

Freight rates—passenger rates too, for that matter—were never very low upon the river in the days of the packets; but neither were costs low for wagon or railroad transportation. In the middle thirties, for instance, coal worth $1 per ton at Pittsburgh regularly brought more than $3 at Cincinnati and if a prolonged spell of dry weather lowered the water on the Ohio sufficiently to impede traffic and cause a shortage in the latter market, the price might jump to $10 or even more. Small wonder that the housekeeper continued to use wood for cooking and heating as long as a supply existed even reasonably near her back door.

Even so, 1840 saw Ohio produce 3,500,000 bushels of coal, Illinois, 424,000, and Indiana, 242,000; although no reliable figures exist it may be assumed that Pennsylvania west of Pittsburgh and the Virginia section of the Ohio Valley considerably exceeded the total from the north shore of the river; Kentucky's coal production was not yet significant.

As a matter of fact the shipping of coal downstream on the Ohio and Mississippi seems never to have been a very profitable business in the nineteenth century. Most shipments were by flatboats or barges but even such theoretically cheap transportation, in the days before dams were constructed to maintain a standard level of water and the depth of the river depended upon the uncertainties of natural rainfall, might run the shipper into terrific expense for wages to a stranded crew or for the loss of an entire cargo in a boat run aground and wrecked.

Those comparatively few barges which proceeded under power in the last half of the nineteenth century were either towed astern or lashed at the sides of a packet. With the increasing boom in coal production after 1870 and the quick transportation that competition of railroads made necessary, even this method was too slow and the quantity per tow too little for practical purposes. Coal became mainly an item of railroad freight and so continued until, shortly before the beginning of the twentieth century, a new development again gave the advantage to the river.

Stern-wheel steamboats had been an early experiment and they had been constructed, usually for use upon narrow tributary streams, during the nineteenth century. Now this type of boat took on a new importance; small stern-wheel boats began to be used to *push* barges instead of to pull them. The powerful little stern-wheeler pushing a long string of barges in front could be maneuvered up or down the river with ease. The average cargo of a single "tow" (such boats continue to be called "towboats" today, even though they have pushed now for half a century) was 15,000 tons but one famous boat, the *Sprague,* made a trip in February, 1907, pushing an aggregate of sixty coalboats and barges carrying a total load of 70,000 tons!

The stern-wheel towboat moved coal traffic upon the Ohio even through the ups and downs caused by the cycles of drought years that marked the early twentieth century, when the river could not be navigated most of the time, and through a freight rate war or two with the railroads. Virtually alone, the little stern-wheelers kept river traffic alive on the Ohio until (as will be discussed later) the river had been canalized to assure a continuously navigable level. Coal traffic is no longer the matter of long-distance hauls it once was; now mines in the South supply the lower Mississippi and a great measure of the Ohio Valley's coal is consumed by her own industry, her own homes. Even so, coal traffic on the river shows a steady, though fluctuating, increase: 4,500,000 tons in 1915; 9,750,000 in 1930; 18,500,000 in 1945—with an all-time high in the war year of 1942 of 21,514,353—and a still-increasing tonnage today!

That, in addition to the millions of tons carried by rail and truck, shows the importance of coal as an item of commerce in the Ohio Valley.

But salt, coal, and iron were not the only products dug from under the rich topsoil of the valley. There was also plenty of glass sand and that useful limestone upon the upper river, and by the eighteen-twenties Pittsburgh had a flourishing glass industry. The discovery of natural gas and its employment in the manufacture of glass gave the business a further impetus and by 1850 the Pittsburgh-Wheeling area had achieved a foothold in the industry which it never lost.

Glassmaking was one of the earliest manufacturing ventures of the American colonies: scant record of the operations is preserved but there is no doubt that enterprising individuals in the coastal settlements tried their hands at producing clear glass, and colored glass for beads to be used in the Indian trade, from the very first settlement. The London Company dispatched a party of eight Dutchmen and Poles to the Jamestown colony of Virginia to teach its colonists to make tar, pitch, soap, and glass in 1608 and the glasshouse constructed by these "skillful workmen from foreign parts," as the London Company described them, was the first in the New World. Presumably the business did not flourish, however, for in 1621 Captain William Norton received a charter to set up a glasshouse in Virginia and to hold exclusive right to manufacture the article for seven years. By 1639 there was a glassworks at Salem, Massachusetts, and presently glassware was being produced in New Amsterdam.

Really successful operation had to await the arrival of Caspar Wistar, who brought four glassworkers to New Jersey in 1739, and Heinrich William Stiegel who landed at Philadelphia in August, 1750, and began manufacture sometime before 1763.

It was after the Civil War, with the introduction of cheaper methods and quantity production, that the Pittsburgh-Youngstown area began to dominate the American glass industry. By that time glass itself had a much wider market: builders had become less parsimonious in the use of windowpanes as home-heating systems were improved and glass itself became cheaper, but the prime causes of the more extensive demand for glass were the introduction of the practice of canning food products between 1850 and 1870 and the vast and rapid increase in use of the oil lamp for illumination. As in the case of iron and coal, the upper Ohio had an industry established and ready to expand to meet the growing demand.

While interesting and artistic designs in flasks, jars, tableware, and lamps were produced around Pittsburgh before the middle of the nineteenth century, fabrication of Victorian glass again followed the pattern of clay, iron, and other products; utility ware and the raw product became the specialties of the upper river while more artistic work was done in Cincinnati and even as far downstream as Louisville. Much of the pressed glass now sought eagerly by collectors whose purses do not permit indulgence in the masterpieces of the self-created "Baron" Heinrich Stiegel came from the middle rather than the upper river.

The passing years have brought an increase in demands for glass sufficient to take up any slack which otherwise might have resulted from quantity production under constantly improving manufacturing methods. The use of large shop windows in which to display merchandise gained great impetus after 1900; the vastly increased use of interior and exterior lighting made a market for both filament bulbs and neon tubes which must exceed the old per capita consumption of coal-oil lamp chimneys at least two to one; the increasing use of glass in motorcars; the development of glass brick as a building material during the decade just past and the ever-growing consumption of canned goods resulting from the limited storage space of the apartment dweller—all seem to ensure a growing prosperity for glass manufacturing upon the Ohio.

Not least of the early contributions of the glass industry, incidentally, was that of its original workers and their children to the general economic, cultural, and social betterment of the valley; glassmaking and -blowing was probably the first trade upon the river that attracted highly skilled European workmen to the river in significant numbers.

The making of pottery is at least as old an industry as salt-making in the Ohio Valley. Ancient predecessors of the mound-building people began it, and it was continued with results quite remarkable in some stages of the mound culture—even apparently having been carried on as the specialized trade of groups of individuals by people of the Hopewell and Middle Mississippi cultures, as in a modern industry. Indians of historic times also practiced it with varying degress of skill; always to an extent more limited in quality and quantity of production than their predecessors, but still sufficient to keep the continuity.

The early colonial period saw efforts at pottery making on the

Atlantic coast but the industry made little headway. Those dining and cooking utensils produced within the present borders of the United States which were not fashioned of sheet iron, copper, or pewter, and thus within the province of the smiths, were often either turned or coopered of wood. What pottery was manufactured in the early days mainly took form as jugs, crocks, and the crudest forms of clayware.

After 1800 table "china" came into more general use through importation from England and France and within twenty-five or thirty years Connecticut, Vermont, New Jersey, and Pennsylvania had thriving pottery industries which manufactured, in some centers, tableware of considerable artistic merit.

As the iron and coal industry developed upon the Ohio there naturally occurred an awakening of interest in other manufacturing. Clay deposits had been mentioned in early surveys and they now presented a possible field for the investment of capital. Investigation led to the discovery of potter's clay of all grades deposited here and there from one end of the Ohio to the other, though existing in greatest quantities upon the north shore, especially in the state of Ohio. For these, as well as some other exploitable natural assets— even plain sand and gravel—the people of the valley have to thank the cataclysmic changes of the preglacial geologic epochs and the fact that the moving glaciers of the ice age delivered and deposited here much of the material they had picked up in their inexorable march southward from the arctic.

Earliest of the famous potteries on the Ohio were established at East Liverpool (called "the pottery capital of the country," where good clay and natural gas combined with skillful management resulted in enormous production), in the Zanesville district, and at Cincinnati. Today the Weller Pottery in Zanesville and the Rookwood in Cincinnati produce some of the finest American products and enjoy a world-wide reputation.

But clay products were not all of high aesthetic virtue. The first longing of the prospering settler was for a brick house. It was easy enough to produce brick in that early day when the subsoil from the basement could be burned on the premises with wood cleared off the lot into cheap, serviceable, and well-colored brick—always supposing that the brickmaker knew his business—but with rising labor cost such procedure became expensive. After farming became general and all-weather roads were built, tile ditching and culverts became a necessity, and with the growth of cities, sanitary and drainage systems called for sewer pipe. The demand for both clay brick and clay tile increased rapidly.

The manufacture of brick and tile soon became industries in themselves and by the nineties every major town on the river had one or more brick or tile factories. There has been much consolidation of small plants but tile made of either clay or concrete and brick made of clay or shale are still major items of production—while the cement-block manufactories have recently become as numerous as were the small brick plants fifty years ago.

The American oil industry, the world industry in fact, had its beginning in the upper Ohio Valley and oil and natural gas made an important contribution to the economy of the Ohio—although the enormous production of fields in the South and on the Pacific coast have rather dwarfed the recollection of the wealth the older fields produced.

Petroleum (called "rock oil") and gas, which seeped naturally from the ground, were recognized by early travelers but were considered to be at best only a natural curiosity, worth detouring a mile or so to see and worthy of a note in a diary—as George Washington mentioned his "bituminous spring"—but of little conceivable utility.

After the drilling of salt wells began in the early nineteenth century those who ventured in the business were often inconvenienced by striking oil instead of salt water—although this misfortune was considered to be somewhat mitigated by the fact that the presence of oil was a fairly certain indication that salt water could be found by drilling deeper!

Early settlers heard from the Indians that petroleum was considered to have medicinal properities and—influenced by professional old wives and the opinion of subscribers to the Indian Doctor school of therapeutics—they applied it in dozens of prescriptions for inward and outward use. Samuel Kier, of Pittsburgh, undertook to skim oil from seepage on the surface of ponds and creeks and to market it in the forties but, though he priced the smelly nostrum at 50 cents the half pint, he still found it a losing business. He did not give up easily, however; convinced that there was a fortune somewhere in oil he finally designed a crude still with which he cracked petroleum to produce a light oil which could be burned in lamps and a heavier by-product which could be sold to wool processors.

Kier's problem lay in his inability to supply himself with a sufficient quantity of the raw product. That remained for others to solve.

Other enterprising individuals besides Kier began to investigate possible uses for petroleum. That it might sometime be used for illumination was realized by many experimenters, for lamps burning whale oil and lard oil had long been in use, though both were odorous and inefficient. Whale-oil production decreased in the forties but Cincinnati alone was making 1,200,000 gallons of lard oil per year after 1850. Before petroleum came to be available in quantity, however, James Young, an English chemist (or Dr. Abraham Gesner, of New York, if one prefers to honor his conflicting claim), had invented a process for distilling true "coal-oil" from coal to make a more satisfactory illuminating agent than either whale or lard oil. This industry also came to the Ohio; American cannel coal was found to be a good source of the oil and manufacturing plants around Pittsburgh and in West Virginia, Kentucky, and Ohio were in production during the late fifties. One of these, the Lucesco Oil Company, of Pennsylvania, eventually reached a capacity of 6,000 gallons per day.

But even before this *real* coal-oil industry was well started the way was being paved for the introduction of illuminating fluid derived from petroleum—still commonly called "rock oil"—which would replace the product distilled from coal entirely after the last coal-oil plant not converted or abandoned burned down in Beaver County, Pennsylvania, in 1871.

In 1854 George H. Bissell, visiting at his alma mater, secured some crude oil from Professor Dixie Walker of Dartmouth and, becoming interested in its possible utility, joined with Jonathan Eveleth and leased some swampy land located on Oil Creek, a tributary of the Allegheny River in Pennsylvania. The owner of the land was Professor Walker's father-in-law. Bissell and Eveleth undertook to gather oil which had seeped to the surface in ditches which they caused to be dug across the land. The operation was not very successful but they did gather a few barrels of oil, some of which they submitted to Professor Benjamin Silliman of Yale for analysis and study. Silliman's report satisfied Bissell as to the commercial value of the various lighter products that could be distilled from petroleum and were suitable for use as lamp fuel, lubricating agents, and for other purposes.

Bissell decided to try to secure a larger supply of petroleum by drilling a well. He organized a company and subleased to it some of the land he and Eveleth had leased from Professor Walker's father-in-law and on which they had tried the ditching experiment. His proved to be the proper solution to the problem. The company's drillers struck oil at a depth of 69½ feet on August 28, 1859. The approaching Civil War notwithstanding, the boom was on. By the war's end it had reached astounding proportions.

The oil distilled from coal had been called "kerosene" or more commonly, "coal-oil" and the cheaper and more plentiful illuminating fluid refined from petroleum kept the same names; was processed, as a matter of fact, in some of the same plants. Within a few years it came to be the *only* proper illuminating medium for families of even the slightest social pretension, where natural gas was not available. As already mentioned, this new development presently made an additional contribution to the Ohio by encouraging still further the production of glass and pottery lamps—table, bracket, and most elegant of all, "hanging"—and of glass lamp chimneys.

Hundreds of old wells are still pumped but the flush production of the original Pennsylvania oil fields was soon exhausted—though not before such astute Pennsylvanians as Andrew Carnegie had amassed enormous profits from them—and phenomenal production moved west, first to Ohio, West Virginia, Kentucky, and Tennessee, then in even more amazing quantities, to Texas, California, and Oklahoma. Even so, oil strikes are not a thing of the past in the Ohio Valley, as witness those recent ones in the field of which the junction of the Ohio and the Wabash is the center and which now pour wealth into surrounding Indiana, Illinois, and Kentucky. There is plenty of oil business on the Ohio—even involving oil brought through pipes from fields half across the continent away—showing itself in the miles upon miles of refineries and storage tanks that line the river, especially above Cincinnati, and the more than eight million tons of oil and gasoline transported upon the Ohio by tank barge each year!

Of course the Ohio soil and rock harbor other, less prosaic, wealth than salt, iron, coal, clay, stone, and oil; for instance, there's gold. Many tributary brooks hold gold dust in their gravel-lined pools—in just sufficient quantity to engage the lifelong attention of an occasional eccentric individual whose temperament could not bear a day of laboring at any orthodox calling.

Especially for the benefit of the eccentric, also, there probably exists along the Ohio and its tributaries a vast hoard of buried treasure. It is the loot, generally, of river bandits, Spanish miners, lost Roanoke colonists, or others of that ilk—or it is the ill-gotten wealth buried by an outlaw about to be taken by a posse, or the fortune of a rich emigrant and his beautiful daughter about to be slain by Indian or outlaw; or traveler in fear of an approaching enemy in, variously, the French and Indian, Revolutionary, 1812, and Civil Wars. (The retellers of these episodes are mostly uninhibited by historical chronology or geographical fact, so that almost any combination is available.) Only two circumstances remain more or less static in the treasure stories: while coastal treasure trove is usually deep buried in shifting sand or sunk in water, that of the Ohio Valley is uniformly concealed in caves or rock shelters well up on cliffsides—and mostly, except for an occasional cache in southern Indiana and Illinois, it is deposited on the West Virginia and Kentucky sides of the valley.

Taking it all and all, even without free gold dust or buried treasure, there was and is a great deal more gold in the Ohio hills than anyone suspected during the first hundred years in which they were familiar to whites—far more in value than there ever was of the actual metal in those far western hills to which the ungrammatical quotation "Thar's gold in them thar hills!" first referred.

Chapter Nineteen

THE RIVER MAKES
A COMEBACK

Despite the inference that may be drawn from pictures painted by the more alarmist-minded among the soil and timber conservationists, the Ohio was no uniformly deep, smooth-flowing stream even before a single acre of virgin timber was cut, a single inch of topsoil plowed. Sizable American rivers which never flooded, never muddied, never experienced low water were few if not nonexistent except on the coastal plains.

As the earliest boatmen noted, the wind usually blows upstream on the Ohio. The flatboatmen found this circumstance only added

to downstream steering difficulties but the keelboatmen were grateful, for it sometimes enabled them to add a sail to aid manpower
in toiling upstream. The boatmen were right about the wind; its
direction was no folk tale—the wind does blow across from west
to east in the Ohio Valley during the winter and spring and in its
blowing it brings much rainfall. Rain upon the western slope of
the Allegheny Mountains, which feeds the Allegheny and Monongahela rivers and thus the Ohio, is unusually great in volume and
the steep descent of the smaller tributaries from the mountain
country pours water into the upper river very rapidly after every
snow or rain: thus the Ohio has always had and always will have
a flood problem.

Floods were certainly somewhat encouraged after settlement by
the draining of swamps and the felling of timber but even in the
flood of 1763, before a single swamp was drained, the river rose to
a height equivalent to 44 feet at "the point" in present Pittsburgh.
It even caused the Shawnee Indians, downstream, to remove their
main town from the mouth of the Scioto River north to the present
site at Chillicothe, Ohio. That rise at Pittsburgh stood as a record
until December, 1936. As accurately as wandering hunters on the
ground at the time could estimate, the Ohio reached a stage of
about 75 feet at the future site of Cincinnati in 1772 or 1773—and
that stage also had to wait to be exceeded at the place until the
river reached 80 feet on the harrowing 26th of January, 1937.

Man had not disturbed the balance of nature much in 1763 or
1773 and there will continue to be floods upon the Ohio when
the proper combinations of temperature, wind, and rainfall occur
as long as the river flows.

Thanks to the endorsement of communities, tired of the necessity to clean up and rebuild after every flood, and thanks also to
civic and industrial organizations such as the Ohio Valley Improvement Association, the Corps of Engineers, U. S. Army, has been
permitted to make measurable—and in some instances spectacular—
progress in reducing flood damage. Headwater reservoirs gradually
are harnessing tributary streams, but not even the most hopeful of
the practical public-spirited gentlemen who have carried on this
work expect to prevent the Ohio from flooding.

The river was not only ravaged by floods from its earliest day but it was also subject to extremes of low water—that last is *one* ill which man, as represented by the Corps of Engineers, has conquered to an extent verging upon the miraculous, as will presently be demonstrated. The river, in the state in which it existed when it was first thoroughly surveyed, had available for navigation over its worst shoals at lowest water a minimum of only one foot in places above Louisville; even below that city there were minimums of only two feet. The flow of water at the Falls themselves could, in extreme drought, be reduced to a trickle between the serrated rocks. In the river's natural state it was only when a minimum of three feet of water maintained over the worst shoals above and below the Falls that it was considered to be at a navigable stage for steamboats or other river craft of cargo size.

The first opinion of thinking rivermen seemed to favor placing upon neighboring states and cities the responsibility for maintaining the river at a navigable stage of water and free from obstructions: as a result little was done toward removing driftwood or marking shoals even in the first two decades of the nineteenth century; no effort was made to gain permanent control. Virginia (later West Virginia) and Kentucky own the river to the low-water mark on the north shore—they were expected to use wisdom in granting millsites, ferry charters, or permits to build bridges in later days, but there were no reasonable grounds for demanding that they maintain the river as a public thoroughfare. There was, in fact, in the experience of the young nation, no precedent for operating a thoroughfare of such magnitude by any public means.

One of the early suggestions for specific governmental responsibility appears as an opinion expressed by Zadok Cramer in the *Navigator*. He says, as of 1814:

A lock-canal around the Falls would . . . be of immense advantage to the Ohio trade. . . . There has been some talk of commencement. . . . What jarring and clashing interests prevent the undertaking are not easily to be found out. It can scarcely be supposed to be the lack of publick spirit in the Kentuckians or their legislature . . .

Cramer added:

There are many smaller impediments, however, in the river from Pittsburgh to the Mingo Town [below modern Steubenville] which may be as long getting removed as the Falls themselves; these consist of rocks that might be blown to pieces, and ripples that might be easily cleared . . . This is certainly an important national concern, but the people must begin to act first . . . It must be done by grants of monies from the state, aided by subscriptions from the people . . .

Cramer's theory as to the proper mode of attacking the problem did not prove immediately workable. Talk about a canal at Louisville had begun even before 1804 and a company to build it was incorporated in that year, but those "clashing interests" mentioned by Cramer had interfered—rival companies had been formed to build on the Indiana side, and Louisville teamsters and warehouse operators had objected. Traffic on the river continued to increase, however—a million dollars' worth of goods shipped downstream in 1798, five million a year by 1807—and finally the successful Louisville and Portland Canal Company was organized in 1825 and the federal government subscribed to a large block of stock. The work was completed and the steamboat *Uncas* passed through late in 1830. The canal was a complete success as a public utility and an investment—Ohio, Indiana, and Pennsylvania boatowners were still complaining of its enormous profits after half a century—and Cramer's judgment was vindicated: it was indeed a "national concern, but the people must begin to act first," even as Cramer postulated. This sequence of procedure certainly got results at the Falls.

The full realization of the significance of the Spanish Conspiracy, which took place before 1800, and Burr's plotting afterward, put the federal government rather in a mood to cherish the states upon the Ohio River lest their ire with the East be again aroused. By the eighteen-twenties the potential value of the West was evident to all but a few of the most reactionary of seaboard statesmen, and western representatives in Congress were enthusiastically flaunting the glories of the western country at every opportunity. Many an appropriation was currently being granted for improving eastern and southern harbors; some such gesture for the benefit of the western rivers was obviously in order. By an act approved May 24, 1824, Congress assigned superintendence of the Ohio and Mississippi rivers to the officers of the Corps of Engineers of the United States Army and provided an appropriation of $75,000 to be spent—in the language of the original act—in removing some of the "planters, sawyers and snags" from the channel of the Ohio.

Besides the stump pulling, the Corps began the next year to blast some of the more dangerous rocks and to construct dikes which would direct the river's own waters in such a way as to scour out and maintain a channel in places where sand bars tended to form. This sort of work was carried on, as Congress could be importuned by western members into making appropriations from time to time, until 1875.

In 1875 came a new proposal.

A survey of the Ohio was made in that year with the intention of providing a 6-foot depth of channel throughout the length of the river through the agency of a series of movable dams. Sixty-eight, in all, were expected to be sufficient to accomplish this purpose along the river's 981-mile length.

The first of these dams was constructed at Davis Island, a little less than five miles below Pittsburgh. It was completed in the fall of 1885 and proved most valuable in that, for the first time, Pittsburgh had ample harbor space even during low water. Experiments were carried on in redesigning and improving the dams and the plans for their location were changed. The program for canalizing the Ohio was being perfected, as the years passed, but little was accomplished except on paper, for obvious reasons. The period after 1875 was no time to interest either Congress or the majority of citizens in the valley in improving the river for navigation. River commerce was dead, in the opinion of most people: "What's the good of a river," they asked, "with all these railroads?"—and it looked, at the time, as if their view was reasonable enough. The efforts of the Corps of Engineers were therefore confined to dredging the channel in an effort to maintain at least three or four feet in the low-water season.

The seventies and eighties saw a brief renaissance of steamboat traffic, with as many as a million passengers and two and a half million tons of freight loaded in twelve months, but the period also witnessed a renewal of railroad building which doubled and redoubled the mileage of midwestern track.

Any town which expected to hold its post office felt that it must have rail connections with its neighbors at any cost. Tracks were laid to connect every imaginable combination of points. Railroad promoters' wars and the excessive civic pride of Ohio Valley towns (often the pride seemed to run in obverse ratio to discernible assets) resulted in competing lines paralleling each other for miles in competition for a freight tonnage which could have been easily handled by a couple of the Studebaker boys' new South Bend wagons.

By the beginning of the nineties there were railroads for all—many of them in the hands of receivers for the receivers for the receivers—and the cause of progress had been served, apparently to the ultimate. A figure reported for 1903, only a little later, shows 213,000 miles of track in the United States—more than four and a half yards each for every man, woman, and infant in arms in the country!

The railroads had become the symbol of progress and the man who used other means of transportation betrayed himself as green, as reactionary, and as a traitor to his home Commercial Club or town council which had invested local contributions and even public funds in precarious bonds or stocks in order to bring the railroad to his door.

Steamboating was dead; the river hadn't been "brought" to town, it was always there. It was slow; it was not of the twentieth century; it lacked "élan" and it was in no way "recherché," as the Midwest understood these terms after the World's Columbian Exposition at Chicago. The Spanish sideburns of the fifties had given way to pomaded hair parted dead center; the banjo had made way for the mandolin and alabaster shoulders were covered by leg-of-mutton sleeves. Steamboats rotted at the levees and the river was all but deserted except for rickety coal barges, catfish magnates' floating shanties, and an occasional excursion boat, aflame with Floradora girl skirts, shirtwaists, straw sailors, bobtailed coats, brown derbies, and bamboo canes.

So things continued—save for changes in the fashion of costume —until another war in Europe gave an added push to upper Ohio industry. Actually, traffic upon the Ohio dropped still further immediately before the United States entered World War I and tonnage on that stream reached a low of 4,598,875 tons in 1917. War influences were, nevertheless, preparing river traffic for a comeback.

Wartime shipping was mainly from Atlantic ports—New Orleans had to await another war to regain her ancient importance —but the failure of already run-down railroads under wartime stress and subsequent government operation created a demand for supplementary freight service when hostilities ceased. In the prosperity that followed 1918 freight tonnage on the river began a rapid increase.

The rolling stock and tracks of the railroads were in a bad way and some active mind (perhaps that of a long-time river fan) recalled the river traffic of the past and called attention to the fact that a barge could be built more cheaply than a string of coal cars totaling equal capacity, and a towboat as quickly as a locomotive. Soon ancient river pilots—their faith in the comeback of steamboatin' at last justified—were being hoisted out of their front-porch rocking chairs, dusted off, and assisted up to the pilothouses on new and small but powerful towboats.

In 1922 Congress, probably somewhat alarmed by the catastrophe that had almost overtaken the American transportation system during and after World War I, voted an appropriation of $42,000,000 to the War Department for the improvement of rivers and harbors and the next year the appropriation was increased to $56,000,000. The Ohio received its share—to complete the program for canalizing the river, which had been proposed in 1875!

Even the prospect of improvement had its effect. Freight hauling made an immediate jump; supplementing the few new ones, ancient towboats and resurrected pilots must have worked wonders, for in 1923, 1924, and 1925 the tonnage exceeded, progressively, ten, fifteen, and nineteen million tons.

Canalization of the Ohio was completed in 1929 and was

marked by ceremonies held in Cincinnati which featured the dedication by President Hoover of a bronze plaque commemorating the occasion. The plan thus brought to fruition had been somewhat changed and refined from that proposed years before by Major Merrill: a depth of nine instead of six feet had been established and forty-eight dams comprised the new system, although later construction of the Gallipolis Dam eliminated two others immediately upstream and reduced the present total to forty-six.

All but four of the Ohio River dams are "movable," that is, their wickets—made of huge timbers and placed across-river much like a board fence—are hinged near their lower ends so that they may be dropped to the bed when the flow of water reaches a certain depth. When the wickets of this type of dam are down lock operation is not necessary, as boats are able to pass directly over the dam structure.

With the river at ordinary stages, however (or "in pool" by the language of the riverman), boats must be raised or lowered to the next pool through the locks, which are gated concrete structures abutting each end of the dam. Passage through locks is necessary at all river stages in the case of the four "fixed," or nonmovable, dams located at Emsworth, Dashields, and Montgomery, Pennsylvania, and at Gallipolis, Ohio.

A fairly complete study of the most efficient methods man has yet devised for regulating a river to produce navigable water could be carried out by inspecting the construction the Corps of Engineers has completed from end to end of the Ohio; a construction which permits navigation throughout the year with only slight interruption during the ice and flood periods.

Ice, like floods, the Ohio will have always with it: nowadays it is not often serious, for the variable temperature in the valley seldom holds a subfreezing level for long and the dam and weir structures are designed to render an ice jam far less spectacular than it was in the days when Harriet Beecher Stowe's Eliza leaped from floe to floe. Even so, every second or third winter still produces some incident which lends itself to dramatic newspaper treatment. No longer ago than January, 1948, eleven barges loaded with 406 hard-to-get automobiles were icebound in the middle of the Ohio above Evansville, Indiana, for several days—while photographs of them stirred up unrest among newspaper readers still car-hungry in the postwar shortage.

Back in 1929, when the work of establishing the minimum 9-foot depth was completed, the Ohio was already carrying over twenty-one million tons of freight a year. As more and more industries saw the advantage of cheap river transportation for heavy freight, the total grew steadily with only slight declines during depression years: in 1941 it passed 36,000,000 tons, and in 1942 the wartime high 38,280,812 tons was carried—enough, probably, to sink all the side-wheel packet boats that ever plied the rivers of America!

Of course there was a slackening after wartime pressure but the annual total fell only a little in the first postwar years, then the increase began again and in 1947 (last completely reported to date) an all-time high of 41,396,798 tons was established. There is every indication that 1948, when the total is computed, will exceed even that!

New and profitable uses for river transportation are being discovered. Automobiles for the southern market or for export from New Orleans are assembled by plants of two large manufacturers at Cincinnati and Louisville and cars of other makes can be easily driven or trucked from the Detroit area to Cincinnati, Madison, Indiana, or Louisville to be shipped by barge down the river at a fraction of rail or trucking cost. (All of those eleven barges carrying 406 cars which were icebound were being pushed—with a twelfth barge of miscellaneous freight—by one small diesel-engined towboat which, incidentally, went right ahead to complete delivery after a few days' delay.) But even four hundred automobile tows will soon be obsolete. A newly developed multidecked barge will carry, on the several decks of its 600-foot length, six hundred cars of average size.

This great highroad of American commerce, the modern Ohio River, is maintained and operated by two branches of government service. The United States Coast Guard, under the direction of the Treasury Department in time of peace and as a part of the Navy in wartime, is charged with the "construction, operation, mainte-

nance, repair, illumination and inspection of aids to navigation"—
that is, the lights, signals, channel markers, etc.—and acts as a
lifesaving, distress-relieving, and policing agency on the river. An
unobtrusive branch of the service—and as powerful and as positive
in carrying out its duties as such unobtrusive individuals and
services are likely to be—the Coast Guard is always on the job but
never fully appreciated, and seldom the object of any romantic
journalistic attention until disaster strikes.

Equally modest in its demands upon the public notice has been
the Corps of Engineers, whose province it was to make the original
plan for perfecting the river for navigation and which was charged
with the construction and maintaining of channel aids and flood-
control measures.

Far too little credit for the present importance of the Ohio
River as a national asset has been given to the Corps of Engineers.
Its sane planning, scientific study, and wise expenditure of often
very limited funds nursed the project through more than a century
before the aim of a constant minimum of water necessary for
navigation was achieved: the efficiency and vision of its personnel
continues to maintain and to plan greater improvement for the
future. Few casual tourists in the valley—not many of even the
actual dwellers upon the river—realize who is to be thanked for
both the advantages of easy navigation and the protection of
property from flood damage. Major General J. C. Mehaffey, Division
Engineer, describes the responsibility of this proud branch of the
service in a statement which is in itself far too modest:

The Ohio River Division's major mission, as a civil works agency
of the Government, is the investigation, construction, maintenance and
operation of all navigation, flood control and related improvements
within parts of 14 states comprising the Ohio River watershed.

The Division maintains administrative supervision over four District Engineer Offices located at Pittsburgh, Huntington, Louisville, and
Nashville. Each of these is responsible for its own geographical portion
of the 203,900 square miles of territory included in the Division proper.

In connection with our responsibility for navigation and flood
control developments, I would emphasize one point:

The Corps of Engineers is the servant of the people as their desires are expressed through the Congress. We recommend to the Congress what the people want when the desired improvement can be economically justified, not what we think they should have. We may recommend a certain improvement or a given type of construction, but the residents of the city or other political subdivision concerned have absolute veto power. They have exercised this more than once, on many occasions proposing alternative plans, which, upon investigation, we have been happy to approve.

It is this responsibility to the people that has kept the record of the Corps of Engineers a proud one over the last 125 years since we first assumed custodianship of the Ohio River and its tributaries.

Although the statement fails to mention the fact, the work of the Corps of Engineers upon the river has been of far more than simple peacetime importance. It played its part in every war after 1824—a truly enormous one in the Civil War—but the river it had opened to year-round navigation really came into its own in the nineteen-forties. Then it was that the Ohio became an item of top-drawer importance to all the peoples of the world, for now the port of New Orleans, in the protected Gulf of Mexico, was the door to the fully developed production of the Middle West and the only dispatching point equally accessible to the Atlantic and Pacific theaters of action.

Perhaps the United States War Department was uninhibited by a knowledge of some of the early unfortunate efforts to build oceangoing ships on the Ohio—"They have built some ships," said Edmund Montule of the Ohio in 1817, "*one* of which was fortunate enough to descend the Ohio without accident"—or perhaps the War Department recalled the success of James B. Eads, formerly of Lawrenceburg, Indiana, in applying protective armor to river boats during the Civil War. In any case, with coastal shipyards working a seven-day overtime week to produce thousands of Liberties, C-2s, and other large cargo vessels, to say nothing of even heavier aircraft carriers, orders were placed with Ohio River firms to build not only such ships as harbor tugs and vessels, but mine layers, mine sweepers, even LSTs and naval dry docks. One Pittsburgh-built LST spearheaded the invasion of Leyte, being lost on that faraway beach only four months after it was launched seven miles below "the Point" at Pittsburgh. Various Ohio ports successfully launched such ships and dwellers on the river were progressively more amazed by the procession of strange vessels that passed downstream, as the war effort gained momentum, bound for every sea on the globe.

The subject of the Ohio's part in World War II, however, seems worthy of an official statement. The Corps of Engineers, in a postwar report upon the the record of the river said:

Actually, the Ohio River served a twofold purpose to the country's war effort—it shared . . . in the transportation of strategic materials and made possible the growth of manufacturing processes that normally would have been forced into congested coastal areas.

The most outstanding movements of traffic . . . related to the supply of petroleum products. In the early years of the war, the submarine menace on the east coast and the transfer of many tankers to the naval service seriously curtailed the usual supply of petroleum on the eastern coast. . . . Fleets of oil barges . . . were towed upstream to refineries on the Ohio River. Inland cities such as Pittsburgh now became supply instead of receiving points for these products . . .

From a strategic viewpoint . . . second in importance to gasoline and crude oil was . . . sulphur, necessary for explosives . . . barge loads were towed up the Mississippi and Ohio to industrial centers . . .

Steel centers such as Pittsburgh, and others . . . required vast quantities of coal . . . One single plant in the Pittsburgh area consumed 40,000 tons of coal each day, all of which arrived by water. Eight hundred railroad cars would have been required each day to supply that plant alone . . .

But tons of commodities shipped . . . tell only a part of the story. Existing coastal shipyards were filled with repair and construction jobs. . . . By slight modification the Marine ways along the Ohio River were altered to accommodate the hulls of all types of landing craft and the smaller sea-going vessels with a draft of 11 feet or less. Over 1,000 craft destined for deep water service were thus constructed near the sources of material. . . . Passage of these vessels to tidewater presented some problems in the normal 9-foot channel of the Ohio River. Those that were ready for delivery during open river stages in excess of 11 feet experienced no particular difficulty, but during lower stages it became necessary to mount some on pontoons, while others were passed from dam to dam by holding higher than normal pools and releasing the excess water at intervals that would create a series of artificial waves. Even ships that would have exceeded any possibility of passage through the Ohio River channels were assembled at coastal shipyards from sections that were built at these inland sites. Nearly 44,000 tons of such sections were contributed from these plants.

The Corps of Engineers and organizations like the Ohio Valley Improvement Association—composed of business, labor, and civic leaders representing the cities and states on the river—are by no means content to rest upon laurels already earned.

Plans call for a future increase of the navigable depth to twelve feet and for the opening of a standard channel varying from 500 to 750 feet in width from Pittsburgh to slightly above the junction with the Mississippi. The maximum tow upon the Ohio today consists of a small but powerful diesel-engined towboat pushing 20 to 22 barges, *the tow covering an area approximately 1,200 feet long by 105 feet wide:* that, for comparison, comes fairly close to a quarter of a mile long by a third of an average city block in width —or only a little less than three acres in area! The deeper channel as proposed will greatly increase safety factors and facilitate maintenance operations. The advantages of a 12-foot minimum depth are obvious: it would permit greater loading of present barges, encourage the building of larger and more efficient ones, and make for even more economical boat operation.

In view of all these facts accomplished and plans made for the future one gets the impression that it is safe to state that navigation of the Ohio is here to stay.

It is considerably less expensive, rivermen love to point out, to dredge an occasional hundred yards of sand bar than it is to relay an equal distance of railroad track or of hard-surface road. There's now a far quicker way to carry mail or passengers, upstairs, than either railroads or motorcars can ever achieve and railroad men are gloomy over rising costs. On the water diesel power is cheaper than on rails and crews are small; in competition for heavy shipping when the hurry's not too great the difference in cost is enormous— $6.02 per ton of oil from Baton Rouge to Pittsburgh by river in 1948 as compared to $12.62 per ton for the same haul by rail; this does not cheer railroaders to any noticeable extent, but rivermen and river-boat fans find it quite encouraging!

Chapter Twenty

COLORED WATERS

Two centuries have passed since Céloron de Bienville moved his caravan from Canada to the Ohio's head and thence floated downstream to claim the river for France's glory—and her hoped-for profit—by burying those leaden plates on the eroding shores and nailing the painted arms of France to trees along the banks.

Today the river has been free of European rule for but a little less than those two hundred years and not much more than a century has passed since William Henry Harrison, first president elected from the Ohio's shores, swept an election which gave to him the office and to the valley recognition as a power within the nation.

Changes have come upon the people of the valley—their number, wealth, and way of living—no doubt of that, and the River

herself has changed in superficial ways; both her waters and the
air above them are rather dirty in places and she has been corseted
and molded to a shape to please the fancy of contemporary man.
Nowadays there never comes a season in which her waters sink so
low—thanks to the U.S. Corps of Engineers—that she cannot muster
the full minimum nine feet of water necessary to maintain her
traffic and even the annual spring tantrums by which she used to
express her impatience with the ending of winter have been tamed
somewhat in the process of refining the old girl.

The population on the river is more numerous—say one thousand now to one in Céloron's day—which means that her admirers
and those who gain a living through her bounty are counted now
in millions rather than thousands—but certainly there can be no
doubt but that the standard of living maintained today by the least
prosperous roustabout upon the landing in her most unprepossessing
port includes comforts unknown two centuries ago to Céloron de
Bienville himself.

Each of the three greatest cities on the Ohio's banks is larger
now than any French metropolis save Paris in the day France made
her claim and each year they, and the smaller communities along
her course, produce a quantity of goods greater than did all France
and all Great Britain in the day of Céloron. Every town, and almost
every homestead between towns, employs the wonders of the
harnessing of nature—power from steam, oil, electricity—and uti-
lizes all the once rare manufactured products such power has now
made commonplace and put within the reach of even the least
of men.

Modern commerce has built great cities on the river—and in so
doing has left many a town, fully as promising in its beginnings,
to stagnate and decay; Ohio River traffic has developed to prodigious
tonnage; population has multiplied—but has there really been much
change along the shores since, say, the days of Harrison, except in
size of settlements and improvements in the mechanics of living?

Does not life on the river and in the towns maintain an
obvious link between the present and the past—as the professional
vocabulary of the modern diesel towboat captain harks backward
to that of the keelboat boss?

There is, indeed, an obvious link. The modern water fronts, the ferryboats, the earthen levees and the design of barge and towboat, skiff and packet, are easily recognizable developments of earlier day. (Why, the towboat *R. J. Nugent* even had a little mutiny in the summer of 1948!) There still are eddies, pockets, backwaters, and bayous on the river which have remained unchanged for centuries; in which the traveler may be hard put to decide whether, through some mischance of time, he may not have arrived in the day of the river pirates, at a quiet moment in the War of 1812 or, frequently indeed, in the midst of the "War Between the States"—as many residents on the southern shore still prefer to call the misunderstanding of the eighteen-sixties.

The small screw-driven, oil-powered towboat cannot command the majesty of the gigantic frosted wedding-cake packet of the glorious fifties but, itself little larger than a trailer truck on the highway, it shoves a "tow" of barges loaded with freight enough to sink a dozen of those earlier, gaudier monsters or to load several dozen railroad freight cars! Even though these Davids of the modern river handling their Goliath-size loads of freight are less spectacular than were those elegant packets, it still takes purpose and concentration beyond the common lot to keep a river dweller— or a stranger, for that matter—from dropping whatever work he may have at hand to watch a towboat's passing out of sight.

Commercial powercraft have surely changed, as has the kind of freight they haul, but that does not apply to lesser vessels. Those narrow, flat-bottomed, square-end skiffs tied up to willows by the thousands from one end of the river to the other must look exactly as did those in use when early settlers first brought tools to rive or saw the planks that shape their simple pattern.

And shanty boats? Shanty boats are as elegant, at best, as that which Aaron Burr had built to serve as headquarters for whatever move it was he planned; and at their worst they remain as crude in design and as disreputable in ownership as were the first of the early "store boats" that floated meager stocks of pewter buttons, shoddy stroud, and brass pins down from Pittsburgh.

Driftwood looks much the same, as it comes bobbing down floodwater—for it is hard to tell, from shore, whether a frayed

plank was split by wedge and mallet or sawed by electric power—
and drift still lodges, until it is pulled or blasted out, to make a
navigation hazard which the pilot of a diesel towboat fears as much
as did the owner of an 'Orleans-bound flatboat.

Small boys still paddle up and down the stream on rafts and
angle for catfish; raw sewage and the waste from mills has driven
off most of those other species which Constantine Samuel
Rafinesque-Schmaltz saw, or thought he saw—but sewage doesn't
show much, on the surface, and catfish thrive on it and boys
survive anyway. This sewage-waste disposal matter is to be solved
on the Ohio, for West Virginia, last of the great states on its shores
to give thought to the problem, finally agreed in 1948 to guard the
waters of the river against pollution. The time is no doubt imminent
—say within the next century or two or three—when the fine words
of the Ohio River Compact, recently endorsed with considerable
fanfare by the governors of all the affected states, may actually
come to have some measure of meaning in preserving the purity
of the water.

Not that the river will ever become crowded with game fish;
great rivers, polluted or not, just do not suit their finicky tastes—it
is the swifter, colder tributary streams that always were and will
continue to be the favorite haunt of bass, pike, pickerel, and all their
relatives. In spite of stories to the contrary, the Ohio was never a
haven for the sporting fisherman. Christian Schultz, traveling down
the unpolluted, still comparatively untouched Ohio of 1807, said,
in regard to his efforts:

> The Ohio, as yet, has not produced any fish . . . I am inclined to
> believe it is not so well stocked with fish as it is represented to be . . .
> I have been assured, by a respectable gentleman, that many [catfish]
> are frequently caught which weigh from fifty to ninety pounds. I have
> seen a few cat fish, buffaloe fish, perch, chub, suckers and herrings;
> but no sturgeon or pike . . .

The people in the riverside villages, where connection with
the world is still shared only between a riverside highway and the
river itself, are a bit more leisurely than their neighbors ten miles
back from the stream. You bide the river's own good time, when

you live on the river; you don't plan anything to conflict with the date when the ice may go out or when the spring rise is due—you wait to see whether the ice is going out peaceably or with one of those grinding, screaming breakups of which the Ohio is occasionally guilty; whether the spring rise is to be a rise only, flooding bottom lands and jamming up driftwood, or whether it will continue up into the second story of the Odd Fellows Hall on Main Street.

Much of the architecture on the Ohio is in harmony with the historic past. Houses and buildings date from all periods since shortly after the end of the Revolution, and some near the river carry a high-water mark from the last preceding rise, as they have ever since the first spring season when their builders discovered they'd selected sites too close to the river. Many towns retain evidence of their original character. If they were settled by respectable people whose chief aim was to make a home in which to live and rear and educate their children, they are still that kind of towns; if they developed from less auspicious beginnings, *that* mark shows too.

This last is a peculiar and interesting fact: visit the Ohio towns and list the ones which appear the least salubrious in the matter of cleanliness, with the most obvious evidences of moral turpitude and ignorance on the part of the inhabitants. When the list is completed skim through the descriptions of river towns by early travelers and you will note that the majority of what were reported to be the more disreputable and depraved settlements in the eighteen-twenties and -thirties give every evidence of maintaining their early standards today! Certainly that is not invariably true, for frequently some God-forsaken little river hamlet, which appears at first glance as though its only conceivable contribution to society might be a particularly bounteous crop of juvenile delinquents, turns out to be the childhood home of dozens of distinguished citizens, and several highly respectable communities have grown on most unpromising soil. But there remain enough still-disorderly towns with equally shady pasts to make the general rule hold good.

For instance, there is Ironton, Ohio. Not greatly admired in the previous century, it received sordid publicity as the center of a white-slave ring in 1947; there is the section of Covington, Kentucky, in which gambling establishments flourish in an atmosphere of Old Kaintuck hospitality and German-American comfort with a touch of the triggerman tradition, Chicago style; there is sad, moldering degeneracy on the river front of Jeffersonville—described as less than promising a dozen decades back—where businesslike gambling establishments (sans Covington hospitality and comfort) and dingy marriage parlors jostle a superabundance of taverns. Adjoining Clarksville, like its original proprietor George Rogers Clark after his fall, supports some handsome old façades backed by rat-eaten and rotting interiors. Brandenburg, Kentucky, while staid and respectable, still has a glint in its communal eye which reminds the visitor that in the sixties it was the headquarters of General John Hunt Morgan's raiders.

On the other hand, Pittsburgh can easily be identified as the smoky, busy city it was said to be a century and a half ago—only a hundred times more so; Wheeling is immediately recognizable by its strategic location at the junction of the river, the National Road, and railroads which marked its early day; a ten-minute stroll through Marietta's residence district establishes its New England origin; Maysville, Kentucky, looks handsome from the river much as early travelers described it. Cincinnati fairly exudes an atmosphere of financial stability; Carrollton, Kentucky, and Madison, Indiana, still qualify as "thriving," and Louisville's location is as handsome as its early publicity maintained.

Evansville gives every promise of becoming an even more important industrial center than its promoters anticipated. With Owensboro and Paducah, Kentucky, it is forming a new manufacturing and distributing district on the lower river around the very spot from which the Rappites first carried on those branches of commerce in the West.

As a matter of fact, about anything one cares to seek can be found upon the Ohio—within certain climatic, ethnic, and geographic limitations, of course! There is the matter of restaurants, for instance. Cincinnati, Pittsburgh, and Louisville maintain some of the best dining establishments in the country, while a much smaller river town, which may as well be nameless, rejoices in only one eating place for its more than one thousand inhabitants.

It is not Maude's Cafe, but it is pretty close to that, and its exterior is marked by a sign which reads "no Pankin BusStop." Inside there are four dining tables of various styles, a counter, a mission-oak table stacked knee-deep in unopened advertising circulars, and a desk of similar material and design similarly burdened. The walls carry four family portraits, four pictorial calendars, and posters without number while the counter is set off by two large trays of highly colored plaster fruit and six punch boards with valuable prizes prominently displayed.

The staff includes Maude, fifty, plump, and the blue-blackest brunette on record; a grandmotherly lady who does not wear her teeth while on duty, a buxom 16-year-old, obviously a daughter of the rural gentry fallen on evil times and unable to conceal her distaste for the commonalty of diners. There is also one male, a sort of sad-eyed Frankenstein's monster who is mainly preoccupied with toothpicking and hairdressing.

The luncheon menu includes boiled beef, cold mashed potatoes, cold navy beans, cold canned corn, wilted head lettuce fragments, a stewed pear, and a cup of fluid of the color but not the taste or aroma of coffee. The pear is very good—after all, who could misstew a pear?

Someone's baby is on the other side of the 6-foot partition (labeled "Employs Onely") that bars the kitchen from public view. The baby is probably not an Employ and possibly for that reason it occasionally frets audibly. Someone then coos, damp pats are heard resounding on what might be plump baby anatomy, and the baby gurgles. Of course the unseen person who is solacing the baby may not be an Employ either, so there is really no cause for alarm.

On the other hand, many small towns on the river boast restaurants which offer inspired cooking, the best of which usually employs the steak, chicken, ham—even the catfish—of the neighborhood in proper combination with valley butter, eggs, milk, vegetables, and fruit. All are grown, marketed, and processed with know-how passed down from the most accomplished of the early settlers added to some worth-while improvements borrowed from the good-living German immigrants who appeared in force during the eighteen-forties—with an occasional contribution by more recent arrivals from other lands.

In the economic structure that results in modern American plenty, agriculture must always come first. Following it come iron, oil and coal, with gas, clay, stone, aluminum, sand, and salt in about that order of importance. To the production, refining, and fabrication of these products even the most biased proponents of lesser rivers must admit that the Ohio contributes much.

It is a proud thing to take this part in maintaining the American Way—but the fact must be admitted that such an excessive contribution of coal and iron and chemicals as is made by the upper Ohio has a tendency to cast a blight upon the landscape. Quite a few of the cities and towns upon either bank of the river between Pittsburgh and Portsmouth have the appearance, by night and by day, of studies which M. Doré might have made while preparing the illustrations for *Dante's Inferno*.

Most beautiful stream in America the Ohio certainly is—where man has let it alone! Where he has exploited its hidden wealth it remains at least picturesque (a term long ago discovered to be agreeable for use in cases like this) and its products have indubitably added enormously to the wealth, comfort, and welfare of mankind throughout the world.

Modern Pittsburgh still answers the description that the astute and farsighted authority, Zadok Cramer, published in 1814: "It is a place of note and celebrity not only in America but even in Europe. The traveler, however, on entering it for the first time, meets with some disappointment. The town is enveloped in clouds of smoke which even affect respiration . . ." Though, unlike some of its less thriving contemporaries, there is little to indicate the fact, Pittsburgh is quite an old city as things go in the West. In 1760 a small settlement grew to the east of Fort Pitt—its inhabitants, according to contemporary accounts, varying from respectable to the extreme opposite. The town was laid out five years later, after Indian incursion seemed no longer to threaten, and was surveyed on its present plan in 1784. Its growth was not spectacular until

1793 but by 1800 it claimed twenty-four hundred people and Cramer estimated a possible eight thousand in 1814.

Well before the latter date Pittsburgh's future was settled; it had become a manufacturing city and, this fact established, it drew capital from the readily accessible money markets of New York and Philadelphia to develop not only the resources in its immediate neighborhood but also those far down the river. Of course the place had its ups and downs; several disastrous fires—the worst of which burned over 56 acres of the "most wealthy and business portion" of the city in 1845—and the inconvenience occasioned by the wars waged in the reasonably near vicinity during its first century. In spite of these distractions its progress and improvement was steady.

Although Pittsburgh was the home of the first novelist who wrote in the West and saw the country's first significant composer into his young manhood, the place was not particularly given to the pursuit of either learning or culture until the end of the nineteenth century when some of the enormous profits of its industry began to be invested in libraries, museums, and educational institutions. When the matter of culture was finally brought to the attention of the energetic Pittsburghers, however, they took to it as they had taken to industrial promotion and with similar success.

Pittsburgh today certainly lags behind no place of similar size in its enthusiastic and intelligent patronage of the arts and of learning. The city has the University of Pittsburgh, prospering in its 42-story "Cathedral of Learning." Regardless of whether one cares to undertake to verify its apparently rather tenuous claim to having been established in 1787, one must admit that the University of Pittsburgh is one of the nation's great educational institutions today. There are also Duquesne University, a Roman Catholic institution founded in 1878, and Carnegie Institute of Technology, which was planned in 1900 and opened its doors in 1905.

The Carnegie Institute, a group of cultural centers which includes the Carnegie Library of Pittsburgh, the Music Hall, Carnegie Museum, and the Department of Fine Arts, was established in 1895. Andrew Carnegie's money did extremely well by the city in which he rose to power—as, later, did that of the Mellon interests in the matter of Mellon Institute, connected with the University of Pittsburgh; as also did the lesser fortunes of dozens of other citizens.

Try as one will, it is impossible to identify any distinctive atmosphere—except that obvious one of success—in modern Pittsburgh. Upon the point formed by the meeting of the Monongahela and the Allegheny Rivers, the "Golden Triangle" as the citizens would have it, there stand a good many high buildings but they might be a small sampling of those of New York or Chicago; the success of the city's industries has attracted workers from every state in the Union and every continent on the globe so that, while a virile population has resulted, there is no cultural unity (unless its citizens are correct in maintaining that, beneath the cosmopolitan surface appearance, Scotch-Irish Presbyterianism still holds sway). The greatest asset about which Pittsburgh can boast is the efficiency of its industry and, per se, it is not efficient to waste time in boasting! Owing to intelligent regulation, Pittsburgh has even lost one of its earliest distinctions; there are today half a dozen midwestern cities smokier than is "the Smoky City" itself! Pittsburgh and the surrounding area is unquestionably the industrial heart of the nation—but it is the plain truth that it, like the human heart, is not a particularly handsome organ, no matter what its importance!

Through many miles the towns and cities on the Ohio below Pittsburgh partake of the characteristics that mark that place itself. Almost to the Ohio line the river is lined with manufacturing and processing plants, railroad lines, and grimy warehouses, the limits between towns indistinguishable. Most of them have mercantile districts set back from the river and the residential sections spreading up into the hills—so it is through Bellevue and Coraopolis, both residential suburbs of Pittsburgh before industries came to them—and so, also, is the pattern generally followed by industrial communities down to Portsmouth. After that, for some reason, industry has a tendency to locate back from the river; railroads come to but not necessarily along the river, and residential areas often occupy the riverside.

The little city of Sewickley, twelve miles downstream from

Pittsburgh on the north shore, has maintained its individuality to a far greater extent than has any other community in the highly industrialized upper fifty miles of the Ohio. Like many another well-located river town it rests on a shelf above the flood level which had been occupied by mound builders and modern Indians before whites settled upon it after the Revolution. Even proud little Sewickley (it is now famous for its Hunt Club, its horse and dog shows, its fine library, and the homes of Pittsburgh business executives) is willing to admit that the early settlements on the present townsite bore names somewhat suspicious—Dogtown, Contention, Devil's Race Track—but they were organized into one community in 1840 and presumably the rivermen who ruled the town in its early day saw to it that the elements responsible for those names were either reformed or removed, for Sewickley is and has long been a community properest of the proper.

Ambridge, next on the river, is a town as different in origin from Sewickley as might well be imagined, but with plenty of interest in its own right. "Ambridge," as is easily seen, derives from American Bridge, the American Bridge Company having laid out and developed the town. The site is that of Economy, the Rappites' second community in Pennsylvania. As has been already related, that society died with the nineteenth century, leaving its somewhat involved affairs in the hands of a faithful trustee, Dr. John Samuel Duss. With the passing of the last of the members for whose support the property had been maintained, the valuable tract of community land was sold to the bridge company, six acres surrounding the principal Rappite buildings being reserved.

Thus Ambridge, modern industrial city, which boasts the largest fabricating steel plant in the world, has within its limits a memorial to the successful but foredoomed industrial community of Father George Rapp and his followers.

The river passes Aliquippa, a steel city of some thirty thousand inhabitants, which occupies several miles of the south bank below Ambridge and about opposite the site of old Logstown, establishment of the Indian traders in the eighteenth century. There is a bend which directs the current more toward the north until it reaches the town of Monaco on the south bank and opposite, at the mouth of a stream of the same name, lies Beaver. From 1778 to 1785 this was the site of Fort McIntosh. In 1814 Zadok Cramer claimed that Beaver "has nothing in it to invite settlers," then promptly contradicted his pronouncement by adding that "Mr. George Grier, whose beer is esteemed at Natchez and New Orleans" maintained a brewery there.

East Liverpool is the first Ohio city on the river. With a population of more than twenty-five thousand it has utilized nearby deposits of clay, flint, and feldspar in the manufacture of clay products since 1839. The industry today produces electrical porcelain and pottery supplies as well as its earlier specialties of semivitreous porcelain tableware and brick. The city has an industrial payroll of more than twenty million dollars per year and supplies a large trading district with merchandise. Newell and Chester, across the river in West Virginia, are a part of East Liverpool's metropolitan area.

The river is running only a little south of west when it crosses the state line from Pennsylvania to Ohio but actually it is at the point, at last, of making the decision to flow to the Mississippi rather than to Lake Erie. Just past East Liverpool it turns to begin the longest southward stretch of its course, defining one side of the narrow panhandle of West Virginia, which juts up to enclose almost half of Pennsylvania's western border.

Wellsville and Yellow Creek are the next towns below East Liverpool on the Ohio side, still in the clay country.

Toronto, which flourished in the early nineteenth century under the name of Newburg Landing, occupies a level shelf high up on the Ohio shore. Under its present title, Toronto specializes in the manufacture of sewer pipe and casts a hopeful eye toward the light-steel fabricating industry, which has been spreading into new locations in its recent efforts to decentralize.

First place of importance on the West Virginia shore of the upper Ohio, and newest city on the river is the city of Weirton, which, with something over thirty-five thousand population, stretches back from the Ohio into the hills. Weirton's business is tin plate and steel—men in the location have dealt in steel, or rather iron, since Peter Tarr erected an iron-ore furnace on the site in

1790, although there was a considerable break between the termination of Tarr's venture, some time after the War of 1812, and 1909 when the Weirton Steel Company purchased the land, set up its plants, and laid out the town. Other communities grew up around it and it was not until 1947 that these joined to form the present corporation.

Being the newest city on the Ohio, Weirton should be, and is, among the cleanest—certainly *the* cleanest steel town—and the passer-by gains the impression that its citizens are rather more prideful of their community than is common. Apparently they have just reason for their pride; Weirton is unusually well supplied with recreational facilities, for its neighborhood, and it is able to claim that more of its children complete the full twelve years of public schooling than those of any other city in West Virginia.

Then, on the Ohio shore, comes Steubenville. Slightly larger than Weirton, it is also dirtier, but if it is a bit less prepossessing than its neighbor it also enjoys the distinction of having greater importance.

Nothing is left to suggest the fact to the visitor—except for a few sooty plaques and statues—but Steubenville's beginnings were early indeed for the trans-Allegheny country. According to local historians, one Jacob Walker bought the land on which the town stands in 1765 (although just who could have given a merchantable title to it at the end of Pontiac's War is a question) and he presumably settled on it. In 1786-1787 Fort Steuben was built on the site preparatory to the opening of the Territory North-west of the River Ohio and a settlement known as La Belle sprang up as the fort was building. Claims are made that La Belle rather than Marietta was the first town settled by Americans in the Old Northwest—but the Mariettians are a race much more interested in maintaining their own priority than are the people of Steubenville, and there is probably little to be gained by backing La Belle's claim at this late date.

Manufacturing began very early—in fact, Steubenville may be considered the first factory town in the state of Ohio—with pottery, coal mining, ironworking, boatbuilding, and the weaving of woolens occupying the populace about equally until shortly before the Civil

War when a rolling mill was opened. From that time on coal and steel dominated the scene, although glass and tin manufacturing enjoyed a boom and there is a comfortable variety of small manufacturing. Steubenville has a rather impressive business district, reflecting, perhaps, the superior agricultural prosperity of the Ohio back country over that upon the West Virginia side and, farther down, a good bit of the Kentucky shore. The city—as does Weirton —takes particular pride in its schools, its recreational facilities, and its residential districts. It claims to have had "the first classical school in the West" but since the school was established in 1814, that claim can scarcely stand; Kentucky had a half dozen or so before then. Perhaps Steubenville should be pardoned in view of the fact that so few "firsts" were left to be appropriated after Marietta had made her boasts.

It is sometimes difficult, when there is no breeze and the smog hangs heavy enough to obscure the highway signs, to distinguish one town from another in the coal and steel belt; traveling on the river gives no advantage because the shore is often enough only a slightly darker gray to right and left. Within the next few miles, however, the river flows between Wellsburg, West Virginia, and Brilliant, Ohio—which is a community as inappropriately named as any in the world!

Immediately below Steubenville, identifiable or not, lies Mingo Junction, Ohio. The Mingoes, who once had a village here, would now neither recognize the place nor care to return to it. They would find Yorkville, twelve miles below, even less desirable as a campsite, for Yorkville is one great smoking slag heap which would certainly offend the sensitive noses of a race which named an area "Chicago— The Place of the Stink" only because it was redolent of wild onions!

Martins Ferry, still on the Ohio side of the river, is larger, slightly cleaner, and considerably more prosperous than its near neighbors. It, too, had a claim to being the first settlement by whites in the Territory North-west of the River Ohio but the claim has been withdrawn under the haughty stare of Marietta's more militant citizens. Actually Martins Ferry does not seem to be sure of the name of the predecessor upon its site which is said to have been christened in 1785. Some local authorities refer to it as Mercertown, some as Norristown—under either title the inhabitants occupied the land illegally and were chased away after a year or so. The present city was not laid out until 1835. William Dean Howells was born in Martins Ferry and Zane Grey chose the place as the locale of a historical novel before he found his true spiritual home in the Purple Sage and Buckaroo country.

Wheeling, across on the West Virginia side, is a community which has outgrown its lively—and in spots somewhat disreputable —early reputation to become a modern industrial city of seventy thousand population. While its railroads are routed along the water in the conventional upriver pattern, its business district begins within sight of the river.

U.S. Highway 40, the National Road of history, crosses the Ohio here and through the first three or four decades of the nineteenth century vast quantities of goods arrived by this road from Maryland and Virginia ports to be transshipped by boat to the West. The city was also second in importance only to Pittsburgh in its facilities for building emigrant boats and outfitting emigrants. The business of warehousing and wholesaling merchandise continues, although goods are now largely distributed to the South, rather than the West. Wheeling is also a considerable retail shopping center and the mercantile business as a whole gives the place a somewhat different aspect than have those cities in the neighborhood which depend solely upon manufacturing. It proclaims itself "Playground of the Central West," which would appear to take in considerable territory but it is indeed well supplied with parks and playgrounds.

Wheeling's industry is based chiefly upon coal and iron and its extensive factories are as smoky as any in the valley, but the fact that breezes from the open river have access to the business district helps to keep it looking clean and prosperous. Wheeling no longer occupies the important place it once did upon the river but it is still a community of substantial and enterprising people who take a justifiable pride in their city's past and present.

Opposite Wheeling on the Ohio shore, Bridgeport, Bellaire, and Shadyside—with a combined population of around twenty-five thousand—string along the river over an 8- or 9-mile stretch. There is some independent industry but mostly the inhabitants are dependent upon Wheeling for a livelihood.

The more concentrated industrial manifestations upon the Ohio's headwaters disappear temporarily as the move downstream continues. Several coal-mining towns and several burning heaps of slag are still ahead but for the most part the sun shines again and green is a predominant color. As the mines and oil-storage plants cease, there begins a big sweep of farm land as the river turns west of south for the long reach to Marietta: it is country of little importance to the world economy—nowadays it offers only a minor production of corn, hogs, beef, and apples—but it is certainly a relief to eyes, nose, and lungs to reach it!

Principal city in this long stretch is Moundsville, at the mouth of Grave Creek on the West Virginia shore. Moundsville is named in recognition of the prehistoric burial mound in the shape of a cone almost 80 feet high and 900 feet in circumference—highest on the river. Present Moundsville is the combination of two pioneer villages, Grave Creek and Elizabethtown, and has a population of about fifteen thousand. Industry is varied but the principal product is the famous glass tableware produced by the Fostoria Glass Company. Early residents took considerable interest in the prehistory of the place and some pride in the artistic production of their mound-building predecessors—who, in truth, would have made exceptionally competent employees for the modern city's chief industry.

New Martinsville is located twenty-five miles below Moundsville and, after another ten miles, is Sistersville.

Sistersville has a good deal to talk about besides its rather unusual name. It was laid out in 1815 by two of the daughters of Charles Wells, who had owned the land before his death in that year. The sisters, Delilah and Sarah, were not Charles Wells's only progeny; he had twenty-one children in all—the twentieth of whom, a daughter, was named appropriately enough, Twenty Wells. The Wells family was not particularly notable for numbers in *that* neighborhood, however, for when it arrived there was already a group of three families which boasted a total of seventy-three offspring!

Sistersville did not thrive particularly until the late thirties, when some industry developed, and it had to await 1892 for real prosperity. In that year oil prospectors brought in the Polecat Well and a boom resulted which drove the more staid element among the citizenry well-nigh crazy. Sisters seem to guide the destiny of Sistersville, for it was another pair, Mrs. O. B. Ramey and Mrs. Joshua Russell, who persuaded their husbands to organize the company that drilled the first successful oil well.

To the surprise of early oil men, the Sistersville field failed to "play out" and continues in production. Even the original Polecat Well still produces after more than half a century and offices of eight major oil companies are located in the town.

There follows another long stretch of the river occupied only by an occasional village and then, on the Ohio shore, comes Marietta.

Marietta insists upon calling itself the "first organized settlement west of the Alleghenies" or, only little more conservatively, the "oldest city in the Northwest Territory"—overlooking in the first case some dozens of thriving communities in West Virginia, Kentucky, Indiana, Illinois, Michigan, and Missouri and, even in the second and more modest claim, ignoring such well-known municipalities as Detroit and Vincennes. The tendency to make such claims may be only an inherited manifestation of the same spirit which prompts the New Englander to ignore any settlement of America prior to that of the Pilgrim Fathers, but it is obviously in a rather extreme degree of error. How, though, can outlanders argue with the citizens of a community of whom it was said in 1814 that "their industry is proverbial as their system of life is economical, moral and religious" and, in 1850, that they "are noted for their morality and intelligence"?

The story of Marietta's founding and early history has already been told; today the city has a population of about seventeen thousand and a substantial economy based on varied manufacturing, a substantial wholesale and retail trade, and fortunes providentially laid by in the past.

There are plenty of handsome buildings, a few dating from the early day, many more from Late Victorian prosperity. While Marietta has always exhibited an immense interest in its past, that interest was not quite strong enough to influence the convenience of the moment, so that few historic sites remain in their original state. There are, however, blocks of large, comfortable, elm-shaded residences—which give the impression that one out of, say, seven was built around 1900 by a retired farmer or the president of the building and loan society.

Marietta is, in short, a beautiful and apparently a soundly prosperous city—though a very, very smug one!

Parkersburg, West Virginia, is a dozen miles below Marietta on the river and as directly opposed to the latter in the spirit and interests of its thirty-three thousand inhabitants as might be imagined. Parkersburg is a manufacturing city making rayon yarn, oil- and gas-processing and utilizing equipment, and glass. Founded in 1820, it is supremely unaware of its early history and interested but dimly in the future.

Meanwhile, though, it appears to do quite well in the present: if it lacks something in civic virtue, it at least produces needed goods efficiently and thus makes its contribution.

There follows a long stretch of the river with only two Ohio towns worth noting—Pomeroy, which had, sometime, a great coal and salt business and city fathers with vision enough to lay out a park overlooking the river along the main street, and Middleport, which profited from salt and coal to a lesser degree. Once an important victualing point for passenger packets on the river, a steamboat- and barge-owning town, Middleport later took a flier in steel and a hand in the transportation business as the home of railroad shops. Industry, nowadays, has practically disappeared and Middleport seems to find recovery from the occasional floods that ravage its river front to be increasingly difficult.

Fourteen miles downstream the Great Kanawha enters the Ohio on the West Virginia side and forms the triangle upon which Point Pleasant stands. One of the oldest settlements on the middle river, the point was the site of the famous battle between Washington's business associate, General Andrew Lewis, and his Virginians and Cornstalk's Shawnees on October 10, 1774, which ended Lord Dunmore's private war on the Ohio.

Five miles down and across the river lies Gallipolis, the tragic story of whose beginning has been told. Nothing remains even to hint at its unhappy origin; the little city has a population of about eighty-five hundred which appears to prosper from the agriculture behind and above and below it in Ohio. Except for a few French-sounding names in the telephone directory (and they might well be twentieth- rather than eighteenth-century émigrés) evidence of Gallipolis' beginning in 1790 is to be seen in only an ancient building or two. There are, however, plenty of attractive nineteenth-century houses including "Gatewood," made famous by the writings of the late O. O. McIntyre.

In its next forty miles the river passes only a few hamlets on either side before reaching Proctorville, an average scruffy village on the Ohio side, and the eastern limits of Huntington, West Virginia, opposite. Huntington is the commercial metropolis of the Huntington-Ironton-Ashland-Portsmouth area (once the Hanging Rock coal and iron center) and it presents a truly metropolitan sky line from the river. With a population of almost ninety thousand it covers about sixteen square miles along the Ohio and up the Guyandot, which joins the former here.

Huntington claims to be the shopping center for half a million people with a retail sales market of over a quarter billion dollars annually and a well-diversified manufacturing and marketing program within its immediate limits: and Huntington, the investigator notes with some relief, looks as if it might have at least all it claims.

Here is a community which, after the boom days of Hanging Rock iron and coal were over, was able to hold remnants of those industries but, neither depending upon them for sole support nor sitting down to bewail the loss of glories past, went out and gathered other manufacturing (the city now produces textiles, stone and clay products, glass, electric goods, food, and printing) to supplement the metal business which, incidentally, also adapted itself by specialization.

The Big Sandy River empties into the Ohio from the south

just below Huntington to form the West Virginia-Kentucky line and immediately below it lies first the ancient village of Catlettsburg and the city of Ashland, Kentucky. Ashland prospers mightily, especially from its oil-refining industry, but it is not very prepossessing to the eye. The traveler gets the impression that a city with good balance between manufacturing, retailing, wholesaling, and service industries—such as Huntington—is likely to present an attractive appearance, while one which depends upon factories alone—such as Ashland—inclines to be otherwise, no matter what figure its annual per capita income may reach. Ashland produces iron and steel products, oil and gasoline, sulphur and coal and coke—the oil products leading—and ships close to ten million tons of its commerce on the river. Seventy-five per cent of its families are said to own their own homes and its thirty-five thousand citizens are well equipped with parks, schools, and churches. The American Folk Song Festival, held annually near the city, is equally attractive to the singers, guitar, fiddle and dulcimer players and folk dancers who come from the Kentucky and West Virginia hills to perform and to the outlanders who come to see them.

Ironton, across the river just below Ashland on the Ohio side, was once the capital of the Hanging Rock area (where, as some early promoter somewhat tritely put it, "coal meets iron"). Apparently it profited by the example of neither Huntington, which diversified, nor Ashland, which seems to have added only a few basic industries but to have made that addition effective. Ironton has improved since it hit its all-time low in the nineteen-thirties but more improvement is obviously in order. The Ohio side of the river in this stretch is generally much less attractive than the West Virginia and Kentucky shores. Though both are built up nearly solid with oil tank "farms," warehouses, and factories, those on the Ohio side are more often abandoned or run down and manage, someway, to appear universally more disreputable. Even the justly celebrated literacy of the state of Ohio breaks down in the region, one of the outhouses in a state park near Ironton being labeled "LADYZ" while the inscriptions upon the inner walls of this and its associated building exhibit even less orthodox spelling than is common to that school of letters.

Russell and Greenup follow on the Kentucky side and the town of Hanging Rock is next below Ironton on the north. Then comes New Boston, Ohio, which is strung out along the river to join Portsmouth, the largest city on the north shore between Cincinnati and Pittsburgh. A gigantic flood wall is being constructed from New Boston to Portsmouth which will eventually carry the east-west highway on its crown; in the meantime the entrance to the city is by way of a narrow shelf crowded between the river and the hills and packed solid with steel mills, manufacturing plants, and neighborhood shopping districts which have the look, against the backdrop of Ohio hills, of western mining towns.

It specializes successfully in the manufacture of steel and steel products, shoes and cement, clay products and chemicals. The country surrounding Portsmouth was the site of great activity on the part of the mound builders and an important Shawnee Indian town occupied the land on which the city is built until the Indians finally wearied of being driven out by floods every second or third spring and moved up the Scioto to Chillicothe. Whites settled the spot in 1803, and put up with the floods because of the advantages of the junction as a place to trade and of the fertile bottom farm lands along the Scioto. The modern city contains a few handsome old houses as a reminder of early prosperity but they have mainly been crowded out by architecture of the Late President Arthur or the Sun-kissed California Bungalow school—all pretty well be-grimed by coal smoke.

Portsmouth spreads back north as it approaches the Scioto valley but, reaching the wide bottom land of that stream, the city stops abruptly with the cornfields in the Scioto bottom practically under its windows. Beautiful as is the Scioto, normally its water appears to be but a small trickle in its fertile valley, even at the point of junction with the Ohio.

At Portsmouth the Hanging Rock industrial region stops—as does, in fact, the entire coal-iron-glass-oil economy of the upper river. From this point to its mouth at Cairo the banks of the Ohio are clear of industry except within the limits of the cities and in most of them some effort has been made to preserve the river's natural beauty while still utilizing its waters. From Pittsburgh to

Portsmouth there are some beautiful stretches of river and a few towns and cities which are reasonably clean as well as prosperous; from Portsmouth to Cairo the river is always beautiful, the cities and towns usually attractive, and the prosperity, though stemming from different sources, is as great.

In the course of the next fifty miles after Portsmouth the Ohio flows through a beautiful valley two to three miles wide which is bordered on either side by worn hills topped by bald rocky peaks. There are no towns of importance until Manchester, Ohio, is reached.

And Manchester's importance, by appearance, was mainly in the past. It was the fourth town organized in the state of Ohio but it has suffered much from floods since its most prosperous day and occasional buildings in its lower part have succumbed to them, or to neglect, and have been removed to give a sort of missing-tooth appearance to the streets, although the higher section of the old town is clean and attractive.

A dozen miles farther downstream lies Maysville, Kentucky, once called Limestone, and one of the important shipping and receiving points in the western country, the gateway—from the river—to Kentucky's Bluegrass. The place had an excellent landing, where a point jutted into the river at the mouth of Limestone Creek, and it was welcomed as a harbor by people who came downstream to settle interior Kentucky after the American Revolution. Although it has not grown large (its present population is about eleven thousand) Maysville has never ceased to prosper and is now a leading tobacco market and a busy shopping center.

From the river Maysville looks, today, exactly as the old prints show the most prosperous towns of the fifties. Since the prints were made to sell, and undoubtedly idealized their subjects, one may assume that Maysville, now, appears as the old towns *hoped* they looked. Here is a community which has seen fit to preserve, through all the years, the solid commercial buildings and handsome residences of the early day and to modernize and adapt them to present requirements. This good taste and sensible planning has left Maysville the handsomest little city on the Ohio—end to end—and has evidently added to its material welfare besides.

Ripley, Ohio, is ten miles downstream; it claims to be the place where the original Eliza fled across the river to inspire Mrs. Stowe. Just below Ripley are the few remaining buildings of Utopia, the vicissitudes of which have been described elsewhere.

Augusta, Kentucky, across the river and a little above Utopia, is a clean, attractive little community. Founded in 1795 upon the site of an immense Indian burial ground, it was the scene of a Civil War engagement in which General Basil W. Duke, on his way to cross the Ohio and attack Cincinnati, was delayed by Federal Home Guards who fired from houses and buildings. Duke burned them out but the delay they created is generally credited with deflecting the attack from the city across the river.

There stand a few small villages, then the outskirts of Cincinnati begin: after Coney Island (which is, naturally, an amusement park) and the River Downs race track, follow miles of alternating neighborhood shopping districts and groups of nineteenth-century warehouses and factory buildings; many quite picturesque, almost all well kept. Over the river the Kentucky residential suburbs around Cold Spring and Fort Thomas spread through the hills. After six or seven miles of this the river and its accompanying highway enter downtown Cincinnati.

The story of Cincinnati's beginning—as Losantiville—has been told. There have been glimpses of its growth through the nineteenth century, when it was known, variously, as "Porkopolis," because of its vast business in processing products derived from the lowly hog; as the "Literary Emporium of the West," because of its work in writing, editing, publishing, printing, and bookselling; as the "Queen City," because of its pre-eminence in commerce and finance. The place still preserves enough of the qualities that resulted in each of those nicknames—plus conservative and law-abiding stability, industry, and a love of good living—to make it one of the most attractive cities in the country. Cincinnati ranks with San Francisco, New Orleans, Charleston, Lexington, New York, Boston, St. Augustine, and Chicago, as one of the American cities that have a personality unmistakably their own.

The population of Cincinnati proper is now near half a million but it is blest with a vast number of prosperous suburbs upon both

Ohio and Kentucky shores which double that figure and not only furnish comfortable residence for commuters to the city but also make their own substantial contributions to the community as a whole.

Cincinnati has one hundred and twenty-eight parks, all within a few miles of the heart of its business district, and its public library with thirty-eight branches, the University of Cincinnati, somewhere very near the largest municipal university in the country, the Art Academy, Fine Arts Institute, eighty-one theaters and the famous Cincinnati Symphony Orchestra, outdoor summer opera, the May Festival of Music, the Conservatory of Music and College of Music, provide educational opportunity, cultural activity, and entertainment for large sections of Ohio, Kentucky, and Indiana.

The industry of Cincinnati is as varied as its cultural interests. It still packs pork and prints and binds books, carries on a great wholesale and distributing business (over a billion and a half dollars' worth annually), and is the retail trading and banking center for a large area. Its Rookwood Pottery still produces fine ware, each piece a distinguished original, and there are still skilled craftsmen in leather, wood, and metalwork, but today Cincinnati also leads the cities of the world in the production of soap, machine tools, playing cards, and electrotypes. Cincinnati beer is still held in high esteem and the manufacture of printing equipment, stoves, and laundry machines is important; but with the twentieth century the making of radio receivers, automobile parts, office furniture, chemicals, and sporting goods has moved into a still more significant place.

No city in America takes more interest in its governmental management and in planning its future.

Cincinnati's Kentucky satellites, Newport and Covington (undistinguishable were it not for the fact that the Licking River flows between them) and Ludlow, withdraw rather rapidly from the river as the Kentucky shore becomes rough and broken but on the Ohio side, even though the shelf of second bottom is quite narrow, the riverside west through Addyston, North Bend (William Henry Harrison's old town), Cleves, and Elizabethtown is built solid to the Indiana state line, at which the Great Miami enters the Ohio from the north.

The "Great Miami," important as it once was, is great only by comparison with the *Little* Miami. A dozen miles above its mouth it is not more than twenty or thirty yards across. Interspersed with rapids and chutes it flows shallow and swift down a channel paved with small boulders through a valley which, in its southern reaches, is seldom more than a mile from hilltop to hilltop. Boat traffic, from M. Céloron's day to the farm-produce flatboat era in the nineteenth century, must have depended largely upon the flash floods which the steep descent of its tributaries and its narrow valley make possible.

Below the mouth of the Miami the Indiana hills retreat from the Ohio to leave a second-bottom plain thirty or forty square miles in extent and broken only by a ridge or two and an occasional bayou.

Lawrenceburg, behind a gigantic levee, is a prosperous, spotless little town with handsome churches of the past century and a business district in which buildings of the eighteen-twenties and -thirties are either well kept or have been remodeled with modern fronts which leave only the roof lines and an occasional Dutch gable end to indicate their actual age. Lawrenceburg's chief industry is distilling whisky and, from the size of the sprawling modern plants on the hillside behind the town, its capacity must be sufficient not only to cure the nation's snakebites but also to drown its snakes. The distilleries draw workers from twenty miles around.

Aurora, four miles downstream, is fully as busy as Lawrenceburg but there ceases the similarity. Aurora's building boom came at a less happy period of American architecture, the seventies and eighties, and little has been done either to conceal its Byzantine horrors or to maintain them in repair. In spite of these the town has at least two architectural glories: Verestau, the country home which Jesse Lynch Holman, first Indiana and Kentucky novelist, built on a bluff overlooking the river, and the handsome old Gath home, purest examples of steamboat architecture—assuming that the term "pure" may be applied to that school of design—on the Ohio.

Aurora also has a much more diverse and probably more stable industry than Lawrenceburg.

Two or three miles below Aurora, Lochry Creek crosses a second-bottom plain on the Indiana side and enters the Ohio. In a deep cut it is, even today, choked with driftwood from above and flotsam pushed in by the Ohio at every high water. Stagnant even at normal stage, it looks more like a great drainage ditch than a stream and should have been dismal and forbidding enough to warn off Lochry's force by its very appearance.

From the Great Miami to well below Lochry Creek the widest second bottom of the Ohio lies on the Indiana side. It is comparatively low above normal water level and is planted in endless miles of corn every year. Here crop rotation is not necessary to preserve the soil from being "corned to death," for the river deposits a completely new layer of top soil with every flood.

Rising Sun, Indiana, named to compete with its neighbor Aurora, is a clean and fresh-painted village which appears in its older, riverside section probably very much as it did in 1840, except for parked automoblies and neon signs. Its people are friendly—as are those of most towns of two or three thousand population on the river below the upper industrial belt; "mannerly folks" they would have been called a century ago, and mannerly they are!

In the smaller towns all up and down the river a favorite leisure pastime is sitting on the bank watching the waters. Many towns have small parks with benches and most have levees which are equally convenient for those addicted to this pleasure. Rising Sun's park is sixty feet or so above the normal water level but those who happen to be occupying the benches there will show you the marks twenty feet higher still on neighboring houses where the river rose in 1937—81 feet above its normal stage!

The Rising Sun park also gives a fine view of Rabbit Hash, Kentucky, which is, in defiance of those who christened it, a rather attractive little hamlet of a dozen or so white frame houses.

Five miles below Rabbit Hash is located Big Bone Lick, back a mile or two from the river. The historical and zoological importance of this spot has already been discussed; today the brackish water only seeps from the earth at the bottom of its bowl among the hills and, although there may still be plenty of evidence of their passing buried beneath the surface, no bones of prehistoric beasts now dot the landscape. The town of Big Bone is remarkable for its dilapidation. There are two mercantile establishments, half a dozen houses, and a church which, by the apparent iniquity of Big Bone's inhabitants, must be considerably less than 100 per cent efficient in the business of soul-saving. Of the ten citizens of the place in evidence on a Saturday afternoon in the fall of 1948, five were obviously in an advanced state of intoxication, one was less than ten years old, and two were apparently in their late eighties, leaving a minority of two eligible adults sober—but of course it was fairly early in the afternoon. Big Bone is apparently a place of trouble, for men as well as mammoths.

Twenty miles from this point is the next town of importance, Warsaw, Kentucky. It is a clean, thrifty little place—the first example of three which follow on the south shore. In spite of the broken and impoverished hill country that presses behind it, Warsaw prospers from the business of the rich river-bottom farms. Tobacco again begins to be a source of income and signs posted in Warsaw advertise for sale both aged country hams and "Blue-tick coon-dog Pups"—an indication that the community is definitely Deep South in spirit if not in geographical location.

Ten miles farther is Ghent, slightly less prosperous but attractive enough, across the river from Vevay, Indiana, and connected by the *Martha A. Graham,* a diesel-powered side-wheel ferry.

Vevay (you pronounce it *Veh*-vee, if you wish to gain the approbation of the older inhabitants) is a charming little town which was first settled by J. F. Dufour and his associates in 1801. Dufour was a French-speaking Swiss who longed to establish vineyards in the Ohio Valley. He had already tried land on the Kentucky River unsuccessfully before he and his friends bought the land around Vevay. His vineyards were not permanently successful on the Indiana side either, but he wrote an interesting little book on the subject of vinegrowing—one of the earliest of midwestern works on any branch of agriculture. Citizens of Vevay played a prominent part in establishing Indiana's literary tradition, it having been the home at one time of Julia L. Dumont, early writer of verse, and of her pupils, the Eggleston brothers. The neighborhood was the scene of Edward Eggleston's *Hoosier School-master* and *Hoosier Schoolboy* and he employed local citizens, by name, as characters. (The present writer trusts that the characters are overdrawn—especially one whose given name was Hank. Incidentally, the gentleman in question was a relative of Eggleston's, though only by marriage.)

Carrollton, at the mouth of the Kentucky River, is the prettiest town on the Ohio—always excepting Maysville, which is large enough to be classed a city anyway. Called first Port William, it has always been prosperous, as the number of handsome old houses in and near its limits will testify. Carrollton has an interesting two-story jail (interesting from the outside at least) which resembles a pioneer blockhouse, one room above another, with slits in the massive stone walls instead of windows. The town is a tobacco market and celebrates the end of the harvest season with a Tobacco Festival and a Singing Convention, in which singers from all the churches of the county participate.

Madison, Indiana, is the only community of importance in the sixty miles between Carrollton and Louisville. With a present population of approximately seven thousand, it is a comfortable little city with some of the finest examples of early nineteenth-century architecture—in public buildings and dwellings—to be seen on the river. While it is set on a shelf only a half mile or so in width between the river and the steep hills behind, its wide streets manage to convey a feeling of spaciousness—an asset not common among municipalities above it on the river.

Madison was Indiana's most important city during the first third of the past century; its commercial metropolis and chief port. Later it became the starting point of the state's first railroad and even through the fifties Madison society set the pace and introduced eastern fashions to Hoosierdom—a smaller Cincinnati in business, with a touch of the Bluegrass in its character as well. In the seventies, when river traffic began its great slump, Madison began to stagnate and the stagnation continued until a revival began about the time of World War I. That long dull period stunted Madison's growth but to it, and the lack of progress during it, Madison owes the preservation of her magnificent old buildings.

Louisville's beginnings are familiar through the early activities of her citizens and the accounts of travelers. Today Louisville is a notably prosperous city of more than three hundred and sixty thousand population—truly, as it proclaims itself, the commerical gateway to the South.

Once Louisville was only a wholesale market and transshipping point, first of goods around the Falls, later of shipments transferred from the river to the railroads that fan out to the southern states. The distributing business continues—at the rate of a half billion dollars per year—but gradually processing and manufacturing interests developed also. The milling of wheat and corn probably was the first large operation (although Louisville is not particularly addicted to grits with its gravy or breakfast eggs it supplies a large portion of that item which is still considered a necessity in the Deep South) and distilling and tobacco processing came next, in that order. Manufacturing of cement began at the end of the nineteenth century and soon there was a thriving business in metal products and textiles. But Louisville did not really begin to come into its own as a manufacturing center until World War I and it enjoyed an enormous expansion in this field during the recent war, when it supplied powder, artillery, synthetic rubber, plastics, and aluminum products. The Louisville neighborhood has two practical monopolies as far as the United States is concerned: it is the largest producer of baseball bats and at Fort Knox, just beyond the city's back door, is buried most of the nation's gold reserve.

Although Louisville has fine libraries, the oldest municipal university in the country, and several other colleges and seminaries, the cultural feature of which it talks most is the Kentucky Derby, held annually since 1875.

Louisville is a city of broad streets and many trees, even through most of its poorest districts. It, like Cincinnati, received a large German immigration and its Germans also built, as they prospered, those intensely red brick houses with well-scrubbed white stone trim which mark Baltimore, Cincinnati, and St. Louis. There are many parks, a fine shopping district, many pleasant and some very handsome residential districts.

Besides its residential suburban towns on the Kentucky side, Louisville may claim the three Indiana cities opposite—Jeffersonville, Clarksville, and New Albany, with a combined population of over forty-six thousand. The Indiana towns have industries of their own but many of their people are employed in Louisville and most of Jeffersonville's shopping is done there—with the result that its own business district is a rather depressing sight. New Albany's street nearest the river is lined with the handsome old homes of families who grew up in close contact with the water.

Time was when the cities on the Indiana side had an even chance with Louisville for prosperity, but they dawdled and bickered over building the canal around the Falls on their side—as no less an authority than Aaron Burr had urged, and really experienced rivermen thought feasible—while the citizens of Louisville promoted and built their own. After that Louisville's leadership was secure.

Past Louisville the river flows almost due south before turning west again. It has many miles to go before it reaches Owensboro, next important city, but on the way it passes Brandenburg, Kentucky. General John Hunt Morgan paused there before raiding Indiana during the Civil War and, except for the addition of a variety of federal bureau offices of one kind and another, time appears to have stood still since. The town has some excellent examples of rural nineteenth-century merchandising establishments in which goods seem to have accumulated in strata since their beginning. Electrical items and plastic goods top the heaps, with World War I items immediately below; what might be found in the dust at the lower levels is an intriguing question.

Cloverport is the next Kentucky town, a clean, cheery little place, then comes Hawesville, of similar description. These small towns below Louisville have strayed from the Greek Revival and American colonial architectural style seen above. All have many frame buildings with up- and downstairs galleries after the fashion of the deeper South. Hawesville communicates with Cannelton, Indiana, by a stern-wheel ferry plying between wharfboats moored on either shore.

Cannel coal was early discovered in the Indiana hills at this point. The mines were developed in 1843 and the town of Cannelton grew up as a coaling point for steamboats. In 1848 a cotton mill was established at the place. It still operates (manufacturing bagging today) in the original turreted buildings.

Tell City is four miles below, a thriving little place of sixty-five hundred people which was, needless to say, founded by Swiss immigrants.

The hills now fall back; presently the river runs in a bed edged by steep banks thirty or forty feet high through a level plain many miles in width. Fine grain and stock farms stretch along the river from here, more or less continuously, to its mouth.

Troy is next below Tell City, in Indiana, near the mouth of Anderson's Creek where young Abraham Lincoln had his first brush with the law in the middle eighteen-twenties. They called it Anderson's *River* then, although it couldn't have made a significant contribution to the Ohio and bigger creeks were a dime a dozen. Young Abraham was picking up an occasional bit by rowing passengers out of its mouth to board passing steamboats. Someone haled him into the Kentucky court of Justice of the Peace Pate (Kentucky owned the Ohio, then as now, to the low-water mark on the Indiana side), charged with operating a ferry without a license. He defended himself with one of those simple pleas that eventually made him a pretty successful lawyer; ferries, he insisted, went *across* rivers—he was only going out to the channel and back; no ferrying there!

Rockport, beautifully situated on a flat table of land above the river, is the next town; then, on the Kentucky side, comes Owensboro, built on what was once called Yellow Bank.

The city of Owensboro has changed in the past century and a half—once a haven for pirates, wreckers, pioneer hoodlums, it is now a well-policed, prosperous manufacturing city in the center of an oil field.

Distilling, wholesaling, milling, tobacco, manufacturing, gas and oil have brought quick and sometimes large individual fortunes and the city has some atmosphere of the boom town. The men who operate these industries work hard—and they relax by eating and drinking well and, naturally in the light of the driving spirit of the place, doing a little gambling for amusement. Unquestionably they gamble in private and the discerning traveler also notes certain unobtrusive establishments outside the city limits which look as though they might harbor individuals willing to accept a bet, to flip a card, to turn a wheel, or to roll the dice, if requested.

Owensboro is by no means a handsome city; neither, on the other hand, is it a particularly ugly one. It simply appears to be a place in which everyone has been too busy making money to become civic-minded: as though most of its people exhausted their initiative in planning business moves like that attributed—probably erroneously—to one of Owensboro's manufacturers:

Faced by a sit-down strike in his plant one morning he telephoned an acquaintance in Louisville and asked for a crew of strikebreakers of a type which he specified particularly. A chartered bus arrived at noon with the crew—a couple of dozen painted ladies whose profession was obvious at a glance.

The manufacturer welcomed them at the gate, ushered them into the plant where his employees were sitting idle at their machines, and then left them to place telephone calls to certain strategic points. Within half an hour news of the invasion from Louisville was all over town; in even less time indignant wives of striking employees began to arrive at the plant to take their husbands home, out of harm's way; by midafternoon the sit-down strike was over and the ladies were on their way back to Louisville.

Thirty miles downstream the beautiful Green River empties into the Ohio and shortly the suburbs of Evansville, Indiana, begin. With a population of 125,000, Evansville shows the effects of civic pride and careful planning. Laid out in 1817, it had its first great boom twenty years later when it was expected to be the Ohio River terminal of the Wabash & Erie Canal. Additions to the town were platted and advertised for sale in newspapers as far away as Washington, D.C. Some of the lots sold, and Evansville took on a more healthy aspect but without much assistance from the Wabash & Erie—that canal had scarcely reached Evansville (in

1856) before its older portions fell into decay and new railroads made its further operation hopelessly unprofitable.

Here the river has played a trick, changing its course to leave a sizable chunk of Kentucky land on the north side of the river at Evansville's very feet: upon it horse and dog tracks and pari-mutuel betting—outlawed in Indiana—have flourished for the convenience of Indianians and the profit of Kentucky. The city is well supplied with parks and has a special feature which should be copied by other river communities—a handsome plaza facing the river along the main part of the shopping and hotel district.

The decentralizing of northern and eastern industry, a strategic location between southern and northern markets and resources, plus a World War II boom, have all contributed to its rapid twentieth-century improvement but Evansville's prosperity also stems from the manufacture of refrigerators (it is the world's largest producer, with four companies engaged in the business), automobile bodies, furniture, plastic products, packaged food, textiles, agricultural and contractors' equipment, structural steel, and cigars. Coal, gas, and oil are produced in the immediate neighborhood and to the north lie the vast Indiana-Illinois corn, soybean, wheat, and oats producing lands.

Residential Henderson, across the river and ten or twelve miles by highway, is Old Kentucky itself. Founded in 1797 by the Transylvania Company of Richard Henderson, it is slow, easy-going—everything that Evansville is not. Here Audubon played at sawmill operating and storekeeping and worked seriously at painting and describing birds and animals between 1810 and 1819.

Downstream again, and last town before the Wabash River enters the Ohio and forms the western boundary of Indiana, is Mount Vernon. High on a bluff above the Ohio, it preserves its rather different past as fully as does Henderson: tough river town it was fourteen decades ago; tough river town it is today with, now as then, a better element in its community which subdues but cannot entirely still the rowdies down by the river.

Mount Vernon has diversified small manufacturing industries which produce everything from grits to threshing machines. At present it is the shipping center for the oil field that surrounds it.

Daily shipments in 1948 averaged 55,000 barrels, making Mount Vernon the largest inland waterway shipping terminal of petroleum.

Uniontown, Kentucky, lies below Mount Vernon, then the Wabash enters the Ohio from the north, forming the Indiana-Illinois state line and beyond that point sleeps Shawneetown. The approaches to the Shawneetown ferry landing are probably as dusty in summer and as muddy in spring and fall as they ever were but the glory of Shawneetown is no more: that, perhaps, is just as well. By present appearance it has sunk from its honored place as the toughest spot on the Ohio above Cave-in-Rock, and later as the financial capital of Illinois, to a state of unquestioned if shabby rectitude.

Shawneetown fought the river for a century and a third but the river's assault of 1937 won. About half the town was moved bodily back from the river three miles or so (by grace of the RFC, WPA, or other benevolent governmental agencies), where in its reborn form it looks like any barren and none-too-prosperous suburban shopping center except that it is distinguished by one bar named "PEARL'S HARBOR"! Many decrepit buildings were moved bodily from the old site to the new, where some of them continue to disintegrate.

In the old town a few inhabited streets remain with a half dozen distinguished buildings, including that of the old Bank of the United States, which ruled Illinois finances with an iron hand in the early day and refused loans to finance such dubious risks as a project for developing a place called Chicago. Business now consists of four-well-nigh deserted and highly respectable taverns, a filling station or two, a grocery, a couple of hardware stores, three mercantile establishments of unidentifiable purpose, a popcorn stand, two of the inevitable catfish ateliers, and a poolroom. Leadership of this section of Illinois appears to have moved to thriving Harrisburg, twenty-three miles back from the river—which is far more adept at southern hospitality than is most of the South. Many a southern town which advertises that nebulous quality should send an agent to Illinois to take notes.

Cave-in-Rock is below Shawneetown, the cave itself being now the central attraction of an Illinois state park. The adjoining village of the same name—which had no part in the nefarious doings of a hundred and fifty years ago since it wasn't there at the time— is as friendly a community as might be found in a month of Sundays; it bears no taint of the blood spilled in the neighborhood.

Towns are few indeed on the Kentucky side in this area but Elizabethtown and Rosiclare grace the Illinois shore and are the center of vast deposits of fluor spar which account for 80 per cent of that mineral mined in the United States. Golconda is next, in Illinois, and, in Kentucky, Smithland is built at the mouth of the Cumberland River.

Only about a dozen miles farther downstream the Tennessee also empties into the Ohio and below its mouth is Paducah, city of forty thousand population, and the capital of the lower Ohio country. Its site was the rendezvous for young General George Rogers Clark's forces before he advanced into what are now Illinois and Indiana for the glory of Virginia and America in the Revolution. Virginia granted him the land on which he had encamped as a reward. Like most of George Rogers Clark's rewards, it did him little good in his lifetime but his brother William (of the exploring firm of Lewis & Clark) eventually gained a clear title, laid out the town in 1827, and named it for the Chickasaw chief, Paduke.

Modern Paducah rode to fame on the white linen coattails of Irvin S. Cobb, although it already had plenty of interesting activity in its past before he began to publicize it. The city's earliest business was lumber—the flat alluvial valley where the Ohio, Cumberland, and Tennessee meet was one great forest—but the timber was exhausted, in the American way, and new sources of revenue had to be found. Today the farms of that former forest country produce tobacco and enormous crop of strawberries, peaches, apples, grain, and mules, all of which are marketed through the city.

Manufacturing got a good start after 1900 and now Paducah produces shoes, stockings, automobile parts, radios, chemicals, and batteries. Its largest single employer is the Illinois Central Railroad in whose shops locomotives are repaired and rebuilt. Paducah's prosperity is evident in its appearance; there are fine parks, good schools, a modern business district, and above-average residences— but in spite of Irvin S. Cobb it has no perceptible "southern atmosphere" except, by report, a diminishing residue of militant Daughters of the Confederacy. If that statement be heresy it still must stand.

The plain on which Paducah is built being low, the city suffered almost yearly damage by floods of the Ohio and Tennessee from its beginning: as in the case of so many other Ohio River communities the 1937 flood, which covered half the surface of the place, finally brought action. Paducah now has a flood wall which looks to be high enough to guarantee future safety.

A high bridge—one of the most impressive in size upon the Ohio—connects Paducah with the Illinois shore and below it half a dozen miles is the little city of Metropolis. The founders of Metropolis evidently had high hopes for its future; they might be disappointed in its size, could they visit it today, for a certain upstart town on Lake Michigan in the northern part of the state has long since passed it in population, but the founding fathers could find nothing but praise for its neat appearance and obviously good administration. The town has a well-balanced industry—one contributing item being what appears the largest collection of mules gathered in one sales lot outside the Tennessee mule country. The site of Fort Massiac is a public park adjoining the city limits and commanding a fine view up and down the river. The foundations of the old fort are outlined and its former eligibility as a place of defense is obvious.

There remain no more settlements of importance upon the Ohio until Mound City and then, three miles farther, Cairo, last city on the river. Very few buildings in Cairo date from much before the Civil War, which is not remarkable considering that Cairo, until she finally built her truly stupendous flood walls (they kept her dry even in 1937) had fought both Mississippi and Ohio floods since her founding. The city (with a population of fifteen thousand today) presents an interesting appearance; magnolia trees flourish among northern hardwoods and cotton, tobacco, corn, and wheat are all hauled to market from the same farms in the surrounding country, for Cairo, though in Illinois, is as far south as Norfolk, Virginia.

Cairo's great day in history came during the Civil War, when

General U. S. Grant organized the western forces for the Union from his quarters in Room 215 of the St. Charles Hotel, while Admiral Foote assembled the Western Gunboat Flotilla in the river as engineer James B. Eads armored requisitioned river boats on the marine ways at nearby Mound City.

Shortly below the southern point of Cairo the Ohio and Mississippi meet and there is an end to the long line of Ohio River towns, Pittsburgh to Cairo.

The whole length of the Ohio Valley gives an impression of conservative stability and permanence. Even though the river itself can change overnight from a smooth, jade-green ribbon to a brawling, tearing, unmanageable torrent, the color and texture of simmering roast-beef gravy, it maintains in either phase a majestic beauty which has been invariably remarked by those who have seen it through the centuries. The people upon its banks today hail originally from all the world and vary in about the usual proportions—good, bad, and indifferent. The Ohio Valley cities are progressive, handsome enough in their best districts and conventional eyesores in their slums, and its towns range from pleasant and cultured to shabby and dissolute. Perhaps it is from these very extremes and their in-between average that the impression of permanence stems.

Perhaps the Ohio, connecting as it does the North, South, East, and West, its people springing, as they do, from every major immigrating strain, American and foreign, is, as its people like to think, *really* the heartland of the nation.

If this be true, God grant that the impression of stability and permanence is valid.

Acknowledgments

THE NOTES that have gone into this book began to accumulate in my desk, in my bookshelves, and in odd corners of my house almost twenty years ago. There was no reason for their preservation, except my interest in the early days in the Ohio Valley, until I was given the opportunity to write *The Ohio*. Thus the origin of many of the notes—who called them to my attention, how I happened to notice them in the first place—is lost in comparatively recent but no less obscure antiquity.

I am perfectly clear as to whom I am indebted, once the actual writing began, and I am most grateful to all of them. First to Dr. R. Carlyle Buley, my friend of the Indiana University faculty who put me to work on the project, and to Hervey Allen, Carl Carmer, and Miss Jean Crawford, editors of the Rivers of America Series, who, I fear, have devoted far more time to guiding my faltering steps than the results will justify. There was another friend, too, Dr. Thomas D. Clark of the University of Kentucky, who took time from his teaching, lecturing, and writing to read and criticize the manuscript.

For checking copy and making corrections in specialized fields and—in many cases—for supplying original material, I wish to express my sincere gratitude to Mr. Eli Lilly and Mr. Glenn A. Black in midwestern archaeology; Mr. Paul B. Mason and Major General J. C. Mehaffey, of the Corps of Engineers, U.S. Army, in the field of the development of navigation aids on the Ohio; to Mrs. Helen Hoffman Collar in the geology of the valley and to Mr. Charles F. Willner in the industrial development of the Pittsburgh area.

I owe gratitude to the librarians and chamber of commerce officials of the cities and towns from Pittsburgh to Cairo who cheerfully supplied statistics and answered questions. Much material has come from the collections and publications of the American Antiquarian Society, the New York Historical Society, the Historical Society of Pennsylvania, and University of Pittsburgh, the Upper Ohio Valley Historical Federation, the Ohio State Archaeological and Historical Society, the Historical and Philosophical Society of Ohio, the Indiana Historical Society, Indiana University, Purdue University, Wabash College, the University of Kentucky, the Filson Club, and the Illinois State Historical Society Library.

Last but not least I must acknowledge the encouragement—usually unknowing—of the people I have met in my travels along the Ohio, from deck hands to industrialists; after all, it's their river and they make it a most pleasant subject of study.

R. E. Banta

Bibliography

BOOKS

ABDY, E. S., *Journal of A Residence and Tour in the United States of North America from April, 1833, to October, 1834.* (3 vols.) London, 1835.

An Account of Louisiana, Being An Abstract of Documents, in the Offices of the Departments of State, and of the Treasury. Philadelphia, 1803.

ARFWEDSON, C. D., *The United States and Canada in 1832, 1833 and 1834.* (2 vols.) London, 1834.

ASHE, THOMAS, *Travels in America, Performed in the Year 1806, For the Purpose of Exploring the Rivers Alleghany, Monongahela, Ohio and Mississippi, and Ascertaining the Produce and Condition of Their Banks and Vicinity.* New York, 1811.

ATWATER, CALEB, *Description of the Antiquities Discovered in the State of Ohio and Other Western States. . . .* n.p., n.d.

Autobiography of Peter Cartwright, the Backwoods Preacher. Edited by W. P. Strickland. New York, 1857.

BAILY, FRANCES, *Journal of a Tour in Unsettled Parts of North America, in 1796 & 1797 . . .* London, 1856.

BEATTY, CHARLES, *The Journal of a Two-Months Tour: With a View of Promoting Religion Among the Frontier Inhabitants of Pennsylvania, and of Introducing Christianity Among the Indians to the Westward of the Alegheny Mountains . . .* Edinburgh, 1798.

BEGGS, REV. S. R., *Pages from the Early History of the West and North-west; Embracing Reminiscences of Settlement and Growth, and Sketches of Material and Religious Progress of the States of Ohio, Indiana, Illinois and Missouri . . .* Cincinnati, 1868.

BERNHARD [KARL], DUKE OF SAXE-WEIMAR EISENACH, *Travels Through North America, During the Years 1825 and 1826.* (2 vols.) Philadelphia, 1828.

BIRKBECK, MORRIS, *Letters From Illinois.* London, 1818.

———*Notes on a Journey in America, from the Coast of Virginia to the Territory of Illinois.* London, 1818.

BLANCHARD, RUFUS, *Discovery and Conquests of the Northwest with the History of Chicago,* (2 vols.) Chicago, 1898.

[BLANEY, CAPTAIN WILLIAM N.], *An Excursion Through the United States and Canada During the Years 1822-3. By an English Gentleman.* London, 1824.

BREAZEALE, J. W. M., *Life As It Is; or Matters and Things in General: Containing, amongst other things, Historical Sketches of the Exploration and First Settlement of the State of Tennessee; Manners and Customs of the Inhabitants; Their Wars With the Indians . . . History of the Harps, (two noted murderers) . . .* Knoxville, 1842.

BRICE, WALLACE A., *History of Fort Wayne, from the earliest known accounts of this point, to the present period. Embracing an extended view of the aboriginal tribes . . . including, more especially, the Miamies . . . Life of General Anthony Wayne . . .* Fort Wayne, Ind. 1868.

BROWN, SAMUEL R., *Views of the Campaigns of the North-Western Army, &c. Comprising Sketches of the Campaigns of Generals Hull and Harrison—A minute and interesting account of the Naval Conflict on Lake Erie—Military Anecdotes—Abuses in the Army—Plan of a Military Settlement—View of the Lake Coast from Sandusky to Detroit.* Burlington, Vt., 1814.

———*The Western Gazeteer; or, Emigrant's Directory, containing a geographical description of the western states and territories, . . . a description of the great northern lakes: Indian annuities, and directions to emigrants.* Auburn, N.Y., 1817.

BURNET, JACOB, *Notes on the Settlement of the North-Western Territory.* Cincinnati, 1847.

BUTTERFIELD, CONSUL W., *An Historical Account of the Expedition against Sandusky under Col. William Crawford in 1782 . . .* Cincinnati, 1873.

———*History of the Girtys. An Account of the Girty Brothers and of the part they took in Lord Dunmore's War, in the Western Border War of the Revolution, and in the Indian War of 1790-95 . . . from Authentic Sources.* Cincinnati, 1890.

CALL, RICHARD ELLSWORTH, *The Life and Writings of [Constantine Samuel] Rafinesque* . . . Louisville: the Filson Club, 1895.

CARTER, RICHARD, *Short Sketch of the Author's Life and Adventures from His Youth Until 1818, in the First Part. In Part the Second, A Valuable Vegetable, Medical Prescription, with a table of Detergent and Corroborant Medicines to Suit the Treatment of the Different Certificates.* Versailles, Ky., 1825.

CIST, CHARLES, *Sketches and Statistics of Cincinnati in 1851.* Cincinnati, 1851.

CLARK, DANIEL, *Proofs of the Corruption of General James Wilkinson, and of His Connexion with Aaron Burr, with A full refutation of his slanderous allegations in relation to the character of the principal witness against him.* Philadelphia, 1809.

CLARK, REV. JOHN A., *Gleanings By the Way* . . . New York, 1842.

CLARK, THOMAS D., *A History of Kentucky.* New York, 1937.

———*The Rampaging Frontier,* Indianapolis, n.d.

COATES, ROBERT M., *The Outlaw Years. The History of the Land Pirates of the Natchez Trace.* New York, 1930.

COLDEN, CADWALLADER, *The History of the Five Indian Nations of Canada Which are the Barrier between the English and French in that Part of the World* . . . London, 1750.

COLLINS, LEWIS, *Historical Sketches of Kentucky* . . . Maysville, Ky., 1848.

Col. George Rogers Clark's Sketch of his Campaign in the Illinois in 1778-9 with . . . the Public and Private Instructions to Col. Clark and Maj. Bowman's Journal of the Taking of Post St. Vincents. Cincinnati, 1869.

Conclin's New River Guide or A Gazetteer of All the Towns on the Western Waters . . . Cincinnati, 1850.

COOPER, THOMAS, *Some Information Respecting America.* London, 1795.

COPPEE, HENRY, *Grant and His Campaigns: A Military Biography.* New York, 1866.

Cox, SANDFORD C., *Recollections of the Early Settlement of the Wabash Valley.* Lafayette, Ind., 1860.

[CRAMER, ZADOK], *The Navigator; Containing Directions for Navigating the Monongahela, Allegheny, Ohio and Mississippi Rivers; with an ample account of these much admired waters, from the head of the former to the mouth of the latter; and a Concise Description of their Towns, Villages, Harbors, Settlements, &c.* . . . Pittsburgh, 1814.

[CUTLER, JERVAISE], *A Topographical Description of the State of Ohio, Indiana Territory, and Louisiana* . . . Boston, 1812.

DANA, E[DMUND], *Geographical Sketches on the Western Country designed for Emigrants and Settlers:* . . . Cincinnati, 1819.

DAVIS, MATTHEW L., *Memoirs of Aaron Burr. With Miscellaneous Selections from His Correspondence.* (2 vols.) New York, 1836-1837.

DAWSON, MOSES, *A Historical Narrative of the Civil and Military Services of Maj.-General William H. Harrison, and a Vindication of His Character and Conduct as a Statesman, a Citizen and a Soldier* . . . Cincinnati, 1824.

DAY, SHERMAN, *Historical Collections of the State of Pennsylania* . . . Philadelphia, n.d.

DE HASS, WILLS, *History of the Early Settlement and Indian Wars of Western Virginia* . . . Wheeling, 1851.

DILLON, JOHN B., *A History of Indiana, From Its Earliest Exploration by Europeans to the Close of the Territorial Governments, in 1816* . . . Indianapolis, 1859.

DRAKE, BENJAMIN, AND MANSFIELD, E. D., *Cincinnati in 1826.* Cincinnati, 1827.

——*Life of Tecumseh, and of his brother, The Prophet; with A Historical Sketch of the Shawanoe Indians.* Cincinnati, 1841.

DURRETT, REUBEN T., *John Filson, The First Historian of Kentucky. An Account of His Life and Writings, Principally from original sources* . . . Louisville: The Filson Club, 1884.

EDGAR, MATILDA (editor), *Ten Years of Upper Canada in Peace and War 1805-1815; Being the Ridout Letters* . . . Toronto, 1890.

ELLICOTT, ANDREW, *The Journal of . . . Late Commissioner on Behalf of the U. S. . . . 1796-1800: For determining the Boundary Between the U. S. and the Possessions of His Catholic Majesty . . . Containing Occasional Remarks on the Situation, Soil, Rivers, Natural Productions, and Diseases . . . on the Ohio, Mississippi, and Gulf of Mexico . . .* Philadelphia, 1814.

ELLIS, AGNES L., *Lights and Shadows of Sewickley Life . . .* Philadelphia, 1893.

ENGLISH, WILLIAM HAYDEN, *Conquest of the Country Northwest of the River Ohio 1778-1783 and Life of Gen. George Rogers Clark.* (2 vols.) Indianapolis, 1897.

ESAREY, LOGAN, *A History of Indiana from Its Exploration to 1850.* Indianapolis, 1918.

EYRE, JOHN, *The Christian Spectator: Being a Journey From England to Ohio, Two Years in that State, Travels in America, &c.* Albany, 1838.

FAUX, W., *Memorable Days in America. Being a Journal of a Tour to the United States, Undertaken Principally to ascertain, by positive evidence, the condition and probable prospects of British Emigrants; Including accounts of Mr. Birkbeck's Settlement in the Illinois . . .* London, 1823.

FEARON, HENRY BRADSHAW, *Sketches of America. A Narrative of a Journey of Five Thousand Miles Through the Eastern and Western States of America . . .* London, 1818.

FERRALL, S. A., *A Ramble of Six Thousand Miles Through the United States of America.* London, 1832.

FIELD, J. M., *The Drama in Pokerville; The Bench and Bar of Jurytown, and Other Stories.* Philadelphia. 1847.

FIELDING, MANTLE, *Dictionary of American Painters, Sculptors and Engravers.* Philadelphia, n.d.

FILSON, JOHN, *The Discovery, Settlement And present State of Kentucke: and An Essay towards the Topography, and Natural History of that important Country . . .* Wilmington, 1784.

FINLEY, REV. JAMES B., *Autobiography of . . . or, Pioneer Life in the West.* Edited by W. P. Strickland. Cincinnati, 1854.

——*History of the Wyandott Mission, at Upper Sandusky, Ohio,*

Under the Direction of the Methodist Episcopal Church. Cincinnati, 1840.

———*Life Among the Indians; or, Personal Reminiscences and Historical Incidents Illustrative of Indian Life and Character.* Edited by Rev. D. W. Clark. Cincinnati, n.d.

FLINT, JAMES, *Letters from America, Containing Observations on the Climate and Agriculture of the Western States, the Manners of the People, the Prospects of Emigrants* . . . Edinburgh, 1822.

FLINT, TIMOTHY, *A Condensed Geography and History of the Western States, or The Mississippi Valley.* (2 vols.) Cincinnati, 1828.

———*The First White Man of the West, or the Life and Exploits of Col. Dan'l Boone, The first settler of Kentucky; Interspersed with incidents in the Early Annals of the Country.* Cincinnati, 1856.

———*Recollections of the Last Ten Years. Journeyings in the Valley of the Mississippi, Pittsburgh and the Missouri, to the Gulf of Mexico.* Boston, 1826.

[FRENCH, JAMES S.], *Elkswatawa; or, The Prophet of the West.* (2 vols.) New York, 1836.

FULKERSON, H. S., *Random Recollections of Early Days in Mississippi* . . . Vicksburg, Miss., 1885.

FUNKHOUSER, W. D., AND WEBB, W. S., *Ancient Life in Kentucky.* Frankfort, Ky., 1928.

GADDIS, MAXWELL PIERSON, *Foot-Prints of an Itinerant.* Cincinnati, 1855.

GALLAHER, JAMES, *The Western Sketch-Book.* Boston, 1850.

GARRAGHAN, GILBERT J., *The Jesuits of the Middle United States.* (3 vols.) New York, 1938.

GERHARD, FREDERICK, *Illinois As It Is—Its History, Geography, Statistics, Constitution, Finances, Climate, etc.* Chicago, 1857.

GILLMORE, PARKER, *Prairie Farms and Prairie Folk.* (2 vols.) London, 1872.

GREEN, JONATHAN H., *The Reformed Gambler; or, The History of the Late Years of the Life of Jonathan H. Green* . . . *to which is added A Complete and Full Exposition of the Game of*

Thimbles; Diamond Cut Diamond . . . etc. Philadelphia, n.d.

GREEN, THOMAS MARSHALL, *Historic Families of Kentucky. . . .* Cincinnati, 1889.

——*The Spanish Conspiracy. A Review of Early Spanish Movements in the South-West . . .* Cincinnati, 1891.

HALE, JOHN P., *Trans-Allegheny Pioneers. Historical Sketches of the First White Settlements West of the Alleghenies. 1748 and after.* Cincinnati, 1886.

HALL, CAPT. BASIL, *Forty Etchings From Sketches Made With the Camera Lucida, in North America, in 1827 and 1828.* London, 1829.

——*Travels in North America, in the Years 1827 and 1828.* (3 vols.) Edinburgh, 1829.

HALL, DR. FRED'K W., *Letters from the East and from the West.* Washington, n.d.

HALL, JAMES, *Legends of the West.* Philadelphia, 1833.

——*Notes on the Western States; Containing Descriptive Sketches of Their Soil, Climate, Resources, and Scenery.* Philadelphia, 1838.

——*The Romance of Western History; or, Sketches of History, Life and Manners, in the West.* Cincinnati, 1857.

——*Sketches of History, Life and Manners in the West.* (2 vols.) Philadelphia, 1835.

——*Statistics of the West, At the close of the year 1836.* Cincinnati, 1836.

——*The West: Its Commerce and Navigation.* Cincinnati, 1848.

HANNA, CHARLES A., *The Wilderness Trail or The Ventures and Adventures of the Pennsylvania Traders on the Allegheny Path With Some New Annals of the Old West, and the Records of Some Strong Men and Some Bad Ones.* (2 vols.) New York, 1911.

HART, ADOLPHUS M., *History of the Valley of the Mississippi.* Cincinnati, 1853.

HART, GERALD E., *The Fall of New France 1755-1760 . . .* Montreal, 1888.

HARVEY, HENRY, *History of the Shawnee Indians, From the Year 1681 to 1854, Inclusive.* Cincinnati, 1855.

HECKEWELDER, JOHN, *A Narrative of the Mission of the United Brethren Among the Delaware and Mohegan Indians . . .* Philadelphia, 1820.

HILDRETH, S. P., *Biographical and Historical Memoirs of the Early Pioneer Settlers of Ohio, with Narratives of Incidents and Occurrences in 1775 . . .* Cincinnati, 1852.

Historical Account of Bouquet's Expedition Against the Ohio Indians, in 1764. With Preface by Francis Parkman, and a Translation of Dumas' Biographical Sketch of General Bouquet. Cincinnati, 1868.

HODGSON, ADAM, *Remarks During A Journey Through North America in the Years 1819, 1820 and 1821, in A Series of Letters . . .* New York, 1823.

HOPKINS, REV. T. M., of Bloomington, Ind., *Reminiscences of Col. John Ketcham, of Monroe County, Indiana, By his pastor . . .* Bloomington, 1866.

HULBERT, ARCHER BUTLER, *The Ohio River. A Course of Empire.* New York, 1906.

———*Waterways of Westward Expansion. The Ohio River and Its Tributaries.* Cleveland, 1903.

IMLAY, G[ILBERT], *A Topographical Description of the Western Territory of North America Containing a Succinct Account of Its Climate, Natural History . . .* London, 1792.

Incidents and Sketches Connected with the Early History and Settlement of the West. Cincinnati, n.d.

Indian Atrocities. Narratives of the Perils and Sufferings of Dr. Knight and John Slover, Among the Indians, During the Revolutionary War, With Short Memoirs of Col. Crawford & John Slover . . . Cincinnati, 1867.

Indiana Authors and their Books. 1816-1916. Biographical Sketches of Authors who published during the first century of Indiana Statehood with lists of their books. (R. E. Banta, editor.) Wabash College, Crawfordsville, Ind., 1949.

The Indiana Gazetteer, or Topographical Dictionary of the State of Indiana. Indianapolis, 1849.

[JACKSON, ISAAC RAND], *A Sketch of the Life and Public Services of William Henry Harrison. Commander in Chief of the North Western Army, During the War of 1812, &c.* Hartford, 1840.

JACOB, JOHN J., *A Biographical Sketch of the Life of the Late Capt. Michael Cresap.* Cincinnati, 1866.

———*Lieut. Boyer's Journal of Wayne's Campaign.* Cincinnati, 1866.

JACOBS, MAJ. JAMES RIPLEY, *Tarnished Warrior.* New York, 1938.

JANSON, CHARLES WILLIAM, *The Stranger in America; containing Observations made during a Long Residence in that Country* . . . London, 1807.

JOHNSON, OVERTON, AND WINTER, WILLIAM H., *Route Across the Rocky Mountains.* Princeton, 1932.

Johnson's New Illustrated Family Atlas, With Physical Geography, and with descriptions Geographical, Statistical, and Historical . . . A Geographical Index, and a Chronological History of the Civil War in America . . . New York, 1864.

JOHNSTON, J. STODDARD, *First Explorations of Kentucky. Dr. Thomas Walker's Journal of an exploration . . . in 1750 . . . Also Col. Christopher Gist's Journal of a tour through Ohio and Kentucky in 1751* . . . Louisville: The Filson Club, 1898.

JONES, REV. DAVID, *A Journal of Two Visits Made to Some Nations of Indians on the West Side of the River Ohio, in the Years 1772 and 1773* . . . New York, 1865.

KERCHEVAL, SAMUEL, *A History of the Valley of Virginia.* Woodstock, Va., 1850.

KETCHUM, WILLIAM, *An Authentic and Comprehensive History of Buffalo, with some account of Its Early Inhabitants Both Savage and Civilized, comprising Historic Notices of the Six Nations or Iroquois Indians, Including a Sketch of the Life of Sir William Johnson, and of Other Prominent White Men, Long Resident Among the Senecas* . . . (2 vols.) Buffalo, N.Y., 1864.

KING, RUFUS, *Ohio, First Fruits of the Ordinance of 1787.* Boston, 1888.

KIP, REV. WILLIAM INGRAHAM, *The Early Jesuit Missions in North America* . . . (2 vols.) London, 1847.

The Life of Major-General William Henry Harrison Comprising a Brief Account of His Important Civil and Military Services, and an Accurate Description of the Council at Vincennes with Tecumseh, as well as the Victories of Tippecanoe, Fort Meigs and the Thames. Philadelphia, 1840.

LILLY, ELI, *Prehistoric Antiquities of Indiana.* Indianapolis, 1937.

LLOYD, JAMES T., *Lloyd's Steamboat Directory, and Disasters on the Western Waters . . .* Philadelphia [1856].

LOSKIEL, GEORGE HENRY, *History of the Mission of the United Brethren Among the Indians in North America.* London, 1794.

MACGOWAN, ROBERT, *The Significance of Stephen Collins Foster.* Indianapolis, 1932.

MACLEAN, J. P., *The Mound Builders; Being an account of a remarkable people that once inhabited the valleys of the Ohio and Mississippi . . .* Cincinnati, 1879.

MCAFEE, ROBERT B., *History of the Late War in the Western Country,* Bowling Green, Ohio [1916].

MCCALEB, WALTER FLAVIUS, *The Aaron Burr Conspiracy.* New York, 1903.

MCCLUNG, JOHN A., *Sketches of Western Adventure . . . Settlement of the West from 1755 to 1794 . . .* Dayton, Ohio, 1852.

MCCOY, ISAAC, *History of Baptist Indian Missions: Embracing Remarks on the Former and Present Condition of the Aboriginal Tribes; Their Settlement Within the Indian Territory, and their Future Prospects.* Washington and New York, 1840.

MCKENNEY, THOMAS L., AND HALL, JAMES, *History of the Indian Tribes of North America, With Biographical Sketches and Anecdotes of the Principal Chiefs Embellished with one hundred and twenty portraits . . .* (3 vols.) Philadelphia, 1858.

MAGILL, JOHN, *The Pioneer to the Kentucky Emigrant. A Brief Topographical & Historical Description of the State of Kentucky . . .* Edited, with an introduction, by Thomas D. Clark. Lexington, Ky., 1942.

Map of the United States. Drawn from the most approved Surveys. Published by Solomon Schoyer. New York, 1826.

MARRYAT, CAPTAIN, *A Diary In America. With Remarks On Its Institutions.* New York, 1839.

MARSHALL, JOHN, *The Life of George Washington, Commander in Chief of the American Forces, . . . and First President of the United States . . .* (5 vols. and atlas.) Philadelphia, 1804.

MARTIN, PAUL, QUIMBY, GEORGE I., COLLIER, DONALD, *Indians before Columbus.* Chicago: University of Chicago Press, n.d.

MARTINEAU, HARRIET, *Retrospect of Western Travel.* (3 vols.) London, 1838.

———*Society in America.* (2 vols.) New York, 1837.

MEEKER, EZRA, *Personal Experiences on the Oregon Trail Sixty Years Ago.* Seattle, Wash. 1912.

MENCKEN, H. L., *The American Language. An Inquiry into the Development of English in the United States.* New York, 1937.

MICHAUX, F. A., *Travels to the Westward of the Allegany Mountains in the States of Ohio, Kentucky and Tennessee in the year 1802 . . .* London, 1805.

MILBURN, WILLIAM HENRY, *The Pioneers, Preachers and People of the Mississippi Valley.* New York, 1860.

———*The Rifle, Axe, and Saddle-Bags, and Other Lectures.* New York, 1857.

MINNIGERODE, MEADE, *The Fabulous Forties 1840-1850. A Presentation of Private Life.* New York and London, 1924.

Mirror of Olden Time Border Life; Embracing a History of the Discovery of America . . . also, History of Virginia . . . also, History of the Early Settlement of Pennsylvania . . . Abingdon, Va., 1849.

[MITCHELL, S. A.], *Illinois in 1837; A Sketch Descriptive of the Situation, Boundaries, Face of the Country, Prominent Districts, Prairies, Rivers, . . . &c. of the State of Illinois . . .* Philadelphia, 1837.

M'NEMAR, RICHARD, *The Kentucky Revival, or, A Short History of the late extraordinary out-pouring of the spirit of God, in the Western States of America . . .* Albany, 1808.

MONTULE, ED. DE (Chevalier de l'Ordre de la Légion d'Honneur), *Voyage en Amérique, en Italie, en Sicile et en Egypte, Pendant les Annees 1816, 1817, 1818, et 1819 . . .* (2 vols. and atlas.) Paris, 1821.

MORRIS, LLOYD, *Postscript to Yesterday.* New York, n.d.

Murray, Hon. Amelia M., *Letters from the United States, Cuba and Canada.* New York, 1856.

Neuman, Fred G., *The Story of Paducah.* Paducah, Ky., 1927.

Nevin, Franklin Taylor, *The Village of Sewickley.* Sewickley, Pa., 1929.

Northend, Mary Harrod, *American Glass.* New York, 1926.

The Ohio Annual Register, Containing a Condensed History of the State . . . For the Year 1835. By John A. Bryan. Columbus, n.d.

Palmer, John, *Journal of Travels in the United States of North America, and in Lower Canada, Performed in the year 1817 . . .* London, 1818.

Parkman, Francis, *The Discovery of the Great West.* Boston, 1870.

———*History of the Conspiracy of Pontiac.* New York, 1929.

———*The Jesuits in North America, in the Seventeenth Century.* Boston, 1867.

———*La Salle and the Discovery of the Great West. France and England in North America. Part Third.* Boston, 1907.

Parton, J., *The Life and Times of Aaron Burr, Lieut.-Col. in the Army of the Revolution, U. S. Senator, Vice-Pres, of the U. S., etc.* New York, 1858.

Peck, J. M., *A Guide for Emigrants Containing Sketches of Illinois, Missouri and the Adjacent Parts.* Boston, 1831.

Peck, George (editor), *Sketches and Incidents or A Budget from the Saddle-Bags of a Superannuated Itinerant.* (2 vols. in one.) Cincinnati, 1847.

Perkins, James H., *Annals of the West: Embracing a Concise Account of Principal Events Which Have Occurred in the Western States and Territories, from the discovery of the Mississippi Valley to the year 1845 . . .* Cincinnati, 1847.

Perrin Du Lac, *Travels Through the Two Louisianas, and Among the Savage Nations of the Missouri; also In The United States, Along the Ohio, and the Adjacent Provinces, in 1801, 1802 & 1803 . . .* London, 1807.

Perrin, W. H., Battle, J. H., and Kniffin, G. C., *Kentucky. A History of the State.* Louisville, 1887.

Pidgeon, William, *Traditions of De-Coo-Dah, and Antiquarian Researches . . .* New York, 1853.

PIRTLE, CAPT. ALFRED, *The Battle of Tippecanoe* . . . Louisville: The Filson Club, 1900.

PITEZEL, REV. JOHN H., *Lights and Shades of Missionary Life:* . . . Cincinnati, 1862.

PODMORE, FRANK, *Robert Owen, A Biography.* (2 vols.) New York, 1907.

POLK, JAMES K., *The Diary of* . . . *During His Presidency, 1845 to 1849* . . . (4 vols.) Chicago, 1910.

POWNALL, T., *A Topographical Description of Such Parts of North America as are Contained in the (annexed) Map of the Middle British Colonies, &c. in North America.* London, 1776.

PROUD, ROBERT, *The History of Pennsylvania, in North America, from the Original Institution and Settlement of that Province, under the First Proprietor and Governor, William Penn, in 1681, until after the Year 1742* . . . (2 vols.) Philadelphia, 1797.

RANDALL, EMILIUS O., AND RYAN, DANIEL J., *History of Ohio. The Rise and Progress of an American State.* (5 vols.) New York, 1912.

Reports of the Trials of Colonel Aaron Burr . . . *for Treason and for a Misdemeanor in preparing the means of a Military Expedition against Mexico, a territory of the King of Spain, with whom the U.S. were at peace. In the circuit court of the United States,* . . . Richmond, Va., 1807.

ROTHERT, OTTO ARTHUR, *The Outlaws of Cave-in-Rock; Historical Accounts of the Famous Highwaymen and River Pirates who Operated in Pioneer Days Upon the Ohio and Mississippi Rivers and Over the Old Natchez Trace.* Cleveland, 1924.

SAFFORD, WILLIAM H., *The Blennerhassett Papers, Embodying the Private Journal of Harman Blennerhassett, and the Hitherto Unpublished Correspondence of Burr, Alston* . . . *in the attempted Wilkinson and Burr Revolution* . . . Cincinnati, 1864.

———*The Life of Harman Blennerhassett comprising an authentic narrative of the Burr Expedition* . . . Cincinnati, 1853.

SCHOOLCRAFT, HENRY R., *Historical and Statistical Information Respecting the History, Condition, etc. of the Indian Tribes of the United States.* (6 vols.) Philadelphia, 1851-1857.

SCHULTZ, CHRISTIAN, JR., *Travels on An Inland Voyage Through the States of New York, Pennsylvania, Virginia, Ohio, Kentucky and Tennessee, and through the Territories of Indiana, Louisiana, Mississippi and New Orleans; Performed in the Years 1807 and 1808* ... (2 vols.) New York, 1810.

SHEA, JOHN GILMARY, *Discovery and Exploration of the Mississippi Valley: the Original Narratives of Marquette, Allouez, Membre, Hennepin, and Anatase Donay.* Redfield, N.Y., 1852.

——*Early Voyages Up and Down the Mississippi, By Cavalier, St. Cosme, Le Sueur, Gravier and Guignas.* Albany, 1902.

SHETRONE, HENRY C., *The Mound-Builders* ... New York and London, 1930.

SMITH, SOL, *Theatrical Management in the West and South for Thirty Years. Interspersed with Anecdotal Sketches.* New York, 1868.

SMITH, T. MARSHALL, *Legends of the War of Independence, and of the Earlier Settlements in the West.* Louisville, 1855.

SMITH, Z. F., *History of Kentucky.* Louisville, 1886.

SPOONER, WALTER W., *The Back-Woodsmen or Tales of the Borders. A Collection of Historical and Authentic Accounts of Early Adventure Among the Indians.* Cincinnati, 1883.

STONE, WILLIAM L., *Life of Joseph Brant—Thayendanegea: Including the Border Wars of the American Revolution, and Sketches of the Indian Campaigns of Generals Harmar, St. Clair, and Wayne* ... (2 vols.) New York, 1838.

STOWE, HARRIET BEECHER, *The Key to Uncle Tom's Cabin; presenting the Original Facts and Documents Upon which the Story is Founded. Together with Corroborative Statements verifying the truth of the work.* Boston, 1853.

——*Uncle Tom's Cabin; or, Life among the Lowly.* Boston, 1852.

TAYLOR, JAMES W., *History of the State of Ohio* ... *First Period 1650-1787.* Cincinnati, 1854.

THOMAS, DAVID, *Travels Through the Western Country in the Summer of 1816. Including Notices of the Natural History, Topography, Commerce, Antiquities, Agriculture and Manufactures* ... Auburn, 1819.

THOMAS, FREDERICK W., *The Emigrant, or Reflections While*

Descending the Ohio. A Poem. From the original edition . . . Cincinnati, 1872.

——*John Randolph, of Roanoke, and Other Sketches of Character, Including William Wirt* . . . Philadelphia, 1853.

[THOMPSON, MORTIMER M.] Q. K. Philander Doesticks, *Doesticks, What He Says.* New York, 1855.

TONTI, CHEVALIER, *An Account of Monsieur de la Salle's Last Expedition and Discoveries in North America* . . . London, 1698.

[TOULMIN, HARRY], *A Description of Kentucky, In North America: To Which are Prefixed Miscellaneous Observations Respecting the United States.* [London] 1792.

TROLLOPE, MRS. [FRANCES], *Domestic Manners of The Americans.* London, 1832.

——*The Refugee in America: A Novel.* (3 vols.) London, 1832.

VIGNE, GODFREY T., *Six Months in America.* (2 vols. in one.) London, 1832.

VOLNEY, C. F., *View of the Climate and Soil of the United States of America: to which are annexed some Accounts of Florida, the French Colony on the Scioto, certain Canadian Colonies, and the Savages or Natives.* London, 1804.

WALKER, ADAM, *A Journal of Two Campaigns of the Fourth Regiment of U. S. Infantry, in The Michigan and Indiana Territories, Under the Command of Col. John P. Boyd, and Lt. Col. James Miller During the Years 1811 & 1812.* Keene, N.H., 1816.

WALKER, MARY ALDEN, *The Beginnings of Printing in the State of Indiana* . . . Crawfordsville, Ind., 1934.

WALTERS, RAYMOND, *Stephen Foster, Youth's Golden Gleam* . . . Princeton, 1936.

WARBURTON, MAJ. GEORGE, *The Conquest of Canada.* London, 1857.

WEISS, H. B., AND ZEIGLER, GRACE M., *Thomas Say, Early American Naturalist.* Springfield, 1931.

WELBY, ADLARD, *A Visit to North America and The English Settlements in Illinois, With a Residence at Philadelphia* . . . London, 1821.

WILKINSON, GENERAL JAMES, *Memoirs of My Own Times* . . . (3 vols. and atlas.) Philadelphia, 1816.

WISSLER, CLARK, *The American Indian.* New York, 1917.
——*The Relation of Nature to Man.* New York, 1926.
WITHERS, ALEXANDER S., *Chronicles of Border Warfare; or, A History of the Settlements by the Whites of North-Western Virginia; and of the Indian Wars and Massacres in that Section of the State, with Reflections, Anecdotes, etc.* Clarksburg, Va., 1831.
[WRIGHT, FRANCES], *Views of Society and Manners in America; In a Series of Letters From That Country to a Friend in England, During the Years 1818, 1819 and 1820* . . . New York, 1821.
Ye Sneak Yclepid Copperhead. A Satirical Poem. Philadelphia, 1863.

PERIODICALS

American Apollo (weekly). Boston, 1792-1793.
American Pioneer, A Monthly Periodical, Devoted to the objects of the Logan Historical Society; or, To Collecting and Publishing Sketches Relative to the Early Settlement and Successive Improvement of the Country. Cincinnati, 1842.
Cincinnati Literary Gazette. (weekly). Cincinnati, 1824.
Cincinnati Miscellany, Or Antiquities of the West: and Pioneer History and General and Local Statistics compiled from the Western General Advertiser . . . Charles Cist, editor, Cincinnati, 1844-1845.
Family Magazine; or, Monthly Abstract of General Knowledge. Cincinnati, 1836-1841.
Indianapolis Star Magazine. Indianapolis, May 2, 1948.
Olden Time: A Monthly Publication Devoted to the Preservation of Documents and other Authentic Information in Relation to the Early Explorations and the Settlement and Improvement of the Country Around the Head of the Ohio. Cincinnati, 1846-1847.
Western Gleaner or Repository for Arts, Sciences and Literature. Pittsburgh, 1813-1814.

MANUSCRIPTS

BANTA, R. E., *History of the New Harmony Movement.*
——*Life of Edward Allen Hannegan.*
BULEY, R. CARLYLE, *The Old Northwest,* manuscript being prepared for publication.
BURGESS, EDWARD SANFORD, *Annals of American Discovery:* Classical; Medieval, in the possession of R. E. Banta.
Collection of notes on Indiana authors in the Wabash College Library.
JAEBKER, ORVILLE JOHN, *Henry Hamilton; Soldier and Lieutenant Governor of Detroit.* Master's Thesis, Indiana University, 1949.
JONES, JOHN HEWITT, untitled memoirs, in the possession of R. E. Banta.
NEWSOM, NATHAN, untitled journal of experiences in the War of 1812, in the possession of the Ohio State Archaeological and Historical Society.
OWEN, ROBERT DALE, untitled journals, in the possession of the Purdue University Library.
VAN CLEVE, BENJAMIN, memoirs of settlement in Symmes' Purchase, in the possession of R. E. Banta.

INDEX